GREENHOUSE CLIMATE CONTROL

GREENHOUSE CLIMATE CONTROL

an integrated approach

EDITORS

J.C. Bakker

G.P.A. Bot

H. Challa

N.J. Van de Braak

Wageningen Academic
Publishers

ISBN 90-74134-17-3

Keywords:
climate control,
greenhouse climate,
horticulture

Cover design:
Beaufort – Sabine Mannel

Illustrations:
P. van Bommel,
A. de Haas & H. Vahl

Wageningen Academic Publishers,
Wageningen, 1995

Printed in The Netherlands

Preface

At the end of the seventies, following the energy crisis, the focus on energy saving in greenhouse cultivation brought four young scientists from different disciplines together: Gerard Bot (physics), Hugo Challa (plant physiology), Alexander Udink ten Cate (control engineering) and Jan van de Vooren (horticulture). Their cooperation on greenhouse climate and its control sowed the seeds for a research programme initiated in the eighties and named "Greenhouse climate control of the nineties". About twenty researchers of various disciplines, who came from different departments of Wageningen Agricultural University, DLO-institutes and research stations in The Netherlands, participated in the programme.

Through annual plenary meetings, frequent smaller meetings and active participation of graduate and PhD students, the contours of a new, scientifically based, integrated approach towards climate control became visible, and scientists of various disciplines learned to understand each other's language.

In the mean time, related activities took place in various other countries, in particular in France, Germany, Israel, Japan, the UK and the USA. Research groups formed informal networks which have stimulated progress and the exchange of concepts, methods and views considerably. The present book is the result of many years of intensive cooperation among the researchers who were directly involved in the earlier mentioned programme, of international contacts and of the interaction with the advanced Dutch greenhouse industry and with climate control equipment manufacturers.

We believe that the views and the knowledge which have been gathered and generated within the frame work of this programme is of major relevance to all those working in the field of greenhouse cultivation, especially in relation to the greenhouse climate. We hope and expect that this book will be appreciated by a broad readership and that it will stimulate the scientific approach towards various aspects of greenhouse climate. In addition, we hope that this publication will stimulate interdisciplinary and multi-disciplinary cooperation among scientists over the world, which, in our opinion, is of great importance to deal adequately with the complex problems of today and tomorrow.

This book has been written over a period of three years, during which major changes took place in the organisations of most of the authors and editors. This has made the task of all people concerned substantially more arduous. We are very grateful for the excellent cooperation and the valuable contributions of all participants, as well as the support of their organisations. This is also true for the international referees, who dealt with their important task within the narrow time constraints which had to be respected.

The Editors

Contents

1 Introduction

1.1 Aim and approach of this book

J.C. Bakker and H. Challa

Greenhouse cultivation is the most recently developed specialisation of horticulture. It developed as a result of technological advancement and the rise in demand for luxury (out-of-season, exotic) products related to the increasing prosperity of a growing group of citizens. Today, as a result of the large scale of production and product handling, greenhouse products are no longer as exclusive as they were and they provide a wide range of people with fresh food throughout the year, give enjoyment and add lustre to daily life.

Even in developing countries greenhouse cultivation is of increasing importance, because it enables them to enter the world market with products which have a high labour requirement and a high added value and hence to earn badly needed hard currency for their national economies.

The distinctive feature of greenhouse cultivation, as compared to outdoor cultivation, is the presence of a barrier between the crop and the external environment. This barrier creates a distinct micro-climate within the greenhouse, protects the crop against wind, precipitation, weeds, pests, diseases and animals, and enables the grower to control the crop environment to an extent unknown in outdoor cultivation. The protection from the outside world makes it feasible to heat, to add carbon dioxide, and to effectively apply chemical and biological control for crop protection. The high value produced per unit area enables the grower to invest in equipment as well as to improve and facilitate production, by such means as substrate culture, supplementary lighting, control of daylength, screening, cooling, benches, soil cooling/heating, etc. Today's greenhouse cultivation may be considered as the most intensive and sophisticated form of crop production, often referred to as the greenhouse industry, thus emphasising the role of technology in the whole process.

The presence of a cover, characteristic of greenhouses, causes, whether desired or not, a change in the climate conditions as compared with those outside: radiation and air velocity are reduced, temperature and water vapour pressure of the air increase, and fluctuations in carbon dioxide concentration are much stronger. These passive changes in the greenhouse weather, traditionally referred to as greenhouse climate, in combination with fluctuating outside weather conditions, force the grower into an active role with respect to climate conditioning. Originally, when control was still manual, the main issue was to avoid climatic extremes: overheating, especially due to strong radiation, and damage from low temperatures at night and during cold spells. Later, with the introduction of heating and automatic ventilation, the focus shifted towards the question of optimal setpoints for crop growth. Today greenhouse climate control is considered an important issue in protected cultivation, because it enables the grower to control the production process, more or less independently of outside weather conditions. As such it represents an important operational management tool for the optimisation of growth, production and quality.

Climate control has evolved over the last four decades from manual to digital. Initially the incentive for automation was to save labour: the operation of ventilation windows required permanent supervision and time consuming intervention by the grower. With the advancement of control techniques, electronic devices became available that enabled more sophisticated control strategies to be developed. The pioneer work of Strijbosch (Strijbosch & Van de Vooren, 1975) should be acknowledged here, because he was the first to meticulously record the climate control strategy of top growers in The Netherlands. The results of his research were implemented in analogue controllers for climate control in greenhouses and his findings still form the basis of present-day control systems. Examples are: different setpoints for day and night for heating and ventilation, temperature heating set points depending on ambient radiation, gradual increase from night to day set points, wind dependent minimum ventilation setting.

When microprocessors were introduced in greenhouse climate control systems they were "simply" the digital equivalent of their analogue predecessors. The main advantages were increased flexibility through the implementation of more complex control algorithms, without changing the hardware, lower cost when controlling more than one compartment and the great benefit of registration of climate conditions.

Since the introduction of greenhouse climate computers, major advancements in hard- and software have taken place. This process in turn has been reinforced by the vastly improved price/performance ratio of microprocessors. Control algorithms have been improved, for example, through the automatic adaptation of the control parameters to the actual situation. Moreover, an increasing variety in options required by growers with widely diverging crops, traditions and approaches has been incorporated in the systems. The development of improved user interfaces combined with an increase in the use of personal computers has facilitated the transfer and analysis of climate data in relation to other relevant nursery information. Yet, the basic philosophy has remained the same: the greenhouse climate computer can still be considered as an operator dealing with distinct actuators (e.g. mixing valves, ventilators), while control strategies reflect the grower's methods in dealing with these actuators. This situation is a logical consequence of the empirical nature of these strategies, which are a direct interpretation of experiences of growers with respect to (possible) crop responses to control procedures formulated along these lines. A major difficulty in this approach, however, is the complex relationship between actuators, environmental factors inside the greenhouse, short term crop response and the final results in terms of yield and quality (Challa, 1990). Moreover, the aforementioned addition of an increasing number of options in the control system has led to a virtually unmanageable jungle of settings. These features, as well as the increased requirements of modern nurseries, in terms of performance and efficiency, have created the need to reconsider the present control systems (Challa *et al.*, 1988). This process was further stimulated by certain other developments that will be briefly discussed.

Research on greenhouse environment, its control and related crop response was intensified in the late seventies as a result of the energy crisis, resulting in a great deal of new insights and knowledge. Stimulated by the availability of generic crop growth models and a young generation of researchers with an open eye for the great potentials of these new techniques, these insights were integrated into physical and physiological models describing the greenhouse-crop production system. With these tools opportunities were created to predict the response of the greenhouse environment, crop growth and their interactions under varying conditions and hence to design more intelligent, more flexible, and more efficient control systems.

The energy crisis also resulted in a different appreciation of the objectives of climate control. As a result of the dramatic rise in energy prices it became clear that the crucial issue, maximisation of profit, is not necessarily the same as maximisation of total yield. In other words, it was realised that climate control represents an optimisation problem, where the difference between benefits and associated costs appeared to be a useful criterion. It also became clear that there is a major difference between climate requirements in buildings, where human comfort is the central issue, and climate requirements in greenhouses, where control (optimisation) of the production process is the primary objective.

The increasing awareness of the fundamental shortcomings of present greenhouse climate control systems, the increased insight into the functioning of crops, the availability of models for the greenhouse-crop system, the availability of much more powerful hardware at a reasonable price have brought an international group of scientists to reformulate the problem of greenhouse climate control and to find a theoretically sounder approach to this problem. The present book provides a thorough basis for this new approach.

The aim of this book is to describe and analyze crop production in greenhouses in relation to climate control, to redefine the problem of (optimal) control from a theoretical point of view, and to provide a suitable framework for the design of new, scientifically based control systems. Though the principles are generally applicable, they will be discussed against the background of the Dutch glasshouse industry. To provide the reader with some background information, the historical developments and the economic position of the Dutch horticultural industry are briefly reviewed in this introductory chapter.

Since crop production, as influenced by environmental conditions, represents a central issue of this book, this phenomenon and the underlying physiological processes have been elaborated upon rather extensively, bearing in mind how difficult it is to extract such information from the usual, less specific handbooks (Chapter 2). This chapter provides insight into the functioning of the crop and the way the crop can be manipulated by changing environmental factors. It summarises quantitative relations that form the basis for crop growth models.

Once crop performance and requirements have been formulated it is essential to focus on the greenhouse, and to consider the physical processes governing the creation of the greenhouse climate in interaction with the crop (Chapter 3). Moreover, this chapter discusses the theoretical possibilities and constraints for climate control in greenhouses. It also summarises and explains the quantitative relations that form the basis for physical greenhouse models.

The present practical implementation of greenhouse technology within The Netherlands with respect to glasshouse construction and equipment is described (Chapter 4), since references to these topics are not easily accessible and since correct implementation is essential for the behaviour and performance of these facilities. Moreover the internationally recognised high standard of Dutch greenhouse technology justifies some special attention.

At this point all the elements required to design the new climate control system are in principle available, but it makes sense to describe and evaluate current climate control systems, in order to understand the principles and learn from them (Chapter 5).

Finally the theory presented in the aforementioned chapters will be used to introduce the concept of climate control systems, based on scientific principles and an integration of physiological, physical models and control engineering (Chapter 6). The advantages and possibilities, but also the limitations of this still preliminary concept are highlighted in the conviction that they will form the backbone of the next generation of greenhouse climate control systems.

1.2 The history of greenhouse cultivation

J.M. JACOBS

The use of measures to protect plants against adverse climatic conditions is very old. There are indications that already many thousands of years ago civilisations in China, Egypt and India employed means of protection against cold, wind and excessive solar radiation. In later Greek and Roman scriptures we find more detailed descriptions of methods for protecting and forcing plants with the help of windbreaks, hotbeds and even sheets of transparent stone (mica, talc). The purist Seneca strongly condemned these unnatural practices.

The Middle Ages form an uninteresting period in that very little is known about further developments. It is only by the end of the 15th century that we find indications that in Spain and Italy constructions were used with a resemblance to what later developed into the well known orangery: a special building in which containers with plants could be stored during winter. A first more or less systematic approach to plant cultivation was introduced during the Renaissance period. The discovery of new countries and new crops from the Middle East and later from the Far East and the West Indies led to the development of new techniques.

Merchants and gentry considered these exotic varieties not only as a source of trade but also as objects of prestige. The orangery, in which plants such as orange trees, laurels, pineapple plants were kept, was commonly designed with a north facing brick wall with a lighter window-like structure in the south side. This is known as the "lean to type", and formed the basic design for a long period.

In the 18th century efforts were made, within the limits of existing technical possibilities, to optimise growing conditions for the plants. During that period there was a special interest in plants for medicinal purposes. Experiments were performed on the gradient of the slope of the glass screens, heating systems (smoke flues), light reflectors, hotbeds, and insulation materials at Leyden University. In England the Royal Society and the Apothecarie's Garden of Chelsea were great sponsors of these experiments (Hix, 1974).

In the first half of the 19th century a new breakthrough occurred (Hix, 1974). The invention of techniques for making cast-iron and, in a very limited way, plate glass brought new incentives to glasshouse cultivation. In England industrial development was just beginning whilst commercial activity throughout her vast empire brought a wave of prosperity to at least part of the population. This resulted in the construction of huge botanical showplaces, of which Crystal Palace (Hyde Park, London 1850) became the most famous. On the continent of Europe such places followed in their wake later in the century, and several of these wonderful greenhouses still exist in botanical gardens and royal parks.

Though these structures served no real commercial purpose they contributed greatly to later developments, not only concerning greenhouse construction, but also because all kinds of additional problems, such as heating, condensation, ventilation, water storage and supply had to be studied and solved. Meanwhile commercial horticulture showed little progress. Where vegetables and fruits were grown round the towns and cities, some simple forms of protection were in use: frames, hotbeds, glass cloches, straw and reed mats, oil paper, but little else. A condition for the further application on a commercial scale of existing technical skills was the creation of a market with purchasing power, able to absorb more expensive products.

England may be regarded as the cradle of the development of the glasshouse industry. This is not surprising, as the Industrial Revolution began earlier than on the continent: as a larger part of the city population became more prosperous, production in glasshouses became a commercially viable process.

A more or less similar development took place in the later part of the 19th century in the north-eastern part of the USA. The fast developing heavy industry in that area created a luxury market with scope for, by the standards of the time, a large and modern glass area of several hundreds of hectares. All this had little effect on the horticultural scene in The Netherlands, where vegetable and fruit production kept its traditional character of mainly outdoor crops, even though there was some trade to the English market via the port of Rotterdam.

On the European continent the political situation stabilized after the Franco-Prussian War of 1870–1871. Only then did industrialisation get under way, leading to a growing demand for more luxury products in nearby German areas. However, the effect of this on glasshouse horticulture was not very noticeable.

The first census of the area under glass cultivation in The Netherlands in 1904 indicates little development (Table 1.2.1), whilst in that same period England counted more than 200 hectares and the USA 900 hectares of proper greenhouses. The growth of the Dutch glasshouse area dates from a later period. It was only in the 1920's that it reached an appreciable size (Table 1.2.1).

From a technical point of view this growth was not very impressive. A large proportion of the total area was taken up by frames (Dutch lights) and simple unheated grapehouses. The other structures were also of simple design, mostly the so-called Westland warehouse, consisting of Dutch lights supported by a wooden substructure. Only few were of a more sophisticated design, like the English type heated cucumber houses (in Loosduinen) and an occasional steel framed tomato house.

In The Netherlands horticulture was mainly the occupation of large rural families with plenty of labour available but with very little capital. As a consequence greenhouses had to be cheap. The same was true of the next step that evolved from the Westland warehouse in the 1930's during the depression: the Venlo warehouse, in which the loose frames were replaced by a fixed roof, the forerunner of today's modern multispan greenhouse (see Figure 4.2.2).

Nevertheless during that period some important steps were taken, which proved to be decisive for the future expansion of the horticultural branch after the Second World War. These were:

1. the creation of a cooperative auction system, which enabled small growers to obtain a fair price for their products in a large market. As a secondary benefit it taught growers to cooperate where they were weak as individuals;
2. introduction of the cooperative (Raiffeisen) banking system, enabling individual growers to finance their investments;
3. government policy to support the industry with professional training, research and advisory services, offering a healthy basis for growers to acquire knowledge for further development.

When the postwar economic situation in Europe revived in the second half of the 1950's, the Dutch horticultural industry was ready to respond to the challenges of the rapidly growing European market and to take advantage of its favourable geographic position in the centre of that market (Table 1.2.2) because:

1. there existed a large horticultural population with a long tradition of professional skill, used to hard work and long hours;
2. the auction system expanded rapidly, allowing the introduction of useful measures including uniform grading, packing and setting quality standards. The concentration of the products at the auctions enabled the development of large and efficient trade and transport facilities, reaching 200 million European customers as well as many outside Europe;
3. the cooperative banks took a progressive attitude and, backed up by government warrants had a very stimulating effect on investments by growers;
4. the foundation of the European Common Market with its free-trade philosophy for the internal market stimulated the expansion enormously;
5. the scheme of applied and fundamental research, set up by the government and supported by the growers, formed an important basis for the introduction of new techniques and methods;
6. the open structure of the horticultural sector promoted the free exchange of technical and economic information supported by advisory services;
7. the development of an inventive service and supply industry, contributed greatly to technical progress.

From this position the Dutch glasshouse industry changed over a period of some 30 years to become the world leader in specialized knowledge and the main centre of glasshouse crop production. This has partly been the result of continuous improvement in greenhouse structure and equipment. Some of the many technical achievements of the last three decades are summarised below.

1. The emphasis on the relation between plant production and the amount of light has led to the construction of greenhouses with a minimum of light intercepting parts. Since the late 1960's this has been achieved by the application of aluminium roof structures supported by galvanized steel frames.

Table 1.2.1 – Glasshouse-area in The Netherlands (ha) (CBS, 1904–1939).

Year	Frames	Grapehouses	Other glasshouses
1904	178	28	–
1912	495	85	160
1927	833	391	610
1939	1024	c. 1000	c. 1500

Table 1.2.2 – Area of glasshouses and production value in The Netherlands (round figures), (CBS, 1950–1990; auctions).

Year	Glasshouse area (ha)	Production value (billion NLG)
1950	3,300	0.25
1960	5,000	0.6
1970	7,000	1.1
1980	8,800	3.5
1990	9,600	6.4

2. New materials enabled constructors to build greenhouses which are more airtight, meaning the climate can be better controlled. The invention of the movable thermal screen in the early 1980's was a valuable addition to greenhouse equipment.
3. Fuel for heating changed from coal to oil in the early 1960's and from oil to gas in the early 1970's. This was combined with improvements in various parts of the heating systems (boiler, controllable mixing valves, etc.).
4. The first analogue climate control systems were introduced in the mid-1960's, controlling heat and ventilation in relation to temperature and air humidity. These were the forerunners of today's digital computers.
5. Application of CO_2 enrichment (since 1960).
6. Use of artificial light to improve growth and development. The first applications were for day-length treatments (low light levels) in crops such as chrysanthemums and poinsettias. High performance lamps are widely used to improve photosynthesis in valuable ornamental crops and to improve product quality especially in low light winter periods.
7. The greenhouse soil as a growing medium is difficult to control. In the postwar period new ways to confine the root system and control the growth factors at root level were gradually introduced. From spraying lines with nozzles, different sorts of soil containers, trickle irrigation, equipment to distribute nutrients in the irrigation water, peatbags, the system has culminated in today's controllable recirculation systems (mostly on rockwool).
8. Most production improvement has come from breeding new varieties. New breeding techniques (Fl-hybrids in the 1950's), adaptation of the specific glasshouse conditions to different seasons, resistance against virus and fungus diseases and pests has resulted in higher production, longer growing seasons, better quality and an enormous increase in variety of products.
9. In the early 1960's salaries rose steeply with the booming economy and the urgency to replace high labour costs by lower capital costs grew quickly.

Examples of mechanisation and automation are numerous. In addition to the labour saving and labour improving aspects of the developments mentioned above we can include the following examples. Transport in the greenhouse has evolved from the heavy picking basket to the electro trolley which runs on the heating pipes, or to the roller table as part of a robot controlled transport system.

Table 1.2.3 – Production, labour input and energy consumption of some important glasshouse crops (KWIN, various years).

Year	Production	Labour input	Labour efficiency	Gas consumption per m^2 [a]	Gas efficiency
Tomato					
1950	7.7 kg m^{-2}	- h 1000m^{-2}	- kg h^{-1}	43 m^3 gas [a]	0.18 kg m^{-3}
1960	9.5	-	-	54	0.18
1970	20.0	680	30	70	0.3
1980	29.0	720	40	46	0.6
1990	44.0	930	47	65	0.7
Roses					
1950	110 stems m^{-2}	2600 h 1000m^{-2}	38 st h^{-1}	-	- st m^{-3}
1960	160	2400	66	-	-
1970	220	1600	140	70	3.0
1980	240	1200	200	45	5.3
1990	320	815	400	47	6.8
Chrysanthemum (year round)					
1970	90 stems m^{-2}	900 h 1000m^{-2}	100 st h^{-1}	75	1.2 st m^{-3}
1980	130	750	175	40	3.3
1990	180	560	320	30	6.0

[a] For 1950–1970 oil recalculated to natural gas.

Grading and packing of the harvested product has changed from the bumping and shaking hand driven tomato grader to today's sophisticated automatic colour grader. There are comparable examples for all crops.

The combined effect of these achievements has been spectacular in terms of production per m^2 greenhouse, per hour of labour input and per unit consumed (Table 1.2.3).

1.3 The greenhouse industry in The Netherlands

D. MEIJAARD

1.3.1 International context

The cultivation of horticultural crops under cover is practised in nearly every country in the world. The total area of greenhouses at the end of the 1980's was 45,500 hectares, and of plastic tunnels 135,000 hectares (Table 1.3.4). Most plastic tunnels are located in the Mediterranean countries and China and Japan. Cultivation in greenhouses covered with glass is mainly practised in countries with a moderate climate. About 60% of the glasshouse area is located in North-West Europe. The

Netherlands leads with 10,000 hectares, of which 4,700 hectares are devoted to vegetable production, 4,200 hectares to cut-flowers and 1,100 hectares to pot plants (Data 1993).

Most North-European countries are big importers of cut-flowers, pot plants and vegetables such as tomato, sweet pepper and cucumber. More than 80% of imports come from other EC-countries (Table 1.3.5). The main exporters are The Netherlands (vegetables, cut-flowers and pot plants), Belgium (vegetables) and Denmark (pot plants). 80% of these exports goes to the member countries, and to the members of the European Free Trade Association (Norway, Sweden, Finland, Iceland, Switzerland and Austria). Exports to other countries are negligible.

The leading exporters to North-West Europe from outside the area are Morocco and the Canary Islands for vegetables and Israel and Columbia in the cut-flower sector. In general the destination of a country's exports is limited to its neighbouring countries. This is so for imports in North-West Europe as well for the USA.

1.3.2 Socio-economic importance

In North-West Europe the importance of the agricultural sector to the national economy is relatively small. Even in The Netherlands the share of agricultural sector in the national income does not exceed 4%.

Table 1.3.4 – Estimated area of greenhouses and plastic tunnels in the most important glasshouse regions.

	Greenhouse (ha)	Plastic tunnels (ha)
North-West-Europe	32,000	7,000
Mediterranean	4,000	45,000
Middle Europe	5,000	22,000
United States of America	2,500	4,500
Eastern Asia	2,000	55,000
South America	-	2,000
Total	45,500	135,000

Sources: De Groot, *et al.* (1990) and Von Zabeltitz (1992).

Table 1.3.5 – Origin of imports to North-West-Europe as % of total imports of relevant products[1] (NLG).

	Vegetables	Cut-flowers	Pot plants
EC-countries	85	80	96
EFTA-countries	-	-	-
Developing countries	11	16	2
Former state-run economies	2	-	-
Other countries	2	4	2
Total	100	100	100

Source: Data base Exmis, of LEI-DLO, The Hague, The Netherlands (1991).

[1] Cut-flowers, pot plants and tomato, cucumber, sweet pepper and eggplant as vegetables.

Nevertheless the agricultural sector contributes to the diversity of the economic activities in the various countries. In The Netherlands the economic importance of the agricultural sector lies more in its stable and appreciable contribution to the balance of trade. Agricultural exports amount to 20% of total exports.

The Dutch greenhouse industry plays a prominent role in the agricultural sector. 17% of the income in the agricultural sector comes from the greenhouse sector. More than 70% of the production from greenhouses is exported. This export oriented horticulture is strongly linked with other branches of economic activity such as supply of materials, services, trade and distribution. All those activities connected with the primary agricultural production are defined together as the agri-business complex. Thus it is evident that the demand for glasshouse produce generates not only income for the entrepreneurs and their employees, but also for the other participants in the agri-business complex. The share of the primary production in the total complex is 1.7 times the value of the production in the primary sector alone. The total income of the greenhouse complex amounts to approximately 4,500 million NLG (approximately 2,800 million US$).

1.3.3 Structure of production and sales

Most greenhouses are heated. Only 7% are not heated, and this percentage is still decreasing. During the 1980's the area of vegetables under glass with heating stabilized at 4,000 hectares. The main crops are tomato, cucumber and sweet pepper. These crops are mainly cultivated by substrate-culture; the supply is practically year round (Tables 1.3.6 and 1.3.7). 250 hectares of the total area are devoted to plant-raising in specialized nurseries. Other holdings tend to be very specialized, the number of crops being cultivated mostly limited to one or two types.

The assortment in the cut-flower sector is more diverse. The main crops are rose, chrysanthemum, carnation, freesia, gerbera and orchid (Tables 1.3.6 and 1.3.8). In the pot plant sector there is a wide range of crops and the supply is equally distributed over the year.

Foliage plants take up an area of 530 hectares, and flowering plants 400 hectares. The area of flower growing under glass increased during the 1980's by 24%. Within this sector pot plants form the biggest growth group with an increase of 65%.

There are 14,500 holdings with greenhouses. 10,000 of these specialize in greenhouse cultivation. The average area of greenhouses per holding is 8500 m^2. Dutch horticulture can be characterized as a sector with relatively small-scale production units consisting of family-farms, and a large-scale organized marketing sector. Nearly all the produce is sold via the auctions. At the auction

Table 1.3.6 – Pattern of supply (kg $pieces^{-1}$) of the main crops under glass per quarter, expressed as a percentage of total annual supply (1991).

Crop	Quarter 1	Quarter 2	Quarter 3	Quarter 4
Tomato	5	43	40	12
Cucumber	13	41	37	8
Sweet pepper	3	36	43	17
Rose	14	30	35	22
Chrysanthemum	18	28	28	25
Ficus	24	29	28	19
Bromelia	25	23	27	25

Source: LEI-DLO/CBS (1991).

the produce is collected, regrouped and inspected for quality. The price is determined by the auction clock. The concentration in the retailing sector continues and in response to this tendency the auctions are also merging to form larger units. This sales method not only increases the market power of the Dutch industry, but enables the entrepreneur to concentrate his attention more on refining growing techniques. Through this way of selling individual growers are not direct competitors in the market and consequently the exchange of technical knowledge is not hampered. Production in The Netherlands is regionally concentrated which has led to optimisation of the level of supply and services and has ensured rapid dissemination of innovations and knowledge.

1.3.4 Costs and returns

Cultivation under glass in The Netherlands is capital-, labour- and energy-intensive. Total investment (replacement value) in capital assets (excluding land) amounts to NLG 150 per m^2 (1990 price level). The labour-requirement is 1 person per 2,000 m^2, the energy input for heating is $1.30 \cdot 10^9$ joules per m^2 per annum and electricity expenditure is 6 Kwh per annum. Direct costs vary between NLG 41 and 86 per m^2. The main direct costs are labour, (NLG 11–15), energy (NLG 10–12), young plants and seeds (NLG 4–11), delivery (NLG 4–11) and interest (NLG 3). Compared with other branches of agriculture and horticulture in the open, the results of the greenhouse sector are favourable. The results in the

Table 1.3.7 – Area of vegetable crops under glass according to crop and substrate cultivation (ha) in 1991.

Vegetable crop[1]	Ha total	Ha substrate
Tomato	1,570	1,448
Cucumber	796	668
Sweetpepper	749	668
Eggplant	73	72
Other	1,341	86
Total	4,529	2,942

Source: KWIN (1994/1995).
[1] Excluding seedlings.

Table 1.3.8 – Area of cut-flowers under glass according to crop and substrate cultivation (ha) in 1993.

Cut-flower[1]	Ha total	Ha substrate
Carnation	217	54
Orchid	194	194
Gerbera	187	92
Rose	898	355
Anthurium	69	69
Other	3,812	24
Total	5,377	788

Source: KWIN (1994/1995).
[1] Excluding seedlings.

vegetable and cut-flower sector were less favourable than in the pot plant sector (Table 1.3.9). The cash-flow as a percentage of capital assets in the vegetable- and cut flower-sector is 15% and in the pot plant sector 19%.

1.3.5 Business development

Production costs have decreased considerably in the last decade (section 1.2). After correction for inflation there has been an annual decrease of 2 to 3%. The increase in productivity in the cutflower industry has been somewhat less than in the other sectors.

The growth in size of the firms, improved control of the growing process, better utilization of greenhouse capacity through lengthening of the growing period, better logistic and labour organization, have been the main push factors for improvements in production costs.

The costs of the equipment and the labour input per m^2 glass decrease as size of the business

Table 1.3.9 – Income and cash-flow statements for holdings with greenhouse crops per m^2 glass in NLG (average 1987–1990).

		Specialized vegetable holdings	Specialized cut-flower holdings	Specialized pot plants holdings
	Receipts			
(1)	Cash receipt (crops)	62	68	120
(2)	Other cash income	-	-	-
(3)	Gross cash income	62	68	120
(4)	Non money income	1	1	2
(5)	Realized gross income	63	69	122
(6)	Value of inventory change	-	4	2
(7)	Total gross income	63	73	124
	Expenses			
(8)	Cash expenses	41	47	86
(9)	Total expenses	50	60	101
	Income			
(10)	Net cash income (3-8)	21	21	34
(11)	Total net farm income (7-9)	13	13	23
(12)	Off farm income	1	1	2
(13)	Total income before tax (11+12)	14	14	25
(14)	Income tax and premiums	3	3	7
(15)	Total income after tax (13-14)	11	11	18
(16)	Family expenditure	7	8	10
(17)	Savings (15-16)	4	3	8
(18)	Depreciation	9	13	15
(19)	Cash-flow (17+18)	13	16	23

Source: LEI-DLO (1991).

grows. The glasshouse area per holding in the pot plant sector increased in the 1980's from 6,300 to 8,000 m^2 per holding. In the vegetable sector and the cut flower sector the growth was limited to 8 and 4% respectively.

The introduction of substrate culture, especially within the vegetable sector, has contributed to a considerable increase in output. The equipment for and the knowledge about climate control was also improved in the same period. Investments in installations are of growing importance (Table 1.3.10). During the 1980's investments in heating equipment and CO_2-enrichment were relatively stable. The value of other equipment for irrigation, lighting, day-length and environmental control (computers) increased dramatically. The share of these investments as a percentage of total investments doubled.

The volume of production in the Dutch greenhouse industry increased in the last decade by 6% per annum (Table 1.3.11). Growth of the area under production was substantially lower at 1%. (Table 1.2.2). Volume of production per m^2 glass rose by 5% per annum. This has partly been achieved at the expense of the environment. The main environmental problems in the glasshouse sector are the leaching of fertilizers, the dispersion of pesticides, the emission of CO_2 and NO_x into the air by the heating installations and the disposal of waste. The leaching of fertilizers has greatly increased with the introduction of substrate culture. After a decrease in the period 1973-1983, energy use is increasing again. As a result of the increased awareness of environmental problems, the greenhouse industry is confronted with new demands. These demands were voiced in the National Environmental Plan (MLNV, 1990). For the glasshouse sector this plan is specified in a number of concrete objectives.

By the year 2000 production should take place in almost completely closed circuit cultivation systems. To prevent the emission of nutrients 30% of the area under vegetable cultivation must, in 1994, be on substrate with a closed recirculation system for the water and a tank for the remainder of the nutrient solution. In 2000 the total area must be provided with such a system.

Pollution from pesticides should also be reduced. By 2000 only 50% of 1990 levels of the active compound may be used. The reduction can be achieved by increasing the amount of biological and/or

Table 1.3.10. Investments in installations for climate control as % of total investments per annum on glasshouse holdings in The Netherlands.

	1978/1980	1988/1990
Heating and CO_2-enrichment	16	14
Other installations	11	22
Total	27	36

Source: LEI-DLO (1991).

Table 1.3.11 –Turnover as a % of previous year's figures for greenhouse area and volume of the total returns of the Dutch glasshouse industry.

	1981/1986	1987/1991	1981/1991
Area	0	2	1
Volume	6	7	6

Source: LEI-DLO (1991).

integrated control, using more efficient equipment and improving climate management.

Energy efficiency should also be improved. By 2000 the input of energy per kg product should have been improved by 50% compared to 1980 levels.

The waste products are mainly rockwool, plastics and organic material. All these products should be completely recycled. To achieve this objective producers will be confronted with increased investments in new and additional equipment.

Due to the obligations to improve the (global) environment the costs of production, (especially capital assets and labour), will increase per unit product. During the period to 2000 it is estimated that the annual rise in the costs per unit product will be 0.6%. The competitive power of the sector will be weakened by these measures. It is expected that the growth of the area of the vegetable sector will be 2% less because of these environmental demands. On the other hand it is possible to imagine, that increased consumer awareness about production methods will lead to an increase in demand for these environmentally friendly products.

References

Challa, H., 1990. Crop growth models for greenhouse climate control. In: R. Rabbinge, J. Goudriaan, H. Van Keulen, F.W.T. Penning de Vries & H.H. Van Laar (Eds), Theoretical production ecology: reflections and prospects. Simulation monographs 34. Pudoc, Wageningen, p. 125–145.

Challa, H., G.P.A. Bot, E.M. Nederhoff & N.J. Van de Braak, 1988. Greenhouse climate control in the nineties. Acta Horticulturae 230: 459–470.

De Groot, N.S.P., M. Mulder, B. Van der Ploeg & G. Trip, 1990. Ruimtebehoefte Zuidhollandse Glastuinbouw *(Area requirement of greenhouse industry in South-Holland)*. Mededeling 438. Landbouw-Economisch Instituut (LEI-DLO), Den Haag, 134 pp. (in Dutch).

Hix, J., 1974. The Glass House. Phaidon Press Ltd., London, pp. 208.

KWIN, 1994/1995. Kwantitatieve Informatie glastuinbouw *(Quantative information greenhouse industry)*. IKC, Naaldwijk/Aalsmeer, 128 pp. (in Dutch).

LEI-DLO, 1991. Rentabiliteit en financiering van de tuinbouw onder glas *(Productiveness and financing of the glasshouse industry)*. Landbouw-Economisch Instituut, Den Haag. (in Dutch).

LEI-DLO/CBS, 1991. Tuinbouwcijfers 1991 *(Horticultural statistical data 1991)* . Landbouw-Economisch Instituut, Den Haag, 154 pp. (in Dutch).

MLNV, 1990. Structuurnota Landbouw *(National Environmental Plan)*, Ministerie van Landbouw, Natuurbeheer en Visserij, Den Haag, 139 pp. (in Dutch).

Strijbosch, Th. & J. Van de Vooren, 1975. Developments in climate control. Acta Horticulturae 46: 21–22.

Von Zabeltitz, Chr., 1992. Gartenbau/Hausbau im Ausland. Gartenbau Report 18: 29–31.

2 Crop growth

2.1 Introduction

H. Challa and J.C. Bakker

In a treatise on greenhouse climate and its control, the crop represents the central, but also the most complex element of the greenhouse-crop production system. Due to this complexity, and the great diversity of crops cultivated in greenhouses, this chapter will focus only on the most relevant and general issues necessary for understanding crop response in relation to greenhouse climate. In order to deal adequately with this complexity it is helpful to consider what responses are, directly or indirectly, relevant. Relevant outputs for the grower are:

- Total amount produced;
- Onset and time course of production;
- Quality of the produce.

In addition risk control in relation to damage or loss of the crop may play a role.

The major factors characterising greenhouse climate are CO_2 concentration (C_a), air temperature (T) and water vapour pressure (e_a) of the air. Radiation is in most cases imposed by the outside weather conditions and should be considered as a boundary condition, although application of screens and/or supplementary lighting may enable the grower to modify it. Through the processes of transpiration, photosynthesis and respiration the crop will interfere with the mass balances of CO_2 and water vapour in the greenhouse air, as well as with the energy balance (Chapter 3). For this reason the crop plays a double role: it modifies and it responds to its environment. Both aspects are important within the framework of this treatise.The objective of this chapter is to provide sufficient insight into the production process of greenhouse crops as related to environmental factors for the design and implementation of improved greenhouse climate control systems, taking the diurnal dynamics of the crop response to the environment into account and relating short-term responses to long-term performance.

The production process, as will be demonstrated in this chapter, is a highly aggregated process which is one of the reasons why it is difficult to describe its overall response by means of simple and general relations. To handle this complexity adequately and in a generic way, the production process may be divided into subprocesses and related state variables, according to the time scale where changes become manifest (which is partly related to the aggregation level of the processes concerned).

Crop responses that become manifest in the short-term (minutes, hours), are the carbohydrate and the water status. These processes provide energy, primary building blocks and water required for the tissue growth process. The carbohydrate balance, governed by photosynthesis and respiration, which under optimal conditions limits the production process is discussed in section 2.2.1, the water balance, important in relation to transport of minerals within the plant, the energy balance of the greenhouse and for tissue extension is dealt with in section 2.2.2. The interactions between both are discussed in section 2.2.3. The theory concerning the processes involved provides the background to define the short-term requirements (diurnal momentary setpoints) for climate control.
Moreover, these results are needed in order to describe the mass and energy balance of the greenhouse-crop system (Chapter 3).

In the long-term, the production process can be characterized by dry matter accumulation and development (section 2.3.1), dry matter distribution (important in relation to the part of the biomass that can be harvested), dealt with in section 2.3.2, and product quality (section 2.3.3), an important factor in determining the economic value of the produce and also an important objective in relation to the management of the nursery. Theoretical background information needed to explain the response to environmental factors will be treated as much as feasible within the scope of this overview, to provide a generic basis for its understanding.

Based on the theory of this chapter, the environmental requirements for production are evaluated in section 2.4. These requirements have to be considered together with those related to other elements of the production system, such as the control of outbreak and spread of pests and diseases, the interaction with biological control, requirements of bees or bumble bees for pollination, and in relation to labour conditions.

2.2 Short-term crop responses

2.2.1 CO_2 uptake by the crop

H. GIJZEN

2.2.1.1 Introduction

Photosynthetic CO_2 assimilation is a key process in crop production. In the photosynthetic process CO_2 is used as a substrate for the formation of primary building blocks, primarily sugars, amino acids and organic acids. These primary building blocks are transported to the growing parts of the plant, the sinks, where they are converted to structural dry weight (section 2.3.2). Here sugars also serve as a source of energy in the conversion process. In addition sugars are used in maintenance respiration, providing energy to maintain the integrity of the living tissues.

Although many greenhouse crops have a low dry matter content, growth and yield are closely linked to dry matter production (Spitters *et al.*, 1989) and for this reason photosynthesis can be characterized as a key production process for production. The relation between photosynthesis and dry matter production over longer periods of time can be described by

$$dW/dt = C_f(30/44\, P_{gc,d} - R_{m,d}) \qquad \text{(Eq. 2.2.1)}$$

where $\Delta W/dt$ is the rate of production of dry matter (g m^{-2} d^{-1}); C_f is the conversion efficiency of the dry weight formed per g assimilates (CH_2O), $P_{gc,d}$ is the daily rate of gross photosynthesis per unit greenhouse area (in CO_2, g m^{-2} d^{-1}), $R_{m,d}$ is the daily rate of maintenance respiration per unit greenhouse area (in CH_2O, g m^{-2} d^{-1}), and where 30/44 converts the weight of CO_2 to the weight of CH_2O. Over shorter periods of time, however, diurnal variations in storage of reserves play a role, resulting in different diurnal dynamics of dry matter production compared to photosynthesis (P).

The role of greenhouse climate control in dry matter production is largely limited to the effects on photosynthesis and maintenance respiration. C_f is much less affected. It depends on the chemical composition of plant parts and on the distribution of the total dry weight increment over plant parts (Spitters *et al.*, 1989), and does in producing vegetable greenhouse crops not vary much during the growing season.

To describe properly the instantaneous net CO_2 exchange of the crop with the greenhouse air, growth respiration, R_g, also has to be quantified. The instantaneous net rate of crop photosynthesis,

P_{nc} (g CO_2 m^{-2} h^{-1}), is by definition related to instantaneous canopy gross photosynthesis, P_{gc}, according to

$$P_{nc} = P_{gc} - R_m - R_g \quad \text{(Eq. 2.2.2)}$$

where all rates are expressed as g m^{-2} h^{-1}. Photosynthetically active radiation (PAR, 400–700 nm), supplies the energy for crop photosynthesis. The relation between PAR and P, however, is complex: there are interactions with CO_2 concentration, temperature, and water vapour pressure deficit (VPD) of the greenhouse air. Moreover, crop characteristics and the radiation climate play an important role. Also, many of the relations mentioned are non-linear.

In the following the relation between PAR inside the greenhouse and crop photosynthesis is analyzed at the level of subsystems. This is done in order to deal adequately and in a generic way with the complex system of the radiation distribution within the canopy and the relation between PAR and the rate of photosynthesis of individual leaves, as related to other environmental factors. As a next step the behaviour of the system as a whole, i.e. crop photosynthesis, is studied for specific cases. First a characterisation at the level of sub-systems will be made, with an emphasis on the processes that provide an understanding of the photosynthetic response of individual leaves.

2.2.1.2 Components of leaf photosynthesis

The way in which photosynthetic CO_2 assimilation is achieved varies a great deal between different plant species. Types found in higher plants are broadly classified into the following classes: C_3-photosynthesis, C_4-photosynthesis, C_3–C_4 intermediate photosynthesis and Crassulacean Acid Metabolism (CAM). The majority of greenhouse crops use C_3-photosynthesis, and CAM is quite common in some families of ornamental crops (e.g. many orchids, Bromeliaceae and cacti). The discussion of photosynthesis response will concentrate on those of C_3-plants.

The net rate of CO_2 uptake of a leaf, P_n, is by definition equal to leaf gross photosynthesis, P_g, minus the rate of respiration not associated with photosynthesis processes, the dark respiration rate, R_d

$$P_n = P_g - R_d \quad \text{(Eq. 2.2.3)}$$

A summary is given of the most important processes that determine the rate of CO_2 assimilation and that affect the diffusion of CO_2 into the interior of the leaf.

Photosynthetic reactions and related reactions

The photosynthetic reactions comprise both the light reactions that capture light energy and the dark reactions that use the captured energy for binding CO_2 or O_2. In the related reactions, sugars and starch are synthesized from bound CO_2. In Figure 2.2.1 a scheme is given of the most relevant processes. Ribulosebiphosphate, RuP_2, plays an important role in the connection of these reactions to each other.

Light reaction

In the light reaction radiative energy contained in PAR is captured. The captured energy is transferred to the molecules ATP and NADPH. These molecules drive other reactions, including the regeneration of RuP_2.

PAR can be expressed in Joules per unit of time and area, but quanta of different wavelengths contain different amounts of energy. For the purpose of quantifying the PAR response of leaf photosynthesis, it is better to express PAR in mole quanta (or photons) per unit of time and area.

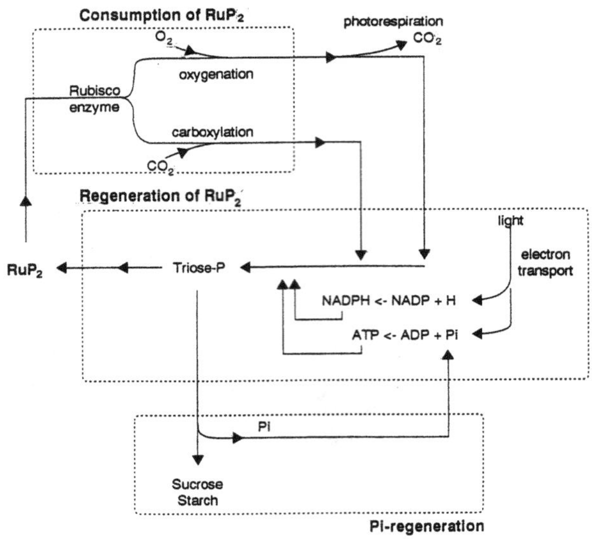

Figure 2.2.1 – Diagram of the most relevant processes contributing to overall photosynthetic CO_2 assimilation. Three main processes can be discerned: consumption of RuP_2, regeneration of RuP_2, and Pi regeneration.

Carboxylation, oxygenation and photorespiration (dark reactions)

RuP_2 can react with CO_2 at the Rubisco enzyme (carboxylation) and split into two molecules of phosphoglyceric acid, PGA. The carboxylation is part of the photosynthetic carbon reduction (PCR) cycle, or Calvin cycle. PGA is further processed in this cycle to yield RuP_2 again, using energy from ATP and NADPH.

RuP_2 can also combine with O_2 at the Rubisco enzyme (oxygenation), yielding PGA and phosphoglycollate, PGlA. The oxygenation is part of the photosynthetic carbon oxidation (PCO) cycle. PGA and PGlA are processed in this cycle to regenerate RuP_2, with energy from ATP and NAPH. In this cycle CO_2 is liberated: photorespiration.

Both CO_2 and O_2 compete to bind to RuP_2, and higher concentrations (or more precisely partial pressures) of CO_2 at Rubisco increase the ratio of carboxylation to oxygenation, and decreases photorespiration. This in turn increases the net rate of binding of CO_2 in the photosynthetic reactions, i.e. carboxylation minus CO_2 release by photorespiration. For each oxygenation of RuP_2 half a molecule CO_2 is released, thus

$$P_g = V_c - 0.5\,V_o \qquad \text{(Eq. 2.2.4)}$$

where V_c is the rate of carboxylation, V_o the rate of oxygenation, and where 0.5 V_o is equal to the rate of photorespiration (Farquhar & Von Caemmerer, 1982).

Orthophosphate regeneration

Orthophosphate (Pi), necessary in the light reaction for the generation of ATP, is liberated during the synthesis of sucrose and starch from triose-phosphate, and can be used again in the light reaction. When the synthesis of sucrose and/or starch is reduced because of too much accumulation in the leaf, the rate of triose-phosphate utilization is reduced and less orthophosphate is liberated. Leaf photosynthesis is then limited due to lack of free phosphate (Sharkey, 1985), i.e. "end product synthesis limitation" (Stitt, 1991). In this way the use of assimilates may regulate the production of assimilates.

Limitation of CO_2 assimilation

The response of P_g at a given level of one environmental factor in relation to other environmental factors is very much dependent on which of the three subprocesses is limiting the overall process: i.e. Rubisco-activity, RuP_2 regeneration or Pi regeneration. When RuP_2 is regenerated faster than it is consumed by carboxylation and oxygenation (normally at high light levels), Rubisco is saturated with RuP_2 and the response of P_g will be largely determined by Rubisco activity ("Rubisco limitation") and CO_2 concentration. When carboxylation and oxygenation occur faster than RuP_2 is regenerated by other processes (normally at low light and/or high CO_2 concentration), P_g is "RuP_2 regeneration limited". The response of P_g will then be largely determined by the rate of light harvesting and ATP and NADPH synthesis, i.e. by the light level. When Pi regeneration is limiting, typically at high light and CO_2, response to light and CO_2 is absent or small, and temperature can have a large effect.

Crassulacean Acid Metabolism (CAM)

In CAM-plants CO_2 is typically assimilated through the prefixation of CO_2 by PhosphoEnolPyruvate (PEP)-carboxylase into C_4-acids, notably malic acid. This can be done at night, so that the stomata need not to open at daytime and water loss can be minimized. When the CO_2-molecules are released from the C_4-acids, they can be processed in the normal PCR-cycle (Ting, 1987). It appears that under well-watered conditions normal C_3-photosynthesis is dominant.

Dark respiration

Dark respiration is the non-photorespiratory CO_2 release of the leaf, and is the consequence of metabolic processes including regeneration of energy, and synthesis of structural plant components. Dark respiration is not directly coupled to the photosynthesis processes, but constitutes the respiration necessary for growth and maintenance processes, not only for the leaf itself, but also for other plant parts, and thus contributes to crop respiration (R_g and R_m in equation (2.2.2)). Dark respiration can continue in the light, but to what extent this occurs is not clear (Kirschbaum & Farquhar, 1987). The rate of dark respiration is about 10% of the maximal rate of gross photosynthesis. It responds momentarily and strongly to temperature. For example, Ludwig & Withers (1978) measured in tomato a doubling of R_d with a temperature rise of 10 °C.

Stomatal and boundary layer conductance

As part of the total diffusion path to the chloroplast stroma where Rubisco is active, CO_2 has to diffuse from the air surrounding the leaf to the substomatal cavity. Two barriers have to be passed that can be characterized by their different conductances: the boundary layer conductance and the stomatal conductance. These conductances will limit photosynthesis more when RuP_2 consumption and, consequently, CO_2 diffusion to the site of carboxylation, is limiting photosynthesis, i.e. at high light intensity. Also, the degree of control that each conductance exerts on CO_2 diffusion is dependent on the magnitude of the other conductance.

Leaf boundary layer conductance

The boundary layer is the layer of still air around the leaf. The low air speeds in greenhouses generate thick layers, and consequently low boundary layer conductances (g_b). Stanghellini (1985) found a value of 0.01 m s^{-1} for g_b, using replica leaves of 5 cm width inside a tomato canopy. With large leaves g_b could be much lower. In field crops with average sized leaves g_b will commonly be in the range 0.025–0.05 m s^{-1} (Monteith & Unsworth, 1990).

Stomatal conductance

By opening and closing the stomata the plant can regulate the uptake of CO_2, and at the same time the loss of H_2O (section 2.2.2). It is commonly believed that the plant optimizes the ratio of CO_2 uptake and H_2O loss in one way or another (e.g. Schulze, 1986). Stomatal conductance, g_s, increases with increasing light intensity and decreases with higher ambient CO_2 concentration. Stomatal conductance of two greenhouse crops with high photosynthetic capacity, cucumber and tomato, reached values of 0.02–0.025 m s^{-1} (Nederhoff & De Graaf, 1993).

It has often been observed that g_s correlates with the photosynthetic capacity and with the photosynthetic rate (Tenhunen *et al.*, 1987). The ratio of intercellular CO_2 concentration, C_i, to ambient CO_2 concentration, C_a, has often been found to tend to a conservative value at full sunlight, i.e. approximately 0.7–0.8 (Jarvis & Morison, 1981; Morison, 1987).

Air humidity and plant water status also affect g_s. High humidities are common in greenhouses in the North-West of Europe and these promote high conductances. Therefore, it is presumed that g_s normally limits photosynthesis to a small extent. However, under summer conditions with high insolation, low humidities may occur which could decrease g_s significantly. Consequently, leaf photosynthesis may be decreased due to the increased diffusional limitation by stomata. When g_s is not decreased sufficiently to prevent desiccation of the leaf, water stress arizes and leaf photosynthesis can be hampered (section 2.2.3).

2.2.1.3 Responses of leaf photosynthesis

Typical responses of leaf photosynthesis to PAR, CO_2 and temperature are discussed which are characteristic of C_3-photosynthesis. See Figure 2.2.2.

Response to PAR

At very low PAR intensities P_n is negative because the CO_2 efflux from dark respiration dominates the CO_2 uptake. At the light compensation point $P_n = 0$. At a low PAR level the regeneration of RuP_2 needed for carboxylation is limited by the rate of ATP and NADPH production in the light reaction; consequently photosynthesis responds maximally to light. The response is near-linear. The slope of the response of P_n to absorbed PAR (in the region 50–150 μmol m^{-2} s^{-1}, Farquhar & Von Caemmerer, 1982) is the quantum efficiency or quantum yield (f_a, mol CO_2 fixed per mol photons absorbed; Bjorkman, 1981). It has also been called light use efficiency (mg CO_2 per J PAR absorbed). With a CO_2 concentration at 350 μmol mol^{-1}, f_a of a thin leaf is about 0.05–0.055 mol mol^{-1}. A higher CO_2 concentration decreases photorespiration and consequently increases f_a and decreases the light compensation point. For example, it was calculated with the leaf photosynthesis model described in Box 2.2.1 that, at 25 °C, f_a increased about 17% when the CO_2 concentration rose from 350 to 700 μmol mol^{-1}. A higher temperature increases photorespiration, and consequently decreases f_a; this, and the increased dark respiration increases the light compensation point. At CO_2 concentration 350 μmol mol^{-1}, f_a decreases about 15% when temperature rises from 15 to 30 °C (Bjorkman, 1981). f_a as would be obtained without photorespiration, e.g., at very high CO_2 levels, is about 0.08 mol mol^{-1} (Farquhar & Von Caemmerer, 1982). f_a can be reduced by previous exposure to high light levels, water stress and phosphate deficiency (Kirschbaum & Farquhar, 1987).

The effect of PAR on the rate of CO_2 assimilation decreases with increasing PAR until reaching saturation, when Rubisco-activity becomes the limiting factor. The rate of leaf photosynthesis at light saturation, P_{nmax}, is dependent on, *inter alia*, the CO_2 concentration and temperature.

The photosynthesis-light response depends much on the average light level to which the leaf is acclimated (Bjorkman, 1981). Leaves acclimated to low light, such as leaves lower in the canopy, generally have lower P_{nmax}. f_a hardly varies with PAR or temperature (Ehleringer & Pearcy, 1983; Evans, 1987). The photosynthesis-light response varies between species, mostly with respect to P_{nmax}. For example, species exhibiting high growth rates generally have a higher P_{nmax}. f_a varies very little

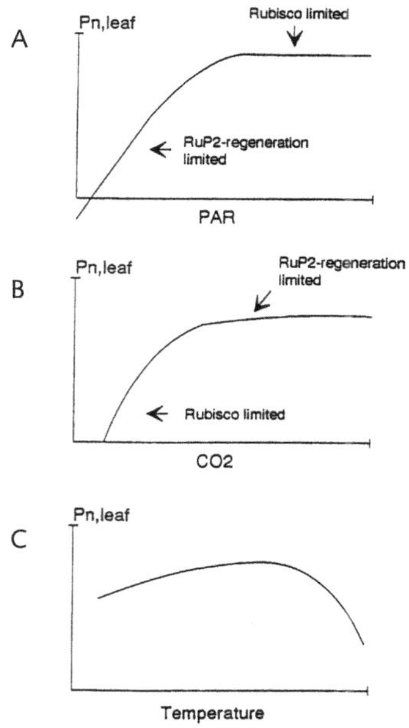

Figure 2.2.2 – Schematized responses of leaf photosynthesis to PAR (A), CO_2 concentration (B) and temperature (C).

between different C_3-species (Ehleringer & Pearcy (1983); Bjorkman & Demmig, 1987). Ehleringer & Pearcy (1983) found an average of 0.052 mol per mol photons at 30 °C and 330 μmol mol^{-1} CO_2. The variation of quantum efficiency is larger when incident PAR is considered instead of absorbed PAR.

Growth chamber grown plants appear to have a f_a that is some 5% higher than that in field grown plants (McCree, 1972; Evans, 1987). This was attributed to the UV- protective epidermis of field plants (Evans, 1987). In glasshouse grown plants, f_a may also be higher than in field grown plants, as glass does transmit littleUV-radiation.

Response to CO_2

P_n increases with the CO_2 concentration. The rate of carboxylation increases due to increased competition of CO_2 with O_2 at the carboxylation site and increased affinity of Rubisco to CO_2. At low CO_2 concentration the response is maximal. At the CO_2 compensation point the total amount of CO_2 that is assimilated is provided by CO_2 released from respiration processes in the leaf. The initial slope has been called the carboxylation efficiency or mesophyll conductance (Thornley, 1976).

The curve slopes off when the increased demand for RuP_2 caused by the increased rate of carboxylation can be less easily met by the rate of RuP_2 regeneration. At high CO_2 concentrations P_n could still theoretically increase, as both O_2 is increasingly replaced by CO_2 at the Rubisco and the affinity to CO_2 increases. However, at a given level P_n does not respond further to higher CO_2 levels, or can even decrease, as a result of end product synthesis limitation (Stitt, 1991). Both the absence of response (Sharkey, 1985) and the reversed response (Harley & Sharkey, 1991) can be caused by Pi regeneration limitation. Lack of response or reversed response was found on several occasions with tomato (Pallas, 1965; Bradford *et al.*, 1983; Stanghellini & Bunce, 1994). Stanghellini & Bunce (1994) found that with tomato the reversed response to CO_2 at high levels disappeared at a higher temperature.

When plants adapt to high CO_2 concentrations the initial effect of an increased rate of photosynthesis may become smaller or disappear, but may also be increased. Apart from morphological changes, the photosynthetic capacity of plants may become acclimated to high CO_2. Stitt (1991) reviewed acclimation responses, and concluded that, when analysing the response of leaf photosynthesis to intercellular CO_2 concentration, in some cases the initial slope and/or the saturation level was increased, but in most cases the initial slope decreased. Upon acclimation, activity of several enzymes in- volved in the photosynthetic and related reactions have been found to decrease (Stitt, 1991). Often the amount of Rubisco decreased, as has been found in tomato (Yelle *et al.*, 1989; Besford *et al.*, 1990) and cucumber (Peet *et al.*, 1986).

Response to temperature

The response of P_n to air temperature generally follows an optimum curve. At low temperature, assimilation increases with T because of the increased rates of the light reaction and of carboxylation. However, dark respiration and photorespiration also increase, which depresses net CO_2 assimilation. These latter effects become dominant at and above the optimum T. At high T the light reactions become less efficient and, in general, enzyme activities decrease. P_n then declines more quickly. The optimum temperature is higher where CO_2 concentration is high and depends on acclimation (Berry & Bjorkman, 1980).

Feedback inhibition

There are many observations of decreased photosynthesis after the removal of "sinks" for carbohydrates, for example in fruits (Geiger, 1976). This decrease is explained as a sink- regulation of photosynthesis, which becomes apparent when the demand for photosynthates is less than the supply, and is aimed at removing the imbalance between source and sink-activity (Herold, 1980). Concomitantly, increased carbohydrate content in leaves has often been observed (Stitt, 1991). Stitt (1991) proposed that feedback could operate via a) direct inhibition, in the short-term, as a result of carbohydrate accumulation, and b) indirect inhibition, in the long-term, by decreasing the levels of enzymes (Rubisco) and other components of the photosynthetic machinery.

It is assumed that under natural conditions synchronous changes in the rate of photosynthesis and the photosynthetic capacity often occur with changes in demand (Geiger, 1976). How often and how strongly periods of decreased photosynthesis occur in practice due to lowered sink demand is not known. In many experiments strong effects of decreased sink demand were observed, but in these experiments the ratio of sink to source activity was often drastically changed. Thus they were not representative of "natural" growth. For example, Marcelis (1991) observed that the rate of leaf photosynthesis of cucumber leaves did not decrease when part of the fruits were removed, but only when all fruits were removed.

Feedback inhibition could also result from acclimation to CO_2 enrichment. Enhanced CO_2 concentrations cause, at least initially, an increased source activity, which can be lowered again when sinks cannot respond adequately (Stitt, 1991).

Air pollutants

Air pollutants that are reported to influence photosynthesis negatively include: ozone, NO_x, CO, and SO_2. Of special importance are the toxic gases that may enter the greenhouse as a consequence of the use of flue gases for CO_2 enrichment (section 4.6.3), i.e. SO_2 and NO_x. In particular, NO_x, a mixture of NO and NO_2, can reach injurious levels (Hand, 1990). Effects of SO_2 and NO_x on photosynthesis are found for levels of less than 1 μmol mol^{-1} (Darrall, 1989; Saxe, 1989). SO_2 appears to be more toxic than NO_x and to affect both stomatal opening and Rubisco-activity, whereas NO_x seems to affect Rubisco-activity more (Saxe, 1989). The stomatal closure induced by high CO_2 concentrations generally reduces the toxicity of air pollutants by diminishing their uptake (Darrall, 1989).

2.2.1.4 Models of leaf photosynthesis

In the literature a large number of models of leaf photosynthesis can be found. Here a "family" of models will be discussed that is used very often in the recent literature: the "Farquhar & Von Caemmerer-type" models. In these models the central issues are the kinetics of the Rubisco enzyme and the regeneration of RuP_2 by the light reaction.

In addition to this biochemical model two classes of response functions for describing whole leaf responses to light and CO_2 will be discussed: 1) the rectangular and non-rectangular hyperbolas, and 2) the asymptotic negative-exponential function.

"Farquhar & von Caemmerer"-type models

A central paper is the one published by Farquhar, von Caemmerer and Berry (1980). Leaf photosynthesis is described as either Rubisco-limited or RuP_2 regeneration limited. The carboxylation rate as determined by Rubisco-limitation is:

$$P_c = V_{cmax} \frac{C_i - \Gamma^*}{C_i + K_c(1 + O/K_o)} \qquad \text{(Eq. 2.2.5)}$$

where C_i and O are the intercellular concentrations of CO_2 and O_2, respectively, Γ^* is the CO_2 compensation point in absence of dark respiration, K_c and K_o the Michaelis-Menten constants for binding of CO_2 and O_2 to Rubisco, respectively, and V_{cmax} the maximal rate of carboxylation. The carboxylation rate as limited by RuP_2 regeneration was described by, as one of several slightly different formulae,

$$P_i = J \frac{C_i - \Gamma^*}{4C_i + 8\Gamma^*} \qquad \text{(Eq. 2.2.6)}$$

where J is the electron transport rate. J has been modelled to saturate with light intensity according to a rectangular (Farquhar & Von Caemmerer, 1982) or a non-rectangular hyperbola (Farquhar & Wong, 1984). The maximal rate of electron transport, J_{max}, is temperature dependent, the optimum being at about 30 °C.

The actual rate of net leaf photosynthesis is determined by the limiting process

$$P_n = 0.044 \min\{P_c, P_j\} - R_d \qquad \text{(Eq. 2.2.7)}$$

In further model developments, Pi regeneration has often been added as a third limiting process (e.g. Sage *et al.*, 1990).

These types of models still leave a number of parameters as estimates. K_c and K_o are biochemical constants and are assumed, for a given temperature, to be species dependent. There is considerable variation in the values of K_c and K_o reported in the literature (Kirschbaum & Farquhar, 1984). Acclimation of leaf photosynthesis to various conditions could be modelled by changing the values of V_{cmax} and J_{max} (Farquhar & Von Caemmerer, 1982).

Note that at high CO_2 concentrations the RuP_2 regeneration and Pi regeneration will be more limiting to CO_2 assimilation, so these subprocesses need special attention when modelling the effect of CO_2 enrichment on greenhouse crops.

Rectangular and non-rectangular hyperbolas

The non-rectangular hyperbola as used, for example, for the light response curve, has the following form

$$P_g = \frac{\alpha I + P_{gmax} - \{(\alpha I + P_{max})^2 - 4\theta\alpha I P_{max}\}^{1/2}}{2\theta} \qquad \text{(Eq. 2.2.8)}$$

where I is the light intensity, α is the initial slope, called the light use efficiency, and P_{gmax} the maximal rate of gross photosynthesis (Thornley, 1976). Parameter θ describes the degree of curvature and is confined to the range 0 to 1. When θ is 0, a rectangular hyperbola is obtained,

$$P_g = (\alpha I P_{gmax}) / (\alpha I + P_{gmax}) \qquad \text{(Eq. 2.2.9)}$$

and when $\theta = 1$, a Blackman-curve is obtained. Parameter θ gives this curve considerable flexibility in describing leaf responses. The non-rectangular form has been applied to the light response of tomato (Longuenesse *et al.*, 1993). The rectangular form has been used to describe the light response of tomato (Acock *et al.*, 1978; Jones *et al.*, 1988; Longuenesse, 1990; Tchamitchian, 1990) and sweet pepper (Acock *et al.*, 1975).

P_{gmax} can be modelled as a product of CO_2 concentration and a conductance to CO_2 transfer, τ. In this case, with the rectangular form, the combined response to light and CO_2 response becomes

$$P_g = (\alpha I C \tau) / (\alpha I + C \tau) \qquad \text{(Eq. 2.2.10)}$$

(Thornley, 1976). The rectangular form, though attractive due to its simplicity, has the drawback that in many cases it saturates too slowly (Jones, 1983), so that it would overestimate the initial slope and the asymptote when fitting data (Acock *et al.*, 1978). However, its simple form enables an analytical solution when integration is carried out over the entire canopy. Equation (2.2.10) was used by Acock *et al.* (1978) to describe tomato canopy photosynthesis. The value of a depends on temperature and CO_2 concentration. Its dependency on CO_2 was modelled by Charles-Edwards & Ludwig (1974) and Thornley (1976).

The negative-exponential response curve

The asymptotic negative-exponential function can give a good to fit to the measured light response of leaves (Spitters, 1986). The curve is characterized by the initial light use efficiency at low light intensity (α_i) and the maximal rate of photosynthesis at high light intensity, P_{gmax}:

$$P_g = P_{gmax} \{1 - \exp(-\alpha_i I / P_{gmax})\} \qquad \text{(Eq. 2.2.11)}$$

Temperature and CO_2 affect the values of α_i and P_{gmax}. With field crops a temperature dependent function of P_{gmax} is often applied. The effects of CO_2 and temperature on α_i and P_{gmax} have been modelled by Goudriaan *et al.* (1985).

Note that parameters of response curves of P_g to a given climatic variable are not interchangeable between various mathematical descriptions.

2.2.1.5 Light interception by the canopy

The amount of PAR intercepted by the canopy is equal to the incident PAR minus the amount that reaches the ground. Part of the intercepted PAR is reflected, and the remaining part is absorbed. Crop photosynthesis strongly depends on the amount of absorbed PAR. However, the distribution of absorbed radiation within the canopy affects the efficiency with which absorbed PAR is used for assimilation. This is caused by the non-linear response of leaf photosynthesis to light. The optimal distribution would be such that all leaves absorb an equal amount of PAR, so that individual leaf photosynthetic rates would be closest to the initial light use efficiency.

Several canopy and greenhouse characteristics influence the amount of and distribution of absorbed PAR, as will be discussed below.

Leaf Area Index (LAI)

The amount of PAR intercepted by the canopy is strongly dependent on the Leaf Area Index (LAI, m^2 leaf area per m^2 ground area) at low indices. With a LAI of 3, theoretically about 90% of the PAR is intercepted for a canopy that is fully closed, with thin leaves that have random leaf angle distribution and that take no preferential position to each other. At higher LAI's canopy photosynthesis increases little with further increases in total leaf area.

Leaf angle distribution

The extinction of PAR also depends on the orientation of leaves. In light interception models a random (also called spherical) leaf orientation is commonly assumed. With tomato (Tchamitchian, 1990) and cucumber (Shell *et al.*, 1974; E.M. Nederhoff, Glasshouse Crops Research Station, unpublished results) a more horizontal leaf angle distribution has been found. With more horizontally oriented leaves light is intercepted more quickly than with more vertical angles. With low light levels and low LAI's this can enhance crop photosynthesis by increasing total light interception, but with higher LAI's and high light levels this effect can be offset by the stronger absorption in the upper layers of the canopy, causing a less favourable distribution of light.

Light scattering

Leaves transmit radiation, but also reflect it due to reflection by the cuticula and as a result of multiple reflections within the leaf. An average thin leaf absorbs about 80 to 90% of PAR. With increased scattering of leaves, radiation penetrates deeper into the canopy, thereby causing a more even distribution within the canopy, but also increasing the proportion reflected by the canopy as a whole, and increasing the proportion reaching the ground. A closed canopy with spherical leaf angle distribution, without interference from the ground, would reflect about 5% of the incident PAR.

Model calculations indicate that variations in the amount of scattering by leaves have little effect on canopy PAR absorption (Goudriaan, 1977).

Extinction coefficient

An important parameter in light interception models is the extinction coefficient for diffuse light, K_{dif}, that appears in the calculations of diffuse light extinction

$$I_{dif} = I_{o,dif} \exp(-K_{dif} L_c) \qquad \text{(Eq. 2.2.12)}$$

where $I_{o,dif}$ is the diffuse PAR above the canopy and I_{dif} the intensity beneath partial LAI L_c. K_{dif} depends on the leaf angle distribution, on the scattering by individual leaves and on whether leaves position themselves preferentially with regard to each other (e.g. to prevent self-shading). For monocotyledonous crops K_{dif} has been found to vary from 0.4 to 0.7, and for dicotyledonous crops from 0.6 to 1.1 (Monteith & Unsworth, 1990). A canopy with random leaf orientation and leaves that scatter 15% of the absorbed light, theoretically has a K_{dif} of 0.74, according to Spitters (1986). For cucumber and tomato, that have canopies with more horizontal leaf angle distributions, K_{dif} should theoretically be 0.85 (Gijzen, 1992). Acock *et al.* (1978) observed for tomato a K_{dif} of 0.6.

Diffuse and direct light

As direct and diffuse PAR have different extinction profiles in the canopy, the amount of and distribution with depth of absorbed light in the canopy will change with the partitioning between diffuse and direct light, and with solar elevation (section 3.3.2). Furthermore, direct PAR causes an unequal distribution of PAR within a given canopy layer, depressing the average rate of photosynthesis in that leaf layer.

Canopy structure

Many greenhouse crops are grown in rows alternated with paths. This results in an altered light distribution in the canopy and a decreased light interception. Row effects were modelled by, among others, Gijzen & Goudriaan (1989) and Tchamitchian (1990). Model studies (Gijzen & Goudriaan, 1989) indicate that, under diffuse light conditions, the effect of path width on crop photosynthesis becomes important when row height is less than the row distance. For example, with row height equal to row distance and with LAI of 3, calculated total absorbed PAR was reduced by 13% when path width was equal to row width. Under direct light conditions, crop photosynthesis can be considerably depressed when the azimuth of the sun (angle with N-S direction) is approaching that of the row. The length of the period of the day when this occurs depends largely on the height of the crop and the path width.

For Dutch cultivation practice it was calculated that, for a fully developed canopy, assuming LAI at 3, and with row height being typically 2.25 m and row width 1.25 m, at a row distance of 1.60 m, daily CO_2 assimilation was on average about 5% less compared with a closed canopy (unpublished results).

In certain greenhouses, notably where pot plants are grown, individual plants can be considered as solitary during a significant fraction of the early growth period. Light interception is then largely determined by the shape of the plant. Interception was modelled by Monteith & Unsworth (1990), by approximating the crowns to cones or spheres.

Ground reflection

The amount of PAR absorbed by the canopy can be significantly enhanced by the presence of reflecting material on the ground, such as white plastic sheets. Bare soil, tiles and concrete have PAR reflectivities in the range 0.1 to 0.2, whereas white plastic sheets, as used commonly in Dutch horticulture, have, when new, a PAR reflectivity of about 0.7–0.8. Experimental data indicate benefits in terms of increased early production (Sondern, 1962).

Sunlit and shaded areas

When the fraction of direct PAR is high, a pattern of sunlit and shaded areas arises on the canopy, due to shade cast by the construction of the greenhouse. This causes unevenness in the distribution of PAR, and hence a lower crop photosynthesis compared to where all direct PAR is assumed to be spread evenly over the crop surface.

For example, when the transmission of direct PAR of the construction and the glass at a given time are 0.8 and 0.9, respectively, 80% of the crop area receives 90% of the level of direct PAR outside the greenhouse, and 20% of the crop receives only diffuse PAR. Using a single transmission factor for direct PAR of $0.8 \times 0.9 = 0.72$ for the whole crop means that all sunlit leaves receive 72% of the direct PAR ouside the greenhouse.

2.2.1.6 Responses of canopy photosynthesis

Not many reports have been made on photosynthesis of whole crops in the greenhouse. Many investigations have been carried out with one of several plants in a small assimilation chamber, so that extrapolation of results to a real closed canopy where light distribution and absorption are completely different, is almost impossible. Crop photosynthesis has been measured, using small cabinets or enclosures, for tomato by Acock *et al.* (1978), Jones *et al.* (1988) and Tchamitchian (1990), for sweet pepper by Acock *et al.* (1975), for rose by Hand & Cockshull (1975) and for chrysanthemum by Acock *et al.* (1979). Crop photosynthesis using whole greenhouse compartments has been measured by Nederhoff & Vegter (1994) for cucumber, tomato and sweet pepper.

Use of a simulation model

As environmental conditions and canopy characteristics varied widely between photosynthesis experiments referred to, it is difficult to obtain a clear picture of specific responses. Therefore, the responses of crop photosynthesis to climatic conditions and crop characteristics are discussed using results of a simulation model of canopy gross photosynthesis (see Box 2.2.1).

An example of the performance of this model is shown in Figure 2.2.3, where simulated P_{nc} is compared with measured P_{nc} of a tomato crop, at the Glasshouse Crops Research Station at Naaldwijk, The Netherlands, on 18 April 1989.

In addition, with the model the daily total of gross photosynthesis at 1 May was calculated with artificially generated climatic conditions, and the effects of changed climatic conditions or crop characteristics on the daily total were assessed. The simulated diurnal patterns of PAR inside the greenhouse and crop gross photosynthesis, P_{gc}, are shown in Figure 2.2.4.

Response to incident PAR

Simulated gross photosynthesis of a canopy with LAI of 3, in contrast to individual leaves, did not show saturation up to a PAR intensity of 2000 μmol m^{-2} s^{-1} photons, as part of the canopy had still not reached light saturation (Figure 2.2.5a). At lower LAI's the response was more strongly curved, as a larger portion of the canopy attained PAR saturation at the higher PAR levels (Figure 2.2.5b).

The response of a canopy with "shade"-leaves was simulated by halving both V_{cmax} and J_{max} ; this decreased maximal leaf gross photosynthesis from 0.9 to 0.54 mg m^{-2} s^{-1} (quantum efficiency remained the same). This canopy responded much less to PAR (Figure 2.2.5c).

Box 2.2.1 Simulation model of canopy gross photosynthesis.

The model was a partly modified version of the model for greenhouse crop photosynthesis from Gijzen, 1992. It consisted of submodels for the calculation of the fraction diffuse in global radiation, of the transmission of the greenhouse cover for diffuse and direct light (according to Bot, 1983), and of canopy light absorption as described by Spitters *et al.* (1989). The submodel for leaf photosynthesis was replaced by a biochemical model (Farquhar *et al.*, 1980). Calculations on the K_m of Rubisco were carried out according to Kirschbaum & Farquhar (1984), the dependence of the rate of electron transport on light intensity, following Farquhar & Wong (1984). Maximal carboxylation velocity, V_{cmax}, and maximal rate of electron transport, J_{max}, at 25 °C were assumed to be 100 μmol m^{-2} s^{-1} and 200 μeq. m^{-2} s^{-1}, respectively. Stomatal and boundary layer resistances (to H_2O) were 50 and 100 s m^{-1}, respectively. Daily gross CO_2 assimilation was calculated for a crop with LAI at 3, and with spherical leaf angle distribution, under a Venlo glasshouse cover with diffuse light transmissivity of 65% (Gijzen, 1992). Simulations were made for 1 May at latitude 52°, with a daily total of global radiation of 25 MJ, 75% of which was diffuse. Greenhouse air temperature was assumed to follow a sinusoidal course, temperature being 18 °C at sunrise and sunset, and 25 °C at noon.

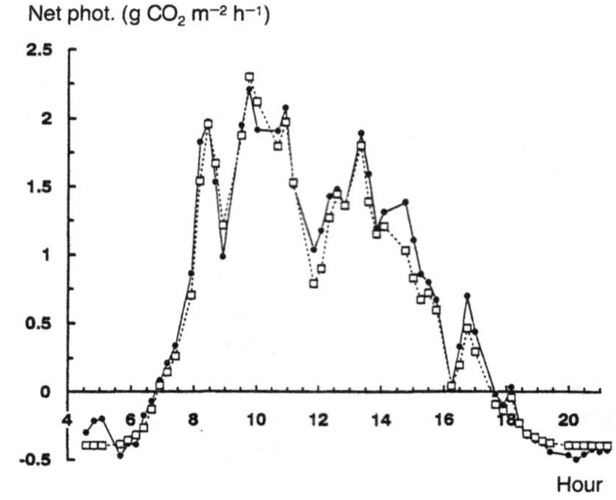

Figure 2.2.3 – The measured (—●—) and simulated (---□---) rate of crop net photosynthesis (g m^{-2} h^{-1}) of a tomato crop at the Glasshouse Crops Research Station at Naaldwijk, The Netherlands, at 18 April 1989. For the tomato crop V_{cmax} and J_{max} were estimated at 150 μmol m^{-2} s^{-1} and 300 μeq. m^{-2} s^{-1}, respectively. On this day the average CO_2 concentration was about 240 μmol mol^{-1}.
The large fluctuations in P_{nc} resulted from strong fluctuations in PAR during the day.

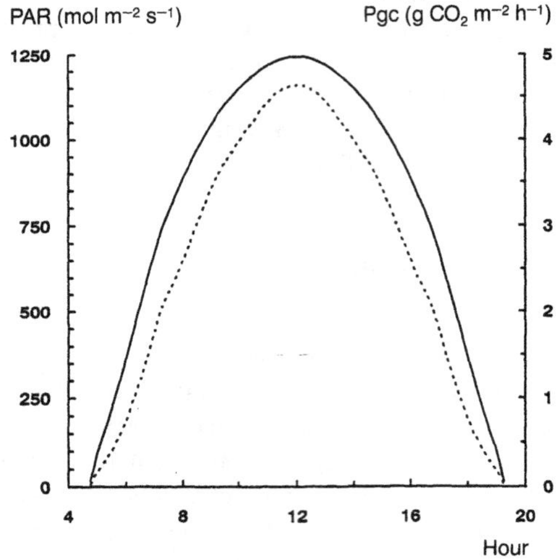

Figure 2.2.4 – The simulated diurnal patterns of PAR (μmol m^{-2} s^{-1}, -------) inside the greenhouse and canopy gross photosynthesis (P_{gc}, g m^{-2} h^{-1}, ———) at 1 May, assuming a total daily global radiation of 25 MJ m^{-2}. See Box 2.2.1 for further details on the simulation.

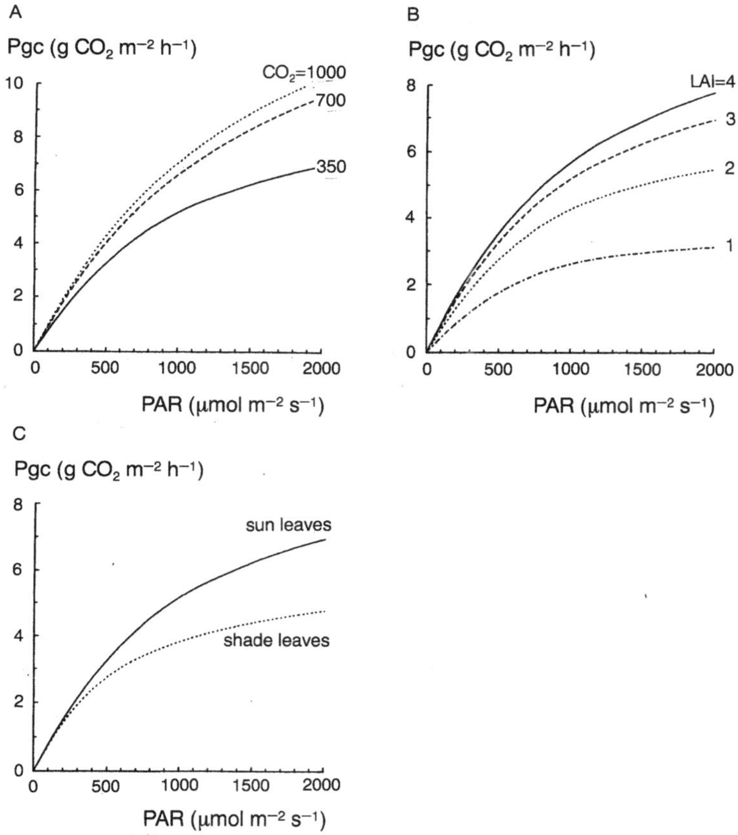

Figure 2.2.5 – The simulated response of canopy gross CO_2 assimilation to incident light intensity (PAR, μmol m^{-2} s^{-1}). K_{dif} was 0.74, temperature was 25 °C, solar elevation was assumed to be 45°, and fraction diffuse in PAR to be 0.50. (A) LAI=3; response for CO_2 concentration of 350, 700 and 1000 μmol mol^{-1}. (B) CO_2 concentration 350 μmol mol^{-1}; response for LAI's of 1, 2, 3 and 4. (C) LAI=3, CO_2 concentration 350 μmol mol^{-1}; response for sun leaves (V_{cmax} = 100 μmol m^{-2} s^{-1}, J_{max} = 200 μeq. m^{-2} s^{-1}) and shade leaves (V_{cmax} = 50 μmol m^{-2} s^{-1}, J_{max} = 100 μeq. m^{-2} s^{-1}).

The quantum efficiency f_a has a relatively large effect on crop photosynthesis, even on bright days, and obviously the more so on cloudy days. Model results indicate that a 10% increase of f_a (0.051 mol CO_2 mol^{-1}) increased daily crop gross photosynthesis by 5% on a clear day, and by 8.8% on a cloudy day (Table 2.2.1). (Note that with another mathematical description of the photosynthesis light-response curve a different effect of the parameter for the initial light use efficiency of leaf photosynthesis on crop photosynthesis will be obtained.)

A canopy with leaves with a smaller photosynthetic capacity lower in the canopy was simulated by assuming that both V_{cmax} and J_{max} decreased linearly with height in the canopy to half their values at the top. Simulated daily photosynthesis was decreased by only 6% (Table 2.2.2).

Table 2.2.1 – The effect of changing climate and crop characteristics on simulated daily crop gross CO_2 assimilation ($P_{gc,d}$, g CO_2 m^{-2} d^{-1}) at 1 May. The clear day was assumed to have 25 MJ m^{-2} of global radiation, resulting in 75% of daily PAR being direct. The cloudy day was assumed to have 7.5 MJ m^{-2} global radiation, being totally diffuse. For other details see Box 2.2.1.
When not indicated otherwise, the reference run was for a clear day, with LAI at 3, K_{dif} at 0.74, zero ground reflection, and with V_{cmax} at 100 µmol m^{-2} s^{-1} and J_{max} at 200 µeq. m^{-2} s^{-1}.

Reference run	$P_{gc,d}$	New run	$P_{gc,d}$	Percentage
All leaves same V_{cmax} and J_{max}	48.3	Decrease in V_{cmax} and J_{max} to 50% at bottom of canopy	45.3	93.7
f_a = 0.051 mol CO_2 mol $^{-1}$	48.3	f_a = 0.056 mol CO_2 mol^{-1}	50.7	104.9
f_a = 0.051 mol CO_2 mol^{-1}, cloudy day	22.7	f_a = 0.056 mol CO_2 mol^{-1}, cloudy day	24.7	108.8
Reflection ground 0.0, LAI=2	40.7	Reflection ground 0.5, LAI=2	45.6	112.1
Reflection ground 0.0, LAI=3	48.3	Reflection ground 0.5, LAI=3	51.9	107.4
Direct and diffuse PAR	48.3	All PAR diffuse	59.6	123.3
Average direct PAR	48.3	Sunlit and shaded crop area	43.7	90.4
Spherical leaf angle*	48.0	Plagiophile leaf angle*	48.0	100.0
Spherical leaf angle, cloudy day*	22.1	Plagiophile leaf angle, cloudy day*	23.1	104.3
K_{dif} = 0.74, LAI = 3	48.3	K_{dif} = 0.6, LAI = 3	51.3	106.2
K_{dif} = 0.74, LAI = 2	40.7	K_{dif} = 0.6, LAI = 2	42.0	100.3
g_b = 0.01	48.3	g_b = 0.005	44.6	92.2
g_b = 0.01, cloudy day	22.3	g_b = 0.005	22.3	98.7

* In these runs diffuse light extinction was calculated for all angles of incidence of individual beams, instead of assuming a single K_{dif} for the diffuse light flux.

Table 2.2.2 – The calculated light use efficiency of crop gross photosynthesis (P_{gc}) at various PAR levels (400–700 nm) and the efficiency of artificial light, calculated as mol CO_2 taken up per mol photons. Artificial light was given at a level of 50 µmol m^{-2} s^{-1} diffuse PAR, at various natural PAR levels. Photosynthetic characteristics of "sun" leaves (V_{cmax} = 100 µmol m^{-2} s^{-1} and J_{max} = 200 µeq. m^{-2} s^{-1}) and "shade" leaves (V_{cmax} = 50 µmol m^{-2} s^{-1} and J_{max} = 100 µeq. m^{-2} s^{-1}) were assumed.

Natural PAR	Canopy with sun leaves		Canopy with shade leaves	
(µmol m^{-2} s^{-1})	Efficiency natural PAR	Efficiency artificial light	Efficiency natural PAR	Efficiency artificial light
0	- *	0.052	- *	0.051
100	0.050	0.048	0.044	0.045
200	0.048	0.045	0.043	0.039
500	0.041	0.035	0.034	0.026

* - = not applicable.

Effects of altered PAR interception and absorption

The effect of leaf angle distribution has been simulated to be smaller. A fairly horizontal leaf angle distribution, called the plagiophile leaf angle by De Wit (1965), which seems to represent that of young leaves of tomato and cucumber has been used in assessing the effects on simulated crop photosynthesis. Under clear weather conditions daily crop photosynthesis is virtually the same for the spherical and plagiophile leaf angle distributions (Table 2.2.1). Under cloudy conditions daily crop photosynthesis increases by 4% for the plagiophile leaf angle distribution; with a LAI of 2 it was increased by 6% (not shown).

The extinction coefficient does affect crop photosynthesis significantly with a LAI of 2, but not with a LAI of 3. By decreasing K_{dif} from 0.74 to 0.6 daily crop photosynthesis increases by 6% and 0.3%, with LAI's of 3 and 2 respectively.

Assuming PAR to be totally diffuse had a considerable effect. The simulated daily crop photosynthesis increased by 23% (in the reference run the fraction diffuse in daily PAR was 0.25). Greenhouse transmission changed very little, so the effect was only on total PAR interception and on PAR distribution within the canopy.

A ground reflection of 0.5 instead of 0.0 enhanced crop photosynthesis by about 12%, for a LAI of 2, and for a fully developed canopy with a LAI of 3, still by about 7% (Table 2.2.1).

The partitioning of the crop into sunlit and shaded areas was calculated to have a significant effect under clear weather conditions. Calculated daily crop photosynthesis decreased by 10% when unevenness was taken into account.

Response to CO_2

Instantaneous crop gross photosynthesis increased more with increasing CO_2 concentration at higher light intensities (Figure 2.2.6a). Also with higher temperatures the effect of enhanced CO_2 increased, up to about 30 °C (i.e. the optimum temperature for the maximal electron transport rate, J_{max}, as assumed in this version of the model) (Figure 2.2.6c). P_{gc} increased by 24% when the CO_2 concentration was doubled from 350 to 700 μmol mol^{-1} at a PAR intensity of 500 μmol m^{-2}, and by 32% at a PAR intensity of 1500 μmol m^{-2} (Figure 2.2.6a). An increase in CO_2 concentration from 350 to 1000 μmol mol^{-1} increased crop photosynthesis by 33 and 43%, at 500 and 1500 μmol m^{-2} PAR, respectively.

Observed increases in production due to increased CO_2 are comparable to these calculated increases. During a more or less steady-state production stage, increased production is probably to a large extent a result of increased canopy photosynthesis. With tomato, Yelle *et al.* (1990) found a production increase of 21% for 4 weeks of fruit growth when CO_2 concentration was increased from 330 to 900 μmol mol^{-1}, and Slack *et al.* (1988) reported, for about 20 weeks of harvest, an increase of 16%, when CO_2 concentration was increased from 350 to 450 μmol mol^{-1}. Total yields of a cucumber spring crop from an 18-week production period were increased by 32% and by 38% by increases in CO_2 concentration from 330 to 660 and from 330 to 900 μmol mol^{-1}, respectively (E.M. Nederhoff, Glasshouse Crops Research Station, personal communication).

Response to temperature

Simulated crop photosynthesis is little sensitive to air temperature (Figure 2.2.7). I.e. simulated crop gross photosynthesis increased only by 6% when temperature was increased from 20 to 30 °C, with CO_2 at 350 μmol mol^{-1} and PAR at 1500 μmol m^{-2} s^{-1}. Only under conditions of high light and high CO_2, a combination not likely to occur often in the greenhouse, was P_{gc} significantly affected by temperature. The optimal temperature was calculated to be higher at greater PAR intensities and higher CO_2 concentrations.

Other factors influencing canopy photosynthesis

Boundary layer conductance

The effect of a low boundary layer conductance (to H_2O) was investigated by halving g_b from 0.01 to

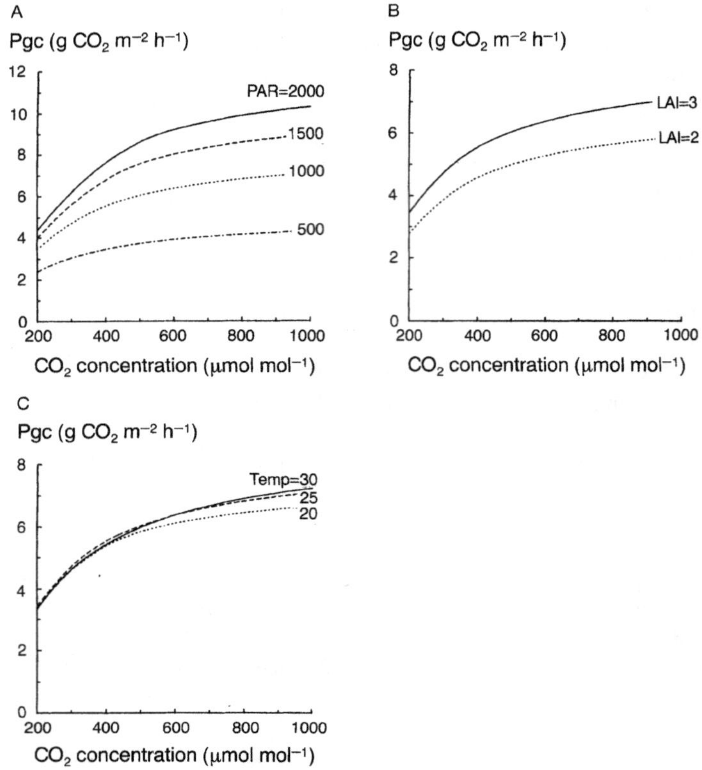

Figure 2.2.6 – The simulated response of canopy gross CO_2 assimilation to CO_2 concentration. Conditions as in Figure 2.2.5. (A) LAI=3; response for PAR at 500, 1000, 1500 and 2000 µmol m^{-2} s^{-1}. (B) PAR of 1000 µmol m^{-2} s^{-1}, CO_2 concentration 350 µmol mol^{-1}; response for LAI at 2 and LAI at 3. (C) PAR at 1000 µmol m^{-2} s^{-1}, CO_2 concentration 350 µmol mol^{-1}; response for air temperature at 20, 25 and 30 °C.

0.005 m s^{-1}. This decreased total (boundary + stomatal) conductance for CO_2 from 0.0046 to 0.0028 m s^{-1}. Daily crop CO_2 assimilation was calculated to have decreased by 8% during a clear day at 1 May, and by 1% on a cloudy day (Table 2.2.1). This indicates that the boundary layer can have a significant effect on crop CO_2 assimilation.

Stomatal conductance

In the present model g_s was assumed to be constant, although g_s increases with light intensity, and a value was adopted that occurs more frequently at high light levels and high humidities (i.e. 0.02 m s^{-1}). A reduction in total leaf conductance of CO_2 equal to halving g_b (see above) would be obtained by lowering g_s from 0.02 to 0.0074 m s^{-1} (i.e. increasing resistance from 50 to 135 s m^{-1}). This change in g_s is not particularly large, which indicates that P_{gc} can be significantly reduced when adverse conditions lead to stomatal closure.

Artificial light

Artificial light aimed at enhancing photosynthesis (section 4.7) is given in low amounts at times when natural daylight is low or absent. At these times its efficiency is maximal, and will be, in most

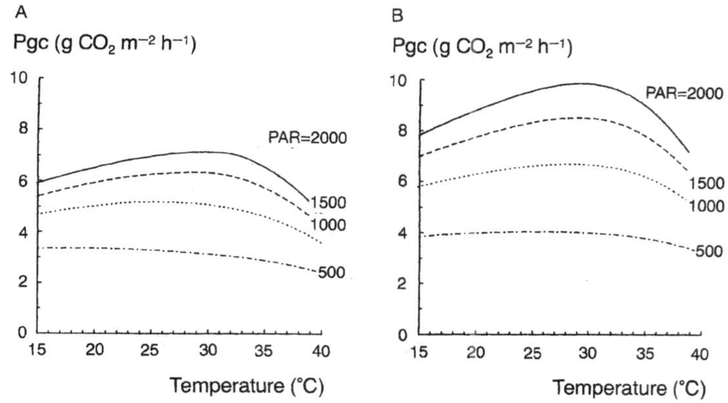

Figure 2.2.7 – The simulated response of canopy gross CO_2 assimilation to temperature. Conditions as in Figure 2.2.5. Response for PAR at 500, 1000, 1500 and 2000 µmol m^{-2} s^{-1}. (A) CO_2 concentration 350 µmol mol^{-1}. (B) CO_2 concentration 700 µmol mol^{-1}.

cases, the same as from natural daylight. However, there exist small variations in the photosynthetic effectiveness of photons of different wavelengths (McCree, 1972; Evans, 1987). Red photons (600 nm) are the most effective; their quantum yield is about 10% higher than of the average photon in white light (Evans, 1987). Tikhomirov *et al.* (1987) found that in plants grown under part of the PAR spectrum, quantum efficiency was changed compared to plants grown under white light. At low light intensities CO_2 enrichment increases both photosynthesis and the efficiency of artificial lighting by decreasing photorespiration.

The efficiency of artificial lighting was estimated by calculating the increase in crop gross photosynthesis at a given PAR level, ΔP_{gc}, when 50 µmol m^{-2} s^{-1} artificial PAR was added as a diffuse flux (ΔPAR). The efficiency of this artificial PAR was calculated as ΔP_{gc} / ΔPAR (i.e. mol CO_2 per mol photons). Thus the efficiency is about equal to the slope of the crop photosynthesis–PAR response curve. It was calculated that the efficiency equalled 0.052 mol mol^{-1}, when given without natural light (Table 2.2.2). This efficiency is equal to the quantum efficiency of individual leaves. At natural PAR levels of 100 and 200 µmol m^{-2} s^{-1}, the efficiency decreased by 7 and 13%, respectively. The efficiency of artificial light increased by 19% when the CO_2 concentration was increased to 700 µmol mol^{-1}. At a time of the year when average light levels are low, leaves may be acclimated to lower PAR levels. The effect of 50 µmol m^{-2} s^{-1} artificial light was simulated for a canopy with shade leaves (V_{cmax} = 50 mg CO_2 m^{-2} s^{-1}, J_{max} = 100 meq. m^{-2} s^{-1}). The efficiency of artificial light without natural PAR was 0.051 mol mol^{-1}, and decreased by 11 and 23% when given at natural PAR levels of 100 and 200 µmol m^{-2} s^{-1}, respectively (Table 2.2.2).

Screening

Screening is sometimes used to diminish the radiation load on the crop at peak radiation levels, in order to improve quality (section 4.5). This decreases the amount of PAR received by the crop and is likely to decrease crop photosynthesis also. It was calculated that the effects on P_{gc} were not large. For example, a bright day at 1 June was simulated (30 MJ m^{-2} global radiation), with peak global radiation intensity of 900 W m^{-2} at noon. Cutting off all radiation above 700 W m^{-2} decreased daily PAR receipt by 9%, and decreased calculated daily crop photosynthesis by 3.5%. However, other factors, such as increased humidity under the screen, may affect overall crop performance and dominate the effects on photosynthesis.

2.2.1.7 Crop respiration

Crop respiration can be divided into respiration for growth and for maintenance of the plant. The CO_2 efflux from growth respiration results from synthesis of plant dry matter from assimilates and nutrients, or from intermediate products. The CO_2 efflux from maintenance respiration results from the combustion of carbohydrates, needed to deliver the energy to maintain ionic balances within the plant, resynthesis of enzymes, repair processes, etcetera. The rate of CO_2 efflux from growth respiration, R_g, is directly coupled to the rate of growth of the crop, and is dependent on the chemical composition of dry matter being formed

$$R_g = \Delta W/dt\, C_{pf} \qquad \text{(Eq. 2.2.13)}$$

where C_{pf} is the amount of CO_2 released per g dry matter formed (CO_2 production factor). C_{pf} is not dependent on temperature. For cucumber and tomato it was calculated to be about 0.4 g CO_2 per g crop dry matter (unpublished results). Note that the daily rate of CO_2 release by respiration is not equal to $(1-C_f) \times (30/44\, P_{gc,d} - R_{m,d})$ (equation (2.2.1)).

Maintenance respiration is correlated with the weight of the crop and its metabolic activity. It is often calculated as the product of dry weight times a maintenance coefficient (g CH_2O needed per g dry matter (DM) per day), being higher for leaves than, for instance, for stems or fruits.

The CO_2 efflux from respiration probably contributes most of the time significantly to the total CO_2 exchange of the crop. This is illustrated by the example of a tomato spring crop grown in a Venlo glasshouse at the Glasshouse Experimental Research Station at Naaldwijk, The Netherlands, in 1989 (unpublished results). Tomato crop growth was averaged 11 g DM per m^2 per day. Maintenance respiration of the crop was estimated to be 5.7 g CH_2O m^{-2} d^{-1} (at 20 °C), based on an average crop dry weight of 400 g m^{-2}, and a maintenance coefficient of 2 g carbohydrate per 100 g DM per day (at 25 °C). Then, from equation (2.2.1), and assuming C_f to be 0.7 g DM per g CH_2O, average $P_{gc,d}$ was calculated to be 31.3 g m^{-2} d^{-1}. With C_{pf} estimated at 0.4, calculated daily CO_2 released by growth respiration was 4.4 g per m^{-2}. Thus, when it is assumed that growth and maintenance respiration continue at night-time at the same speed as at daytime, and daylength equals 12 hours, daytime CO_2 release by respiration is about $(4.4 + 5.7 \cdot {}^{44}/_{30}) / 2 / 31.3 \times 100 = 20\%$ of the daily canopy gross CO_2 assimilation.

The metabolic activity of the crop is probably higher at daytime than at night-time due to higher temperatures and possibly due to higher carbohydrate availability. However, little is known about the diurnal pattern of respiration. Canopy gross CO_2 assimilation is not so much affected by temperature (see above). Then, when $P_{gc,d}$ is assumed to be increased by 5%, and when the rate of daytime respiration is assumed to be twice the night-time rate, crop day respiration would be 26% of daily canopy gross CO_2 assimilation.

These calculations indicate that for validation of models of crop gross photosynthesis, daytime respiration needs to be estimated with a fair degree of accuracy.

2.2.1.8 Conclusions

Crop net CO_2 uptake is the result of canopy gross photosynthesis minus crop respiration for growth and maintenance. Canopy photosynthesis is driven by PAR, and the CO_2 concentration strongly affects the efficiency with which intercepted PAR is used for crop photosynthesis. PAR, CO_2 concentration and temperature strongly interact in their effects on crop photosynthesis.

The response of crop photosynthesis to (air) temperature according to the Simulation model of canopy gross photosynthesis (Box 2.2.1), is relatively small.

Crop characteristics (such as LAI, presence of rows and path, leaf angle distribution and leaf maximal photosynthesis) and greenhouse properties (such as direct and diffuse light transmission and ground reflection) affect the amount of intercepted PAR and the efficiency with which it is used for crop photosynthesis. It has been calculated that the leaf boundary layer conductance may have a significant effect on crop photosynthesis.

CO_2 release by crop respiration contributes significantly to the net rate of CO_2 uptake by the canopy, and needs to be estimated or measured when validating simulations of canopy gross photosynthesis.

2.2.2 Water balance

P.A.C.M. VAN DE SANDEN

2.2.2.1 Introduction

The greenhouse and its climate control offer an opportunity to optimize indoor conditions with respect to the water status of the crop. It is not solely manifest water stress that affects crop growth and productivity. Within the range from full hydration to water stress different physiological processes have their own threshold and sensitivity to changing plant water status (Bradford & Hsiao, 1982) and therefore display a water status dependent contribution to the output of good quality produce. Photosynthesis, for instance, is affected primarily by water status related stomatal conductance (section 2.2.1), although direct effects have also been suggested (Farquhar *et al.*, 1989). Water status influences the extension of tissue and as such the development of leaf area and volume of fruit with impact on plant photosynthetic and transpiration rate and on the distribution of dry matter and fresh weight. Also, the dry matter content of the harvestable product is influenced. The rate and pattern of water flow in the plant influences growth and quality of produce since water acts as a conveyer for the distribution of nutrients such as nitrate, potassium and calcium. For instance, the occurrence of physiological disorders is related to this pattern (section 2.2.2.5).

In order to optimize the greenhouse environment, the anticipated output, in terms of rate or state of plant growth, development and yield of good quality produce with respect to the water status of the crop has to be known. The behaviour of the crop may be quite diverse in different species. For instance, with respect to humidity control Bakker (1991[a]) found high humidity to increase the leaf area of cucumber, while, in tomato, leaf area was reduced. In the case of cucumber its water status at high humidity promoted leaf elongation (Van de Sanden & Veen, 1992), in tomato the water flow related calcium supply to the elongating leaves was reduced (Adams, 1991).

This section will give an introduction to plant water status and water flow and to the relation between greenhouse climate and water status. It will also briefly consider the relationship between plant water status and the processes of stomatal conductance, tissue elongation and root pressure (important for the distribution of water and calcium). For comprehensive reviews the reader is referred to Slatyer (1967), Zimmermann & Steudle (1978), Jarvis *et al.* (1981), Barlow (1982), Lange *et al.* (1982), Tyree & Jarvis (1982), Boyer (1985), McIntyre (1987), Jones (1990).

The water balance

Water is taken up by the root system and lost through transpiring leaves. Evaporation from the leaves is the driving force for transfer of water across the plant (section 3.4.3.2). Only a minor proportion of the water taken up is used for growth. Water moves as a liquid in the xylem and is distributed throughout the plant entering or leaving cells across membranes. To tissues such as apices and fruits water is probably supplied not primarily through the xylem, but through the phloem. The fluxes of water in the plant are related as described by Boyer (1985):

$$U + Tr = H + G \quad \text{(Eq. 2.2.14)}$$

where U is water uptake, Tr flux for transpiration, H the flux to storage or the (reversible) hydration of the plant and G (irreversible) growth. In a steady-state condition or over a longer period of time the storage factor may be ignored and the relation becomes $U + E = G$. Each flux is governed by specific more or less physiologically controlled parameters and driving forces.

2.2.2.2 The status of water in the plant

Water status of a plant or tissue can be described by its water content and its water potential. Although there is some dispute on what measure would be physiologically most meaningful (Sinclair & Ludlow, 1985; Kramer, 1988; Passioura, 1988[b]; Schulze *et al.*, 1988; Boyer, 1989; Jones, 1990; Schulte, 1992), both are relevant to the understanding of plant water relations.

Water content

Water content is generally described relative to a "saturated" water content, the maximum amount of water the plant or tissue can hold, normally occurring in a stationary situation when transpiration rate is close to zero and water uptake rate is not limiting. This relative water content (RWC) is usually measured on (leaf) samples as

$$(W_{fresh} - W_{dry}) / (W_{turgid} - W_{dry}) \qquad \text{(Eq. 2.2.15)}$$

and expressed as a percentage, where W_{fresh} is fresh weight of the sample, W_{turgid} weight after floating on water for several hours, and W_{dry} is weight of oven-dried sample (Barrs & Weatherley, 1962). The RWC of greenhouse crops is usually between 80–100%. The tissue dry weight of greenhouse crops may vary from, for instance, 35% of the fresh weight for rose leaves (De Stigter & Broekhuysen, 1984) to as low as 3% for cucumber fruits (Marcelis, 1992[b]).

Water potential

The water potential (Ψ) is a thermodynamic expression of the energy status of water with units of kJ kg^{-1}. Usually the numerically equivalent unit of pressure, MPa, is used. The water potential of pure water at 25 °C and 0.1 MPa atmospheric pressure is set at zero. The water potential in living plant tissue can be differentiated into two main components, turgor potential (Ψ^p) and osmotic potential (Ψ^s). In special cases more components may be of significance. The components are related according to:

$$\Psi = \Psi^p + \Psi^s \qquad \text{(Eq. 2.2.16)}$$

Within cells Ψ^p is usually positive or zero and is the result of the pressure exerted by the water inside the elastic cell wall. Ψ^p in the xylem vasculature is usually negative, and sometimes positive (root pressure). Ψ^s is always negative and results from the amount of osmotically active solutes in the vacuole of cells, in the apoplast in cell walls or in the vasculature. By approximation (Slatyer, 1967):

$$\Psi^s = - c^s R_{gas} T \qquad \text{(Eq. 2.2.17)}$$

where c^s is concentration of solutes (moles cm^{-3}), R_{gas} the gas constant (8.31 J K^{-1} mol^{-1}) and T absolute temperature. In the apoplast c^s is usually low and therefore Ψ^s often is ignored.

The water potential of the environment of the plant or tissue must also be taken into account. The root environment in water culture for instance has a water potential equal to the osmotic potential of the nutrient solution, with a typical range of -0.03 to -0.3 MPa (a nutrient solution with electrical conductivity (EC) of 1 or 8 mS cm^{-1}). In soilless culture on substrate the actual root environment water potential will depend not only on the nutrient solution added to the substrate and on its water content, but also on the amount and distribution of nutrients already present and on water binding forces (matric potential) of the substrate itself.

The air surrounding the plant may be assigned a water potential using Raoult's Law (see Slatyer, 1967), with a typical value of -71 MPa at 25 °C air temperature and 80% Relative Humidity (RH). In the gaseous phase, however, water does not flow along a water potential gradient but is driven by the partial vapour pressure difference between leaf and surrounding air (section 3.4.3.2).

Höfler-Thoday diagram and tissue elasticity

In the plant tissue water potential and tissue water content are related as shown in the Höfler-Thoday diagram (Figure 2.2.8). Ψ^P diminishes as RWC decreases from fully saturated to the wilting point where $\Psi = \Psi^s$. The rate of turgor loss depends on the elasticity of the tissue. The relation is defined as (see Zimmermann & Steudle, 1978; Tyree & Jarvis, 1982)

$$\Delta\Psi^P = \varepsilon \, \Delta V/V \qquad \text{(Eq. 2.2.18a)}$$

where V is tissue volume of water. ε, the bulk elastic modulus, changes during growth and development of the tissue, but does not seem to respond to drought (Barlow, 1982). Using RWC rather than V, ε is defined as (Schulte & Hinckley, 1985):

$$\varepsilon = (d\Psi^P/dRWC) \times (RWC - RWC_a) \qquad \text{(Eq. 2.2.18b)}$$

where RWC_a is the apoplasmic volume of water in the tissue. ε and RWC_a are usually determined using pressure-volume analysis (Hellkvist *et al.*, 1974). At high RWC the relation between ε and Ψ^P is linear (Hellkvist *et al.*, 1974; Stadelmann, 1984; Schulte & Hinckley, 1985). On the basis of theoretical calculations Nilsson *et al.* (1958) presented the following relation between e and Ψ^P for biological tissue:

$$\varepsilon = 3.6 \times \Psi^P + 2.5 \qquad \text{(Eq. 2.2.18c)}$$

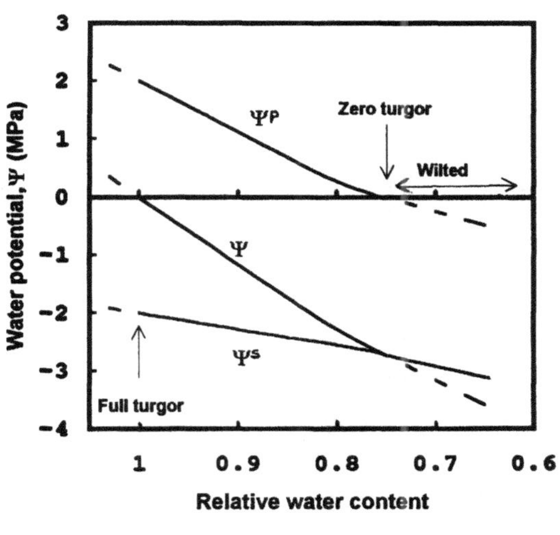

Figure 2.2.8 – Höfler-Thoday diagram illustrating the relationships between total water potential, turgor potential, osmotic potential and relative water content as a cell or tissue loses water from a fully turgid state. The dotted line below zero turgor represents possible negative turgor in rigid cells (from Jones, 1992).

Using a pressure probe Hüsken *et al.* (1978) measured values of ε of tissue cells of sweet pepper fruit from 0.2 to 2.5 MPa, depending on cell turgor and volume. Zimmermann & Steudle (1978) present several values of ε; for example for tomato leaves a typical value of 2.15 MPa is given. Recalculating data from Behboudian (1977[a], 1977[b]) yielded for leaves of cucumber, tomato, sweet pepper and eggplant an average ε of 6.9, 1.3, 2.8 and 1.4 MPa, respectively.

Although basically curvilinear, a linear relation between Ψ_{leaf} and RWC might suffice (Behboudian, 1977[a]; Marcelis, 1989). This relation might shift during the course of the day, due to changing Ψ^s, independent of RWC (Acevedo *et al.*, 1979). Significant cultivar differences may exist, as has been shown for lettuce. These differences are probably related to morphological differences between cultivars (Behboudian & Van Holsteijn, 1977).

2.2.2.3 Flow and distribution of water in the plant

Ohm's Law analogue

Analogous to electrical circuits, plant water flow is regarded as a network of potentials, resistances and capacitances (Figure 2.2.9). The Ohm's Law analogue is used to describe steady-state flow of water through the plant:

$$J_w = \Delta\Psi / R \qquad \text{(Eq. 2.2.19)}$$

where J_w is volume flux density of water and $\Delta\Psi$ water potential difference along a certain path with a liquid flow resistance R. Water will flow from high (less negative) to low (more negative) water potential. In a flowpath without water transfer across membranes, as in the xylem, water flow is driven only by the pressure gradient and J_w is proportional to $\Delta\Psi^p$ instead of $\Delta\Psi$. Following main paths

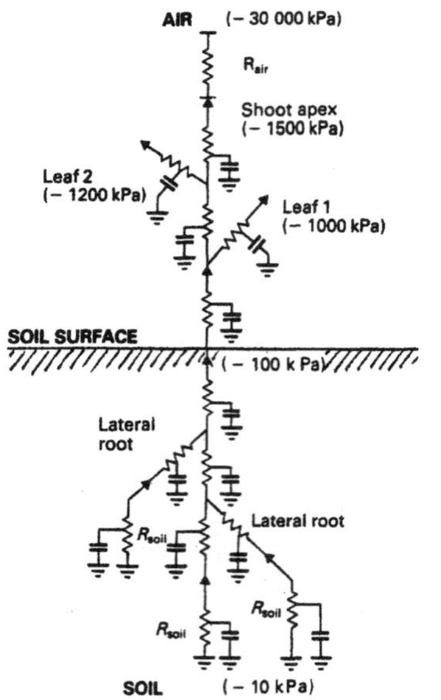

Figure 2.2.9 – Pathway of water movement from soil to air through a plant, showing resistances encountered in soil (R_{soil}), root, leaf and air (R_{air}). The capacitors represent the storage capacities of soil and plant parts. Figures show hypothetical fall in water potential in various parts of the system (from Sutcliffe, 1979).

of liquid flow to and through the plant may be discerned: the soil-to-root, root surface-to-root xylem, the xylem, xylem-to-tissue and the xylem-to-inner leaf evaporating surface path.

Hydraulic conductance

In plants rooted in soil the main source of resistance might be the movement of water towards the roots. In that case soil/root contact is important (De Willigen & Van Noordwijk, 1987; Passioura, 1988[a]). However, in the greenhouse in both soilless culture as well as in well managed soil culture the movement of water towards the roots should not be a limiting factor.

The resistance in the stem and root vascular tissue is considered to be relatively small (Passioura, 1988[a]). From data of Dimond (1966) a total vascular conductance of tomato over 16 internodes may be estimated as about 1 10^{-4} $cm^3\ s^{-1}\ MPa^{-1}$. It seems that the xylem as a whole might not be treated as equivalent. Dimond (1966) found that the driving pressure required to move water to the base of a petiole is considerably less than that which moves water through petioles. Ehret & Ho (1986[a]) and Lee *et al.* (1989) reported a significant hydraulic resistance in the petiole of the tomato fruit. Lee (1989), studying the vascular anatomy of tomato fruit petioles, identified a poor connection of xylem in the petiole as the cause of this high resistance. As observed by Ho *et al.* (1987) and Wolterbeek *et al.* (1987), the relative contribution of the phloem path in water import into tomato fruit is high, as was also found in apple (Lang, 1990). At high flow rate and/or restricted water supply xylem cavitation may occur, reducing plant hydraulic conductance (Jones & Sutherland, 1991). The phenomenon of cavitation is mainly reported for woody species.

From experiments with excised root systems it has been concluded, that the major resistance is within the root where water is transferred from the root surface to the inner stele (Jarvis, 1975), as has been demonstrated in tomato (Jensen *et al.*, 1961). Deposition of suberin in the cell walls of the endodermis of the root (the Casparian strip) effectively forces water to move through the cells across membranes rather than through the apoplast/cell wall pathway. De Willigen & Van Noordwijk (1987) mentioned values for root hydraulic conductance per unit root surface area ranging from 0.3 to 27 10^{-6} $cm^3/(cm^2\ s\ MPa)$ and specifically 5 10^{-6} $cm^3/(cm^2\ s\ MPa)$ for tomato and cucumber grown in rockwool. Root hydraulic conductance depends on root temperature. The temperature coefficient Q_{10} describes the ratio of a rate at a certain temperature to that at 10°C lower. In bean plants a Q_{10} of 4 was measured fot root water uptake below a critical temperature and 1.5 above (Kuiper, 1964). The critical temperature depends on environmental conditions during growth. Jensen & Taylor (1961) found a Q_{10} of around 1.5 for water flow through plant tissue or whole tomato plants.

There are numerous publications concerning the assumed variable nature of root hydraulic conductance. Barrs (1973) found that tomato plant liquid flow conductance per unit leaf area increased from 2.5 to 10 10^{-6} $cm^3/(cm^2\ s\ MPa)$ when transpiration rate increased from 0.4 to 1.6 $g\ m^{-2}\ h^{-1}$, partly preventing Ψ_{leaf} from falling to low levels. He found a similar response in bell pepper and suggested the site of variability to be situated in the roots. A simple mathematical approximation of the generally observed asymptotic response of R versus tranpiration rate is given by Jones (1978) for whole plant hydraulic resistance:

$$R_{soil-plant} = a / (1 + b \times E) \qquad \text{(Eq. 2.2.20)}$$

where E is transpiration rate and a and b are empirical constants. Fiscus (1975) and Fiscus *et al.* (1983) have explained apparent variations by coupling active solute flux and water flux in the roots. Using equations (2.2.19), (2.2.16) and (2.2.17) and $c^s_{xyl} = J_s/J_w$ (internal solute concentration dependent on relative rates of solute and water uptake) and assuming atmospheric pressure in the root environment ($\Psi^p_o = 0$) the expression for J_w across the soil-root interface may be rewritten as

$$J_w = (1/R_{root}) \{(\Psi^s_o - \Psi^p_{xyl}) + R_g T J_s / J_w\} \qquad \text{(Eq. 2.2.21)}$$

assuming the transfer of solute across ideal semi-permeable membranes, where Ψ^s is osmotic potential of the root environment, J_s the net solute influx and T absolute temperature. For tomato a typical value for J_s is 2.5 10^{-6} mol m^{-2} s^{-1} (Marcelis, 1989). Apparent changes in hydraulic resistance result from the transition from osmotic flow to pressure flow into roots, because at higher rate of J_w the weight of the second, osmotic term diminishes.

Part of the curvilinearity in the water potential-transpirational flux relationship might be explained by the fact that water-use for growth was not accounted for, which at very low transpiration rate may not be disregarded. With respect to cultivation in soil, Reid & Huck (1990) proposed a theory in which an increased amount of the root system contributes to water uptake the drier the soil becomes.

In studying root hydraulic conductance an accurate estimate of (effective) root surface area or mass is important, but hard to determine in vivo. According to Kaufmann & Fiscus (1985) it is reasonable to expect that the relative size of the absorbing, conducting, and transpiring tissues remains nearly constant under stable environmental conditions. Leaf area and cross-sectional area of vascular tissue seem to develop concomitantly, stabilising the pressure drop across the pathway (see Jarvis, 1975). So, as an alternative, plant hydraulic conductivity may be expressed on a leaf area basis.

Retarded growth of the roots might decrease their permeability, when, concomitantly, the endodermal suberisation rate is unhampered and the root volume available for uptake along the apoplast pathway is decreased. Not only water uptake but also uptake of essential nutrients, such as calcium, might suffer if root growth is restricted, since the major flux of Ca^{2+} towards the xylem takes place in the zone close to the root tip where suberisation of the endodermis is not yet completed (Drew, 1987).

Capacitance

Boyer (1974) located a source of nonlinearity of flow to potential gradient in the leaves. This apparent change in root resistance is the result of water potential dependent redistribution between storage and xylem which implies the existence of a resistance/capacitance network. He estimated the resistance from xylem to storage in a sunflower leaf at 0.97 10^6 MPa s cm^{-1}, around 30 times higher than its soil to leaf hydraulic resistance.

In plant water relations, capacitance is a measure of water holding capacity of tissue. In a system where there is only resistance, a sudden change in potential will cause an immediate flow response, whereas in a system with capacitance, dynamics tend to be damped. The plant behaves as a capacitance/resistance network (Figure 2.2.9) in which there is interchanging water volume between the main flow path and the tissue. So equation (2.2.19) will hold in steady-state conditions only. Capacitance should be included in dynamic modelling. Models including several resistances and capacitances in series or parallel are described by Cowan (1972), Boyer (1974), Powell & Thorpe (1977), Molz (1979) and several others.

The capacitance of a tissue is defined as (Jarvis *et al.*, 1981):

$$C_{tissue} = dV_{tissue}/d\Psi_{tissue} \qquad \text{(Eq. 2.2.22a)}$$

where V_{tissue} is the volume of water in the tissue. For small changes in volume the following holds (Molz & Ferrier, 1982)

$$C_{tissue} = V/(\varepsilon + \Psi^s) \qquad \text{(Eq. 2.2.22b)}$$

For example, the capacitance of apple leaves is 19 10^{-6} m^3 MPa^{-1} (Powell & Thorpe, 1977). As discussed above elasticity ε is not a fixed characteristic of tissue and hence capacitance varies with water and pressure potential. For several species a constant value, however, would be a reasonable approximation under normal conditions (Jones, 1978). Capacitance will also vary with variation in osmotic potential. Tissue is able to adjust to the water status by physiological control of the amount of osmotica in the vacuoles, which determines vacuolar Ψ^s and hence the capacitance of the tissue.

Using equation (2.2.22a) the flow of water in or out of tissue/storage may be written as

$$J_w = dV_{tissue}/dt = C\,d\Psi_{tissue}/dt \qquad \text{(Eq. 2.2.23)}$$

The response of Ψ_{tissue} in response to a step change in Ψ_{xylem} may be described as (Jarvis *et al.*, 1981):

$$\Delta\Psi_{tissue} / \Delta\Psi_{xylem} = 1 - e^{(-t/R\,C)} \qquad \text{(Eq. 2.2.24)}$$

The half-time for equilibration of Ψ_{tissue} is mostly in the order of 1 to 2 minutes, but the water exchange of some tissues is slower (Zimmermann & Steudle, 1978; Cosgrove, 1986).

2.2.2.4 Variation in water status in relation to greenhouse climatic factors

Dynamic behaviour of leaf water potential

The diurnal variation in Ψ_{leaf} closely reflects the diurnal variation in transpiration and in its main driving forces, radiation and VPD, in field grown (Rudich *et al.*, 1981; Rudich & Luchinsky, 1986) as well as in greenhouse grown tomato (Behboudian, 1977[a]; Marcelis, 1989; Sánchez-Blanco *et al.*, 1991).

A simple model for Ψ_{leaf} is the steady-state equation using the Ohm's Law analogue (equation (2.2.25))

$$\Psi_{leaf} = \Psi_{soil} - E \times R_{soil\text{-}plant} \qquad \text{(Eq. 2.2.25)}$$

which does not include any effects of variable hydraulic resistance, capacitance, coupled solute flow or volume flow for growth. Effects of the greenhouse climate act upon transpiration rate (E) and Ψ_{soil}, while a factor like temperature of the root environment might affect $R_{soil\text{-}plant}$. When E diminishes, Ψ_{leaf} will reach a maximum value, equal to Ψ of the root environment. In growing tissue, however, a small Ψ gradient will be maintained (Cosgrove, 1986). In the field the predawn Ψ_{plant} commonly approaches Ψ_{soil}. It is doubtful whether in a heated greenhouse the same will hold, since night-time transpiration might be considerable (De Graaf & Van den Ende, 1981; Seginer, 1990) due to the energy-input from the heating system. Behboudian (1977[a]) found a maximum Ψ_{leaf} (at $\Psi_{soil} = 0$) of -0.39, -0.43 and -0.15 MPa for young tomato, cucumber and sweet pepper plants, respectively. As the soil dried the drop in Ψ_{leaf} was most pronounced for sweet pepper, intermediate for tomato and the least for cucumber. Figure 2.2.10 from Schulze & Hall (1982) illustrates the relation of Ψ_{leaf} to transpiration rate and possible long-term changes. The relation deviates from linear because of changes in gaseous or liquid path conductance. The intercept, when Ψ_{leaf} approaches Ψ_{soil}, shifts to lower values because of water depletion in the root environment, a phenomenon more representative of soil grown plants, but which might also be relevant in water/substrate culture because of increased salinity of the root environment at a (prolonged) high transpiration rate. The increased sensitivity of Ψ_{leaf} to transpiration rate over a prolonged period of water stress may involve increased soil-root resistance (mainly in soil) or liquid path resistance (e.g. cavitation).

According to Jones (1990, 1992) the behaviour of many species can be approximated by a single resistance and capacitance:

$$d\Psi_{leaf}/dt = (\Psi_0 - \Psi_{leaf}) / (R_{plant}C) - E/C \qquad \text{(Eq. 2.2.26)}$$

Jones gives the mathematical solution of this equation as the time (t) dependent change in Ψ_{leaf} after a step change in E:

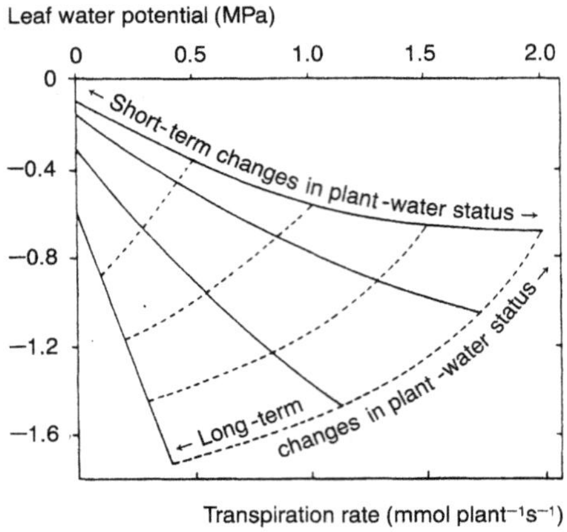

Figure 2.2.10 – Relations between leaf water potential and transpiration with increasing soil drought (after Schulze & Hall, 1982).

$$\Psi_{leaf} = A + B\,e^{(-t/\tau)} \qquad \text{(Eq. 2.2.27)}$$

with $A = (\Psi_0 - E\,R_{plant}) / (R_{plant}C)$
$B = \Psi^{initial} + R_{plant}E - (\Psi_0/R_{plant})$
$\tau = R_{plant}C$.

These equations show the dependence of the dynamic behaviour of Ψ on greenhouse climate as well as the possibilities of the plant to regulate its Ψ, through E (leaf area, stomatal response), R and/or C (Ψ^s and ε). The model might not be sufficient for relatively large indeterminate crops in a greenhouse environment, such as tomato and cucumber, where fruits are distributed along the length of the plant axis. Moreover a "lumped" model is not able to explain internal water distribution and related processes. However, simulating "lumped" Ψ_{plant} gave a reasonable fit to measured data in the greenhouse (Figure 2.2.11) (Bruggink *et al.*, 1988; Marcelis, 1989 & personal communication).

According to equation (2.2.27) the dynamic behaviour of Ψ_{leaf} is dependent on external factors such as Ψ_0 and those governing E, including radiation, temperature, windspeed and air humidity. Behboudian (1977[a]) used multiple regression analysis to derive a relation for the dynamic response of Ψ_{leaf} to radiation and air temperature for normal and stressed tomato and sweet pepper in the greenhouse. In stressed plants the response to temperature seems to become more important relative to radiation.

Not only time variation of Ψ will occur but also spatial variation, especially in the greenhouse, where large vertical gradients occur. Apart from the liquid flow resistance itself causing a vertical Ψ profile with lower values at the top of the plant, vertical profiles in microclimate, especially radiation and humidity, will steepen the profile in Ψ, whereas heating pipes underneath the crop will have the reverse effect.

Dynamic behaviour of leaf water content

The dynamic change in volume of the leaf or other tissue is the result of (net) uptake of water and transpiration (Jarvis *et al.*, 1981; Jones, 1992):

$$dV_{leaf}/dt = J_w - E = (\Psi_0 - \Psi_{leaf})/R_{plant} - E \qquad \text{(Eq. 2.2.28)}$$

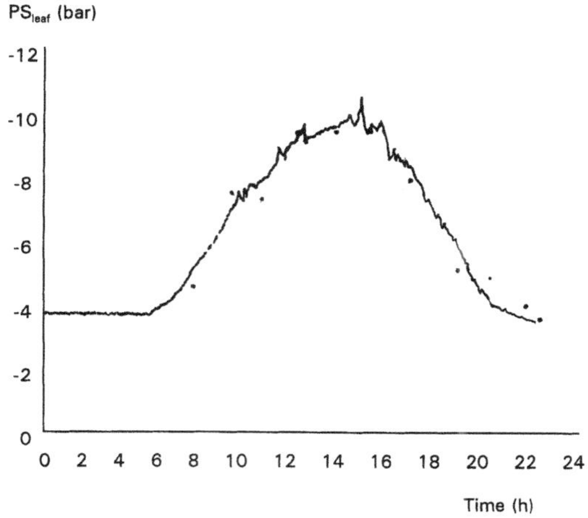

Figure 2.2.11 – Measured (data points) and simulated (line) daily course of Ψ_{leaf} (bar) of tomato plants in the greenhouse (June 14, 1986), (from Bruggink *et al.*, 1988).

where Ψ_0 is root environment water potential. To illustrate this the diurnal course of tomato leaf fresh weight in the greenhouse is shown in Figure 2.2.12. A build-up of water stress over time amplifies the diurnal course as illustrated in Figure 2.2.13 for potato leaves in a controlled environment (Plodowska *et al.*, 1989).

Vapour pressure difference

Interrelationships and control loops can exist of both an hydraulic or chemical nature (Schulze, 1986; Davies & Zhang, 1991). The interdependence of aerial and root environmental factors, such as vapour pressure deficit and Ψ_0, on the one hand and leaf conductance and Ψ_{leaf} on the other has been discussed in the review by Schulze (1986). Schulze distinguishes three possible changes of leaf water potential with variation in leaf-to-air vapour pressure difference:

1. A proportional decline with increasing vapour pressure difference, when leaf conductance does not respond;
2. A gradual decline to a minimum value, when stomata respond to Ψ (feedback control);
3. As 2 but Ψ increasing again at high vapour pressure difference, because of direct stomatal response to humidity (feedforward control) (Figure 2.2.14).

The stomatal sensitivity of four major greenhouse crops to humidity, as reported by Bakker (1991^b), suggests a type (3) response. Barrs (1973), however, found a type (1) response for tomato and bell pepper. In pearl millet grown in a greenhouse Squire *et al.* (1983) found a relation between Ψ_{leaf} and *E*, but not with vapour pressure deficit. They concluded that changes in leaf conductance counteracted effects of vapour pressure deficit on Ψ_{leaf}. The same was concluded by Van de Sanden & Veen (1992) after observing long-term exposure to different levels of vapour pressure deficit of cucumber seedlings grown in a controlled environment. Hoffman (1973) too found only slight effects on Ψ as well as Ψ^s after (long-term) exposure of several crops, amongst others radish and bell pepper, to different levels of vapour pressure deficit. Syvertsen & Levy (1982) presented diurnal curves of vapour pressure deficit and Ψ_{leaf} of citrus grown in the greenhouse and found them to be highly correlated.

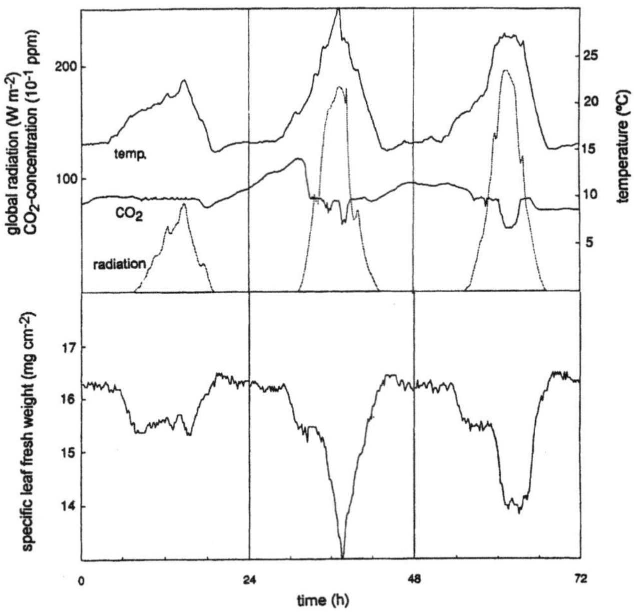

Figure 2.2.12 – Diurnal course over three successive days of glasshouse air temperature and CO_2 concentration together with global radiation (outside) and specific leaf fresh weight of tomato (from Van de Sanden & Gijzen, 1993).

Atmospheric CO_2-concentration

The effect of the (elevated) CO_2 concentration in the greenhouse on plant water status depends on the short term effect of CO_2 on *E*, via stomatal conductance, and on the long term effect on *E*, via the development of leaf area (see Tyree & Alexander, 1993 and section 2.2.2.3). High CO_2 causes a decrease in stomatal conductance in several greenhouse crops, including cucumber and tomato (Shaer & van Bavel, 1987; Nederhoff & De Graaf, 1993) sweet pepper (Nederhoff, Rijsdijk & De Graaf, 1992), eggplant (Nederhoff, 1992), chrysanthemum (Gisleröd & Nelson, 1989) and strawberry (Sruamsiri & Lenz, 1985).

In contrast to in the field, the effect on E in the greenhouse is reported to be small and often negligible, because of aerodynamic resistance and feedback mechanisms in the greenhouse climate (Jarvis, 1985 and sections 2.2.3.2 and 3.4). If this is so, one would not expect any effect on the plant water status apart from alleviation of water stress (Peaz *et al.*, 1984; Tyree & Alexander, 1993) or restricted water uptake because of low Ψ_0 (Mortensen & Gisleröd, 1989; Zeroni & Gale, 1989). Swalls & O'Leary (1976) found CO_2 to lower the transpiration rate of tomato when humidity in the greenhouse was high.

Root environment water potential

According to equation (2.2.25) one would expect Ψ_{leaf} in a steady state, after long-term exposure to salinity, to change in parallel with Ψ_{soil} or Ψ_0 in soilless culture. In fact, Van de Sanden & Veen (1992) found the sensitivity of Ψ_{leaf} to Ψ_0 to be 1 for cucumber seedlings. Hoffman *et al.* (1971) and Hoffman & Rawlins (1971) found the sensitivity of Ψ_{leaf} to Ψ_0 to be close to but a bit lower than unity for cotton, garden beet and onion only when humidity is high, and considerably less then unity when humidity

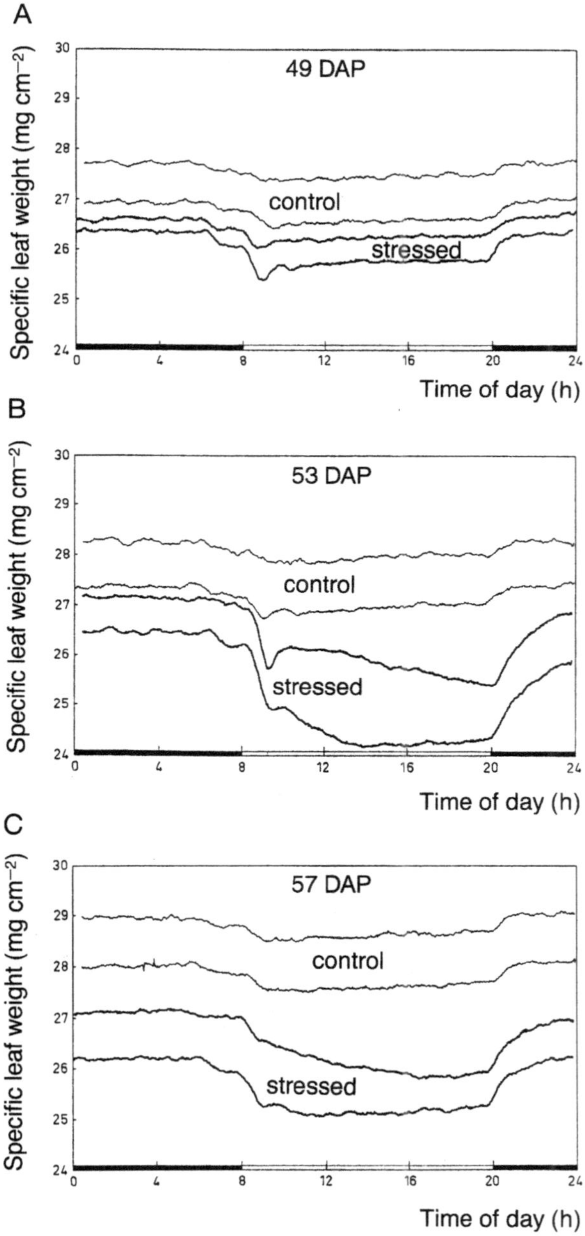

Figure 2.2.13 – Diurnal fluctuation in specific leaf fresh weight of a leaflet (measured as mg cm^{-2}) for two plants without stress (fine lines) and two plants exposed to drought stress (bold lines). The night period lasted from 20.00 h until 8.00 h, as indicated by the shaded parts of the x-axis. The drought stress lasted from 46 – 54 days after planting (from Plodowska *et al.*, 1989).

a. Fluctuation shortly after stress initiation (49 days after planting);
b. Fluctuation at the end of the stress period (53 days after planting);
c. Fluctuation shortly after stress (57 days after planting).

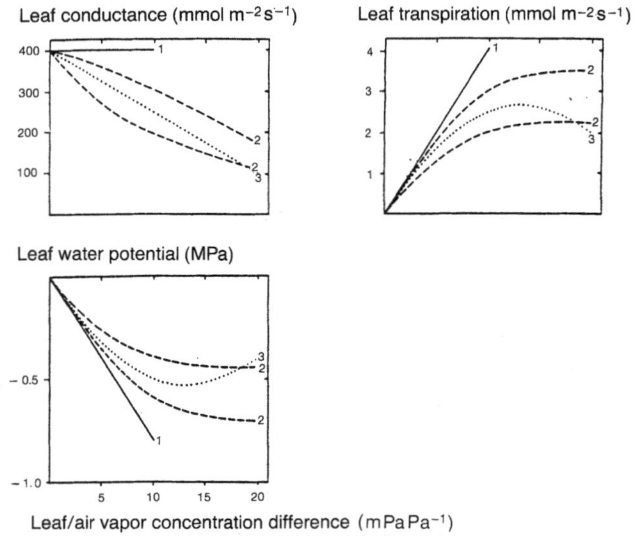

Figure 2.2.14 – Possible changes of leaf conductance, leaf transpiration and leaf water potential with variations in leaf/air vapour concentration difference. 1. no response of conductance to changing humidity; 2. feedback control of leaf water potential on leaf conductance for different minimum water potentials; 3. direct (proportional) response of leaf conductance to changing humidity. Assumption: a proportional effect of transpiration on water potential (from Schulze, 1986).

is low. This might indicate a down-regulation of E at low RH, either through stomatal closure or as a consequence of morphological adaptation, decreasing the sensitivity of Ψ_{leaf} to Ψ_0. In radish Hoffman & Rawlins (1971) found a sensitivity considerably higher than unity. Van de Sanden (unpublished data) found a concurrent change in Ψ_{xylem} of tomato with a change in Ψ_0, but hardly any effect on Ψ_{fruit}. As a result the water potential gradient between xylem and fruit was influenced and consequently also water import into the fruit. With spatially different water potentials in the root environment the water status of the plant might reflect the average root environment water potential or that of the environment of the roots best provided with water.

2.2.2.5 The effect of plant-water relations on some physiological processes

Water status and growth

Cell division is believed to be influenced only to a minor extent by the water status of the plant (Dale, 1988). On the other hand, expansive growth and especially growth of the aerial part of the plant, is one of the processes most sensitive to water deficit. Root growth seems to be less sensitive to water deficit (Sharp & Davies, 1989; Wyn Jones & Pritchard, 1989), an observation confirmed for cucumber response to aerial and "soil" drought (Van de Sanden & Veen, 1992). In this section only a short introduction will be given on the effect of plant water status on tissue extension. For extensive information the reader is referred to Dale & Milthorpe, 1982; Tyree & Jarvis, 1982; Baker *et al.*, 1985; Cosgrove, 1986 and Jones *et al.*, 1989.

The elastic response of tissue to water status has been discussed in section 2.2.2.2. However, growing tissue responds to its water status in an irreversible way as well. Irreversible tissue enlargement results from two processes: import of water into the tissue and the yielding of the cell wall. Yielding of the wall will lower the cell Ψ thus promoting inflow of water. This in turn will give rise to increase of turgor again causing the cell wall to yield, etc. The turgor driven wall yielding is described by the Lockhardt equation

$$dV/Vdt = m(\Psi^{p} - Y) \qquad \text{(Eq. 2.2.29)}$$

where dV/Vdt is relative rate of tissue volume increase, m wall extensibility and Y minimum turgor for extension (the yield threshold). The inflow of water from the xylem is described in analogy with equation (2.2.19)

$$dV/Vdt = L\,\Delta\Psi = L(\Psi_{xylem} - \Psi_{tissue}) \qquad \text{(Eq. 2.2.30)}$$

where L is path hydraulic conductance. Whether inflow of water limits short-term growth depends on the relative magnitude of parameters m and L of the tissue (Tyree & Jarvis, 1982; Cosgrove, 1986; Wyn Jones & Pritchard, 1989). When water transport is not limiting ($L>>m$) volume increase is governed according to equation (2.2.29). When water transport is limiting ($L<<m$), equation (2.2.30) describes the dominant process controlling relative volume increase, and turgor approaches the yield threshold, which might be in the order of 0.1 MPa or less. This might be the case with tomato fruit. Using the isopiestic technique similar to that described by Slatyer (1958) we could not find any detectable turgor in growing tomato fruit pericarp tissue (Van de Sanden, unpublished data). Shackel *et al.* (1991) found small if any turgor in mature green tomato using the *in situ* pressure probe. The existence of highly elastic cell walls (low ε) might be an alternative mechanism buffering tissue turgor against changes in water inflow (Dale, 1988; Wyn Jones & Pritchard, 1989). In tissue with a half-time for water exchange of a minute or so, which is normally the case, inflow of water is probably not limiting (Cosgrove, 1986). Tissue extension will depend on the existence of turgor pressure above the yield threshold. With loss of turgor pressure, water inflow for growth can be maintained by adjustment of Ψ^{p} through Ψ^{s} (equation (2.2.16)), so-called osmotic adjustment or turgor regulation, or by adjustment of m and/or Ψ. The mode of action may be different in response to different sources of water stress (Van de Sanden & Veen, 1992). A change of these wall parameters might occur within minutes and will result in sustained growth, albeit at a lower rate, and smaller cells (Wyn Jones & Pritchard, 1989).

Over time, longer term responses may alleviate negative effects by changed allocation and metabolism of carbon, resulting in an acclimated morphology of the plant, e.g. restricted shoot growth in favour of sustained or even promoted root growth (De Koning & Hurd, 1983; Van de Sanden & Veen, 1992). Changed carbon allocation in response to water availability matches the concept of the functional equilibrium (Brouwer, 1983) according to which the plant adapts its morphology with respect to balance between the performance of the shoot (carbon gain) and of the roots (especially uptake of N and water). Figure 2.2.15 from Geiger & Servaites (1991) illustrates how plants may react to water stress over time; from changed xylem water potential within seconds to, for example, increased root growth within days.

Water status and root pressure

Root pressure results from the osmotic water movement in response to the gradient in water potential across the endodermis membrane (Slatyer, 1967). This movement is wholly passive and follows equation (2.2.21) with the water potential gradient close to zero. The phenomenon, however, is energy-related, since a water potential gradient must be maintained by concomitant active uptake of nutrients. Root pressure is an important feature, because it may refill empty or cavitated vessels induced by high transpiration and because it transports nutrients such as calcium into tissue less adequately fed by transpiration-induced flow (Bradfield & Guttridge, 1984; Ehret & Ho, 1986[a]). Van de

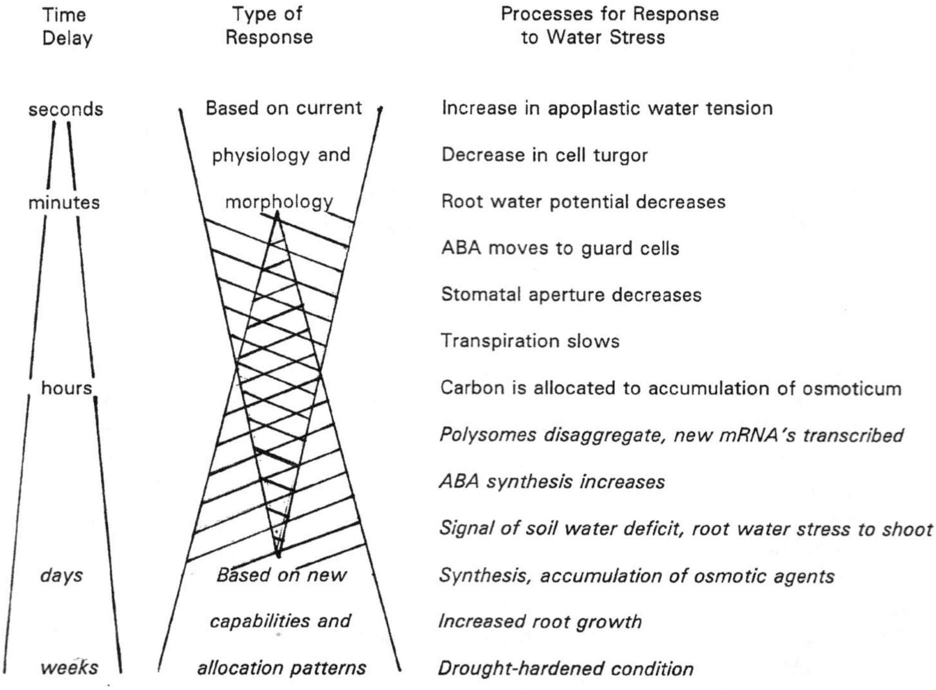

Figure 2.2.15 – How plants respond to stress over time. Progression from current to new capabilities illustrated by responses to water stress (from Geiger & Servaites, 1991).

Geijn & Smeulders (1981), however, dispute the hypothesized dominance of the relation between root pressure and Ca^{2+} distribution. The osmotic potential in the root stele is the net result of (active) import of solute from the root environment and export out of the root via the transpiration stream. So root xylem osmotic potential will be low, and consequently the water potential gradient between xylem and root environment high, when the transpiration rate is low. As a result development of positive root pressure is linked to a low transpiration rate, for example at night and/or at high humidity. Root pressure is, on the other hand, impeded by low water potential (high salinity) of the root environment (Ehret & Ho, 1986[a]). The strength of night-time root pressure might be related to previous day climatic conditions somehow providing energy and solute requirements for pressure build-up during the night.

Water flow coupled nutrient distribution

In general, in well watered conditions, plant water flow has only a slight effect on ion uptake, but the translocation patterns of water and nutrients are closely linked. In conditions of water stress, however, in the case of a lower Ψ_0 of the nutrient solution, the uptake of a nutrient like Ca is restricted, resulting in deficiency symptoms (Van Goor, 1974; Ehret & Ho, 1986[a]; Bakker & Sonneveld, 1988; Adams & Ho, 1989). High EC-salinity of the root environment increases the K content of most tissues in tomato, but NaCl-salinity has the opposite effect; the N content is hardly affected by salinity (Charbonneau *et al.*, 1988; Sonneveld & Welles, 1988; Adams & Ho, 1989; Knight *et al.*, 1992).

Although there does not seem to be an exact proportional relationship between calcium (Ca) and water flow (Bangerth, 1979; Bengtsson, 1982), the supply of this nutrient to specific organs has by far the most been associated with the water distribution in the plant. The plant depends for its supply of Ca, and to a lesser extend Mg, wholly on translocation through the xylem. Furthermore, Ca is not redistributed in the plant. Some organs, such as young leaves, fruits and storage tissue, can be characterized as having a low transpiration rate and a high growth rate. These are less adequately fed by the xylem and tend, in competition with other organs for transpirational water with a high Ca content, to be very sensitive to Ca-related physiological disorders, such as blossom-end rot in tomatoes and tipburn in lettuce and cabbage (see e.g. Wiersum, 1966). The Ca accumulation in grape berries, for instance, is closely correlated with the transpiration rate of the berry itself (Düring & Oggionni, 1986).

There has been a lot of interest in the effect of air humidity on the distribution of not only calcium, but also magnesium, potassium, nitrate and phosphate (e.g. Michael & Marschner, 1962; Gislerød *et al.*, 1987; Bakker & Sonneveld, 1988; Adams, 1991). In general, high humidity during the day causes, apart from a slight decrease in Ca uptake, a relative shift in Ca distribution from young leaves and fruits to older leaves and might result in calcium deficiency symptoms in leaves of tomato, cucumber, strawberry, lettuce and cabbage, and fruit of tomato and apple (see review Grange & Hand, 1987). High night-time humidity (less than 0.2 kPa vapour pressure deficit) seems to have a positive effect on Ca distribution to sensitive tissues, because it promotes root pressure flow, which drives water and nutrients to these tissues, and/or because competition with organs with a transpiration rate is suppressed (e.g. Ho, 1989). Furthermore the maximum concentration of Ca in the xylem occurs at night (Ferrario *et al.*, 1992). So the import of Ca by tomato fruits and meristems is favoured at night (Van de Geijn & Smeulders, 1981; Ho, 1989; Tachibana, 1991). E.g., pre-emerged, non-transpiring strawberry leaves depend for their calcium on water flow arising from root pressure at night, while after emergence calcium is supplied by transpirational water flow, promoted by dry days (Bradfield & Guttridge, 1979). Bangerth (1979) and Grange & Hand (1987) conclude, that an increase in the diurnal transpiration amplitude created by dry days and nights with a low evaporative demand, combined with good water supply or low salinity in the root environment should increase the Ca supply to storage organs and weakly transpiring young leaves. High humidity alone might not suffice to bring about a negative effect. O'Leary and coworkers (O'Leary & Knecht, 1972; Swalls & O'Leary, 1976) did not find any effect of humidity on total salt and on ^{45}Ca uptake in tomato, because nutrients are delivered to the shoot at higher concentration when transpirational flux is low. They did, however, find that a combination of high humidity and high CO_2 concentration to reduce the transport of Ca and Mg to the leaves considerably.

Water status and stomatal conductance

The traditional model describing the effect of water status on stomatal resistance is a feedback mechanism which becomes operative when a threshold leaf water potential has been exceeded (Figure 2.2.16) (Hsiao, 1973). The threshold Ψ_{leaf} depends on previous growth history and on the prevailing temperature. In field grown tomato the threshold value is around -0.7 MPa (Rudich *et al.*, 1981; Rudich & Luchinsky, 1986), which is in agreement with data from Duniway (1971) for greenhouse-, soil-grown tomato. A gradual closure of stomata rather than a threshold response was found in eggplant (Behboudian, 1977^{b}) and was also used in the simulation of tomato water relations by Marcelis (1989), based on the assumption that stomata of greenhouse crops are more sensitive to Ψ_{leaf} than those of field grown crops (Burrows & Milthorpe, 1976). It must be realized that a threshold response of stomatal resistance to Ψ_{leaf} may be transformed into a linear response when considering conductance rather than resistance. The mode of action of water status on gas exchange resistance is still obscure. RWC or epidermal Ψ^{P}, rather than Ψ, might have greater relevance to the description of dehydration effects on stomata. The direct sensing of water stress in the root environment might also be involved, either by way of a chemical signal such as abscisic acid (ABA) transmitted through the xylem (Davies & Zhang, 1991) or by way of a pressure signal transmitted through the phloem

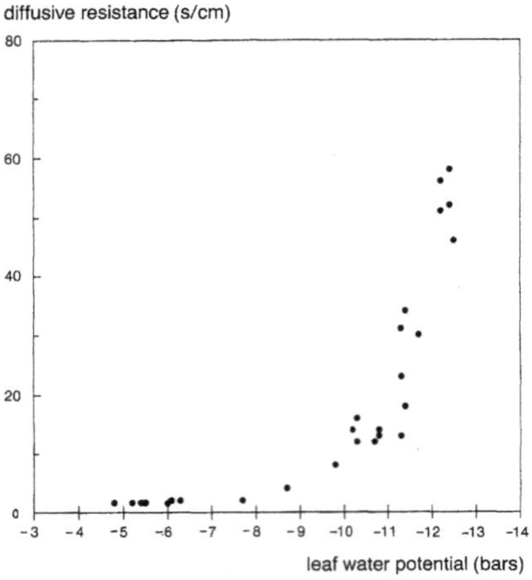

Figure 2.2.16 – Relationship between stomatal resistance and leaf water potential (from Duniway, 1971).

(Schulze, 1991). According to Schulze *et al.* (1987) a direct relation between Ψ and stomatal conductance is disputable. One response of stomatal conductance independent of bulk leaf water status is the stomatal sensitivity to the humidity of the leaf boundary layer, as was also shown in cucumber, sweet pepper, eggplant and tomato (Bakker, 1991[b]). Stomata tend to react to the evaporative demand of the air, thus preventing plant water potential from falling (see above and section 2.2.2.4).

Plant water status might thus be controlled by stomatal conductance (apart from leaf and root area development) rather than the reverse (Jones, 1985; Sharp & Davies, 1989). In line with this view Jones (1985) simplified equation (2.2.25) to

$$\Psi_{leaf} = \Psi_{soil} - a\, 6{,}39 \times g_{leaf} \qquad \text{(Eq. 2.2.30)}$$

with a depending on hydraulic conductance, leaf area and environmental conditions and g_{leaf} as leaf diffusive conductance to water vapour. It is uncertain whether this relation will hold in the greenhouse, since there is a strong "decoupling" between g_{leaf} and transpiration (Jarvis, 1985; Aubinet *et al.*, 1989), apart from situations of restricted water uptake.

Behboudian (1977[a]) presents for tomato, cucumber and sweet pepper a set of linear relations between stomatal diffusive resistance and Ψ_{soil}. Accumulated ABA might be responsible for any restrictive after-effect of water stress on stomatal opening, although plant water status has recovered. The effect of water status on carbon gain, the prime prerequisite for growth and yield, is not only a matter of stomatal conductance. Effects on mesophyll resistance to CO_2, on respiration or on assimilate allocation might be involved as well. Gas exchange properties of cucumber and sweet pepper deteriorated upon drought, while those of tomato did not (Behboudian, 1977[a]). This was attributed to a substantial increase in mesophyll resistance in the former two species.

2.2.3 Interaction between CO_2 uptake and water loss

H. GIJZEN

2.2.3.1 Introduction

Plants need to take up CO_2 from the air by opening stomata. This, however, also entails a H_2O loss. This water loss often represents a cost ("porosity at a price", Mansfield, 1985), although the water flow also enables the plant to transport and distribute nutrients through the transpiration stream and to cool the leaves when radiation levels are high. By regulation of stomatal opening plants can control both the CO_2 and the H_2O flux, although some water loss occurs through the cuticle. Stomata are believed to keep some balance between photosynthetic CO_2 assimilation and transpiration, so as to maximize CO_2 uptake but at the same time acting to prevent possible future desiccation in a variable environment.

In the following an overview is given of responses of stomata to the environment and to plant internal factors, and of the effects of plant water status and water loss on CO_2 uptake.

Stomatal conductance is correlated with the rate of photosynthetic CO_2 assimilation (Schulze & Hall, 1982). This correlation is considered here to form the basic pattern of stomatal behaviour, which can be altered by air humidity and plant water status (section 2.2.2).

2.2.3.2 Stomatal and boundary layer conductance

Stomata and boundary layer

Stomata are one of the resistances in the pathways of H_2O and CO_2 between ambient air and the interior of the leaf (Figure 2.2.17). The start of the H_2O pathway is at the transpiring cell walls within the leaves, for example the walls bordering the substomatal cavity. It is commonly assumed that the air inside the cavity is saturated with water vapour (Ball, 1987). However, in some species internal cutinisation of cell walls may occur (Sheriff, 1984). The end of the CO_2 diffusion pathway is at the stroma of the chloroplasts. Some water is transpired via the epidermal cuticle. This is almost impermeable to CO_2.

Guard cells close the stomatal pore when their turgor is decreasing compared with that of subsidiary cells. Turgor changes in guard cells are brought about by changes in water content and in solute content of these cells. Change of the solute content (notably K^+-ions, along with electrically neutralising anions) is an active metabolic process, which can be followed by passive changes in water content.

Stomatal behaviour is complex and not well understood. Stomatal opening can be affected by current environmental conditions and by various plant internal factors, for example leaf water status and leaf photosynthetic rate. It also seems to be influenced by various metabolites in the transpiration stream, e.g. abscisic acid (ABA), indoleacetic acid (IAA), Ca and cytokinin (Grantz, 1990).
The mechanisms by which stomata respond to internal and external conditions are still largely unclear. In Figure 2.2.18 various feedforward and feedback loops are depicted that could control stomatal opening. They will be discussed below.

The boundary layer (sections 2.2.1.2 and 2.2.2.5) is a thin layer of still air around the leaf, that forms a resistance in the pathways of CO_2 and H_2O to and from the leaf. The boundary layer affects the rates of H_2O and CO_2 exchange of the leaf and modifies the environment of the stomata. This occurs particularly in greenhouse crops, as the low air speeds in greenhouses generate thick boundary layers. Thus, the role of the boundary layer must receive due attention, also because stomatal response is often better understood when related to conditions at the leaf surface (Ball, 1987).

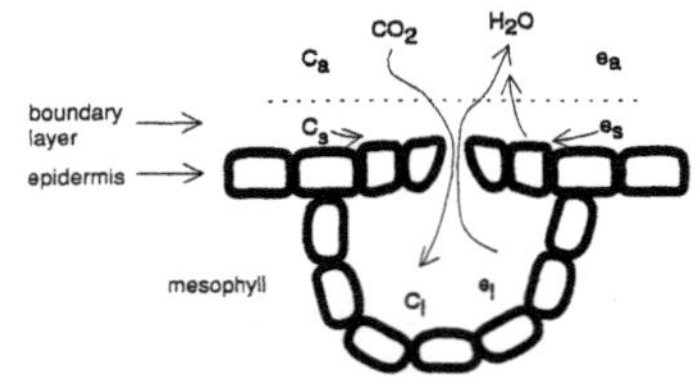

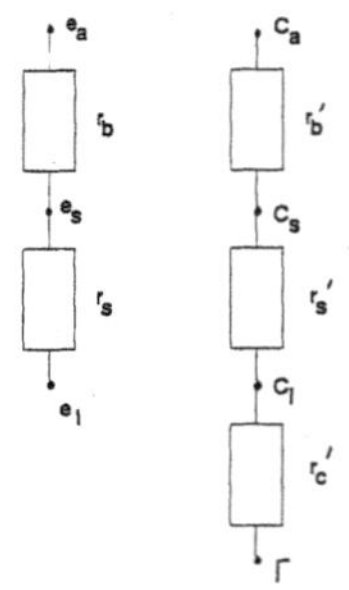

Figure 2.2.17 – Pathways of diffusion of CO_2 and H_2O between stomatal cavity (or intercellular spaces) and ambient air. The resistance scheme for CO_2 diffusion is adapted from Goudriaan *et al.* (1985). C_a and e_a are CO_2 concentration and water vapour pressure in ambient air; C_s and e_s are CO_2 concentration and water vapour pressure at the leaf surface; C_i is CO_2 concentration in the substomatal cavity; e_l is vapour pressure in the substomatal cavity (assumed to be saturated), Γ is the CO_2 compensation point. r_b, r_s and r_c are resistances of boundary layer, stomata and carboxylation, respectively. The prime indicates resistance to CO_2 transfer.
Note that, with $e_{sat,a}$ and $e_{sat,l}$ saturated vapour pressures at temperatures of air and leaf, respectively: leaf-air VPD = $e_l - e_a$; air VPD = $e_{sat,a} - e_a = D_a$; leaf surface VPD = $e_{sat,l} - e_s = D_s$.

Conductances and resistances

Leaf and stomatal conductances are the proportionality parameters relating H_2O and CO_2 exchange to the driving force. Leaf conductance is the sum of stomatal and cuticular conductance. Both the terms conductance and resistance will be used here. The term conductance is useful when considering the magnitudes of transpiration and CO_2 uptake, the term resistance when considering the magnitude of control that the stomata exert on the fluxes, compared with other resistances.

Leaf transpiration, E_l, is proportional to leaf conductance, g_l, and to the difference in water vapour pressure inside the leaf, e_l, and the water vapour pressure at the leaf surface, e_s; alternatively, E_l is proportional to total (leaf+boundary layer; g_{tot}) conductance and the difference between e_l and water vapour pressure of the ambient air, e_a,

$$E_l = g_l (e_l - e_s) k \qquad E_l = g_{tot} (e_l - e_a) k \qquad \text{(Eq. 2.2.31)}$$

where k is a factor for converting pressure to concentration (Jarvis & Morison, 1981; see also section 3.4.3.2).

As e_l can be assumed to be saturated for most species, the difference e_l–e_a is called the leaf-air water vapour pressure deficit (leaf-air VPD), and, similarly, the difference e_l–e_s the leaf surface VPD, D_s. As leaf temperature normally differs from air temperature, air VPD, D_a, differs from leaf-air VPD. Stomata will respond to D_s rather than to D_a, as was found by Bunce (1985).

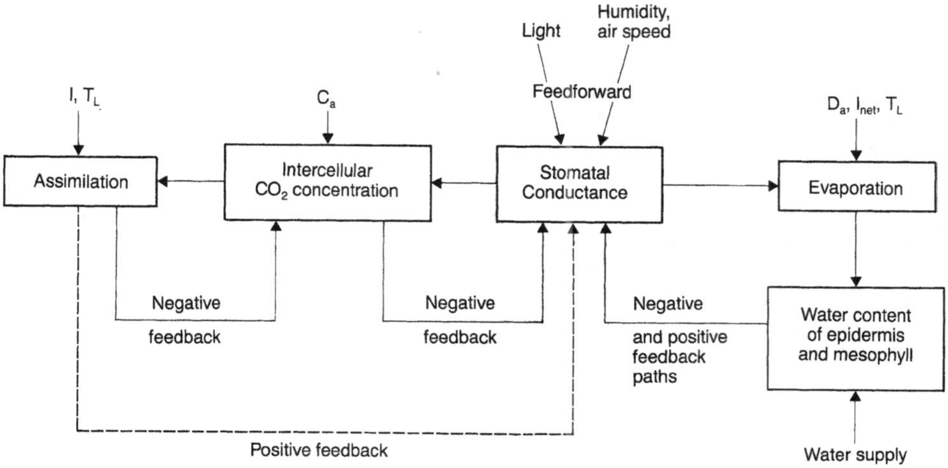

Figure 2.2.18 – Scheme of possible interactions of leaf photosynthesis and water loss, mediated by stomatal conductivity and intercellular CO_2 concentration (from Raschke, 1979). C_a ambient CO_2 concentration, I_{net} net radiation, I photosynthetically active radiation, T_l leaf temperature, D_a vapour pressure deficit of air.

Leaf net photosynthetic CO_2 uptake, P_n, is proportional to leaf conductance for CO_2, g_l', and the difference in CO_2 concentration at the leaf surface, C_s, and CO_2 concentration in the substomatal cavity, C_i; alternatively, P_n is proportional to total (leaf + boundary layer) conductance, g_{tot}', and the difference between CO_2 concentration in the ambient air, C_a, and C_i,

$$P_n = g_l' (C_s - C_i) \qquad P_n = g_{tot}' (C_a - C_i) \qquad \text{(Eq. 2.2.32)}$$

C_i is assumed to be equal to the CO_2 concentration in the intercellular spaces. Note that the resistance chain of CO_2 diffusion contains an important extra resistance, the (chemical) carboxylation resistance for the rate of binding of CO_2 by the Rubisco enzyme. This resistance is normally relatively high as compared with the other resistances, so that the effect of g_s on the rate of diffusion will often be smaller for CO_2 than for H_2O. The resistances of boundary layer and stomata for CO_2 transfer are somewhat higher than for H_2O, i.e., they have to be multiplied by 1.6 and 1.37, respectively (Von Caemmerer & Farquhar, 1981). In the following conductances will refer to water loss.

Stomatal conductance in different species varies from almost zero to highest values of about 0.05 m s^{-1} (i.e. resistances vary from 20 to 5000 s m^{-1}). Stomatal conductance shows a saturating type of response with increasing light intensity. The maximal value varies depending on, among others, the past average light levels and photosynthetic capacity. At ambient CO_2 concentration (350 μmol mol^{-1}) it has been estimated for greenhouse crops to be 0.02 m s^{-1} in cucumber (Bakker, 1991[b]; Nederhoff & De Graaf, 1992), 0.02 in eggplant (Bakker, 1991[b]) and 0.025 m s^{-1} (Nederhoff, 1992), 0.01 (Bakker, 1991[b]), 0.015 (Nederhoff & De Graaf, 1992), and 0.004 m sm^{-1} in tomato (Jolliet & Bailey, 1992), 0.01 (Bakker, 1991[b]) and 0.025 m s^{-1} in sweet pepper (Nederhoff *et al.*, 1992), and 0.005 m s^{-1} for *Ficus benjamina* (Fredrick *et al.*, 1992). Most of these values are quite high and are not often found in field crops. Cuticular conductances are in the order of 0.00025 to 0.001 m s^{-1}. A leaf boundary layer conductance, g_b, of about 0.01 m s^{-1} was measured by Stanghellini (1985) inside a tomato canopy using replica leaves of 5 cm width, and was estimated to be 0.005–0.01 m s^{-1} for *Ficus benjamina* having a leaf width of 5 cm (Zhang & Lemeur, 1992).

Data on g_b of large-sized leaves, as from cucumber and eggplant, are lacking, but they could be much lower. g_b of crops in the greenhouse is normally significantly lower than in field crops. At a windspeed of 1 m s^{-1} g_b will be in the order of 0.05 m s^{-1} for small sized leaves (Jones, 1983).

As cuticular conductance is generally negligible compared with stomatal conductance, in the following stomatal conductance will be equated with leaf conductance.

2.2.3.3 Stomatal conductance and CO_2 uptake

To promote leaf photosynthesis stomata should open maximally. However, stomata do not stay open fully, but respond to light, temperature and CO_2 concentration in a manner that is often related to the response of leaf photosynthesis to these climatic conditions. This behaviour of stomata can be viewed as the basic pattern, which can be modified by responses to humidity and water status, as discussed in sections 2.2.2.4 and 2.2.3.4.

Stomatal responses to light, CO_2 and temperature

Light

Stomatal opening strongly increases with light. At high light intensities the conductance reaches saturation. Photosynthesis has a very similar response: a strong response at low light intensities, as light is a limiting factor for the photosynthetic reactions, and a saturating response at high light intensities, when other factors become limiting to leaf photosynthesis. As light intensity increases, C_i initially decreases sharply (Figure 2.2.19). It then reaches a fairly constant value, due to the parallel

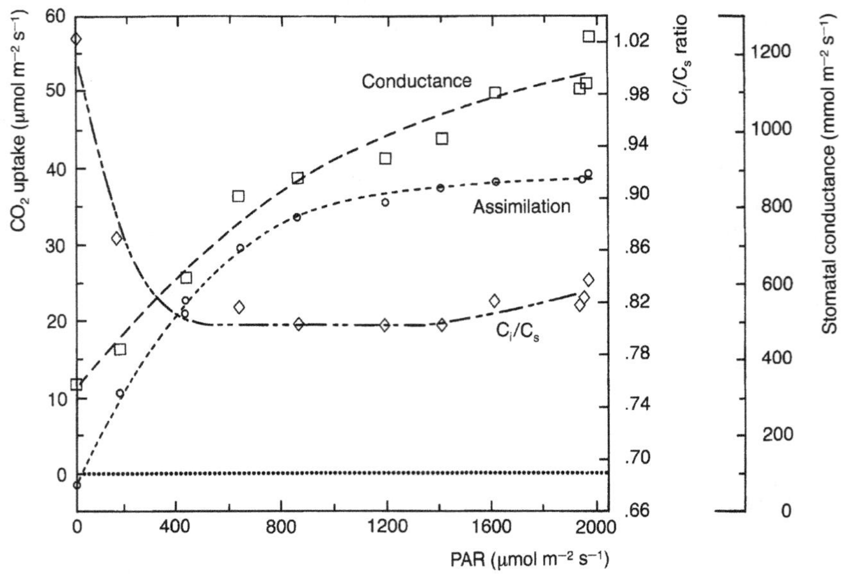

Figure 2.2.19 – Responses of leaf conductance, g_l, net photosynthesis, P_n, and the ratio C_i/C_s to light intensity (PAR, 400–700 nm) for a single attached leaf of *Geraea canescens*. Leaf temperature, CO_2 concentration and leaf-air VPD were kept constant at 20 °C, 330 µmol mol^{-1} and 0.5 kPa, respectively (Ball & Berry, 1981).

responses of stomata and leaf photosynthesis to light intensities above about 200–500 μmol m^{-2} s^{-1} PAR (Morison, 1987). Although stomata respond to C_i (Mott, 1988), the initial response to increasing light appears, for well-watered plants, to be mostly directly to light, and to be less dependent on the decreasing C_i (Morison, 1987).

Note that at higher light intensities the rate of CO_2 diffusion will be more limiting to photosynthesis, and consequently effects of changes in g_s on P_n are more pronounced than at low light intensities.

The light responses of stomata and photosynthesis are considered to be dominant factors in causing the decreased stomatal opening in leaves lower in the canopy.

CO_2

Increasing the ambient CO_2 concentration normally increases C_i, to which stomata respond by closing (section 2.2.2.4). Raschke (1970), however, found that at air temperatures above 35 °C stomata of maize plants well supplied with water became insensitive to CO_2.

In many cases stomata close as C_a increases, in such a way that in steady state situations the ratio C_i/C_a remains approximately constant (i.e., conservative) under full light (Goudriaan & Van Laar, 1978; Jarvis & Morison, 1981), often at about 0.7–0.8. It was found to be 0.7 in tomato at 1.8 to 2.4 kPa leaf-air VPD (Bradford *et al.*, 1983), and as high as 0.9 in cucumber with leaf-air VPD at 1.3 kPa (Peet *et al.*, 1986) or 0.8 kPa (Raschke, 1986). Ramos & Hall (1983) found that in sweet pepper it decreased from 0.82 to 0.61, with leaf-air VPD at 1.4 kPa, when PAR increased from 90 to 480 μmol m^{-2} s^{-1}, but did not decrease further when PAR increased to 1500 μmol m^{-2} s^{-1}. C_i/C_a is not much influenced by leaf age or nutrient status. The ratio C_i to C_s has sometimes been found to be more conservative (Farquhar & Wong, 1984). However, experiments have seldom been performed with low g_b. In the case of greenhouse crops, further investigations on the ratio C_i/C_s would be useful.

Temperature

Controversy exists in the literature about the (air) temperature response of stomata. A clear view is difficult to obtain as response to temperature has often been confounded with the humidity response. Jarvis & Morison (1981) concluded that conductance follows an optimum response, but also noted that stomata frequently are fully open at high temperatures combined with full plant water supply. Several authors reported that conductance in well-watered plants increased with temperature beyond 30 °C, e.g. Hall *et al.* (1976) and Küppers (1988).

Correlation between stomatal conductance and leaf photosynthesis

The leaf photosynthetic capacity (defined as photosynthesis at light saturation and ambient CO_2, i.e. P_{nmax}) can vary during the season when the leaves acclimate to changing growing conditions. Stomatal conductance at P_{nmax}, g_{smax} appears often to be correlated with P_{nmax}. Linear correlations have been found between g_{smax} and P_{nmax} for many species (Schulze & Hall, 1982). Higher P_{nmax}, and correspondingly higher g_{smax}, are generally found with higher average light levels, nutrition levels, and lower values have been found with older leaves and lower ratio of sink to source activity. High ratios of g_{smax} to P_{nmax} indicate little stomatal limitation of leaf photosynthesis and low water use efficiencies (g dry matter accumulated per g H_2O transpired). Schulze & Hall (1982) suggested that high ratios are associated with more humid growing environments. It is likely that they would be common in greenhouse crops.

g_{smax} sets the upper limit to the operational range of stomatal conductances, i.e. the range in which g_s varies during diurnal courses in response to short-term changes in radiation, temperature and humidity. g_{smax} is an important parameter in regression models for stomatal conductance of the type proposed by Jarvis (1976)

$$g_s = g_{smax} f(\text{light}) f(CO_2) f(\text{VPD}) .... \quad \text{(Eq. 2.2.33)}$$

where f( ) indicates the multiplicative effect of a climate or plant variable.

In the short-term, g_s and P_n appear to be correlated to a large extent (Tenhunen *et al.*, 1987; Figure 2.2.20); the correlation was even found at the crop level (Louwerse, 1981). This is related to their parallel responses to light, to the (partly) similar responses to temperature, and the response of stomata to C_i. The correlation is found in the short-term during steady state situations lasting several minutes to hours. Under fluctuating conditions the correlation is less, as g_s can vary to some extent independently from P_n, for example, as a result of changes in humidity or water status. In the very short-term (seconds to minutes), correlation between g_s and P_n is less due to the much faster response time of photosynthetic reactions than stomatal opening and closing.

The correlation between g_s and P_n, both in the long-term and the short-term, is related to the conservative, though not constant, value of C_i. C_i depends on the rate of leaf photosynthesis. It is both the result and, to a given extent, the effecter of stomatal conductance. However, no clear picture exists of the role of C_i in the coupling of P_n and g_s under varying conditions (Morison, 1987). Note that when stomata follow C_i more closely, climatic conditions such as light intensity, CO_2 concentration and temperature interact more in their effects on g_s, via the role of photosynthesis.

2.2.3.4 Response of stomata to water

The humidity of the air, plant water status and water status in the root environment affect stomatal conductance and transpiration. In the control of plant water status 3 types of stomatal response to water have been distinguished (Raschke, 1979; Schulze, 1986; Schulze *et al.*, 1987):

1. A direct response to humidity, not mediated via water status of the leaf mesophyll;
2. An indirect response due to changes in the water status of the leaf mesophyll, and
3. A response to signals from roots experiencing a low water potential in the root environment.

The distinction made between the direct and the indirect response is partly based on the observation that turgor changes of guard cells, causing changes in stomatal opening, are regulated independently of turgor of the leaf mesophyll (Schulze, 1986).

Response of stomata to humidity

Generally, stomata respond to decreasing air humidity (increasing vapour pressure deficit) by closing. Often the closing response is not strong enough to cause a decrease in transpiration, so leaf water content will decline. It is then difficult to decide whether stomata close directly in response to humidity, indirectly via leaf water content, or both (section 2.2.2.4).

Sensitivity of plant stomata in various species to humidity appears to be proportional to stomatal conductance (Schulze & Hall, 1982; Morison & Gifford, 1983), indicating that relative decreases tend to be more or less similar. For a range of wild plants the relative decrease in stomatal conductance has been estimated to be in the order of 50% in the range of 1 to 2 kPa leaf-air VPD for a range of wild plants (Schulze & Hall, 1982). Bakker (1991[b]) found a decrease in stomatal conductance of 65% in the range 0.1 to 1 kPa in cucumber, tomato, sweet pepper and eggplant. Van de Sanden & Veen (1992) found that stomatal conductance of cucumber seedlings was 70% less when comparing plants grown with air-VPD at 0.2 kPa and at 1.4 kPa. Sometimes, however, no stomatal response to humidity is observed.

Temperature influences the humidity response, but in what manner is still unclear (Aphalo & Jarvis, 1991). Ball *et al.* (1987) found with soyabean a decreased response to leaf surface VPD at higher temperatures. The combined effect led them to incorporate a response to relative humidity in their model of stomatal conductance.

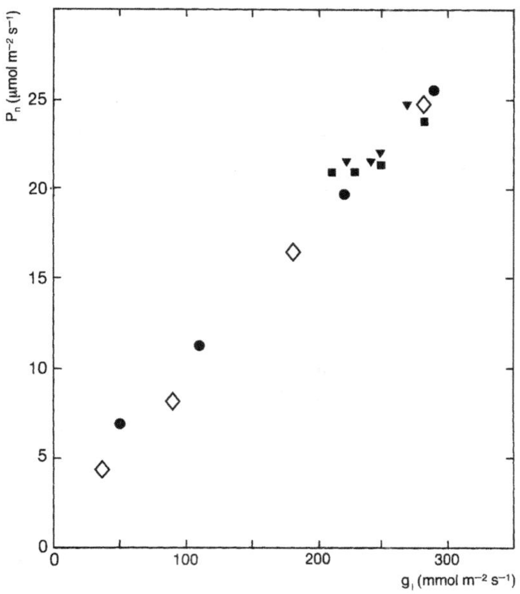

Figure 2.2.20 – Rate of net photosynthesis, P_n, and leaf conductance, g_l, in two leaves of *Phaseolus vulgaris* grown under full sunlight (open symbols) and two leaves grown under 6% of full sunlight (closed symbols). g_l was varied by varying light intensity from 50 to 413 W m^{-2} PAR (Wong *et al.*, 1985).

The growing conditions may affect the magnitude of the response of stomata to humidity. Bunce (1981) found that the response to VPD was higher when plants were grown at low irradiance, or when grown at high temperature.

Humidity partially affects the CO_2 response of stomata. It has frequently been found that well watered plants grown at high humidities lacked any closing response to increasing ambient CO_2 (Raschke, 1986). Bradford *et al.* (1983) observed that stomatal opening in tomato did not respond to C_a when leaf-air VPD was 0.5–1.0 kPa, but did so when leaf-air VPD was at 1.8–2.4 kPa. Morison & Gifford (1983) found in two C_3-grasses that the decrease of g_s with increasing C_a was higher at low humidity, i.e., C_i/C_a decreased from 0.9 to 0.7 when leaf-air VPD increased from 0.4 to 2 kPa. In general, when a response to C_a is present, humidity does not appear to interact with the CO_2 response (Jarvis & Morison, 1981; Morison, 1985), i.e., sensitivity of g_s to C_a (dg_s/dC_a) is independent of humidity. Stanghellini & Bunce (1992) found that leaf conductance of high CO_2 grown tomato plants decreased less at short-term high CO_2 concentrations than that of ambient CO_2 grown plants. At higher leaf-air VPD the difference dis-appeared.

Some possible responses of stomatal conductance and transpiration to humidity as discerned by Schulze (1986), are depicted in Figure 2.2.14 (section 2.2.2.4) No response of stomata is indicated by curves numbered 1, response type 2 could result from a proportional response to decreasing leaf water content (a feedback control). Response type 3 results in transpiration increasing to a maximum level and then decreasing again. The leaf water content improves beyond the maximum. This cannot be effected by feedback control, as in that case stomata would open again. Hence, this is a feedforward type of response (Cowan & Farquhar, 1977).

By which mechanisms stomata respond to humidity is not clear (Grantz, 1990). It has been proposed that humidity is "sensed" by vapour loss from unthickened areas in the outer walls of guard cells (Appleby & Davies, 1983), or that stomata respond to altered water potential of the epidermis, in which process cuticular transpiration could play an important role (Sheriff, 1984). Raschke (1975) and Grantz (1990) suggested that stomata respond to the rate of transpiration, with involvement of signal metabolites, such as ABA, that are carried by the transpiration stream. Although ABA is an inhibitor of stomatal opening, it is present in considerable amounts in the transpiration stream of unstressed plants (Grantz, 1990). This suggestion is supported by Mott & Parkhurst (1991) who found in several species that stomata do not respond to humidity at the leaf surface or difference in water vapour pressure between leaf interior and leaf surface, but directly to the rate of transpiration.

Response of stomata to water stress

Water stress (see also section 2.2.2.5) occurs when metabolic processes such as photosynthesis are hampered. With respect to water stress, greenhouse crops differ from field crops in that water stress experienced by leaves of greenhouse crops should not normally be a consequence of conditions in the root environment. Usually water stress is generated in experiments by not watering, hence decreasing the soil water content. From several studies it became apparent that stomatal closure that developed with decreasing water content in the soil, but high water content of the leaves, could be attributed to root signals (Schulze, 1986). It is believed that ABA plays an important role in this phenomenon (Davies *et al.*, 1991). Also in well-watered plants, as in greenhouse crops, stomata would receive signals from the roots about the water potential in their environment. Although signals of real water stress would be uncommon, gs could still be affected. E.g. it has been found that an increase of the EC (from 1 to 5 mS cm^{-1}) of the nutrient solution in a water culture decreased gs in tomato by 30% (W. Van Ieperen, pers. communication, 1994).

Several authors have concluded from experiments with well-watered plants that stomata are rather insensible to changes in water content of the plant; they would close rather rapidley when a threshold water content of the leaf is reached (Hsiao, 1973; Schulze, 1986) However, this view is still controversial.

ABA is believed to play a major role in stomatal closure when leaf water content is lowering. It appears that when water content is lowering, ABA is released in the transpiration stream by the mesophyll, first by redistribution, later by *de novo* synthesis, and, when arriving at the guard cells, elicits the decrease of stomatal opening (Hartung & Davies, 1991).

Plants often acclimate to repeated water stress by accumulation of solutes in the leaves, which lowers the water potential so that water can still be attracted (Chaves, 1991). Also the amount of ABA in the leaf can increase after water stress. ABA is commonly viewed as a "stress-integrator". A higher ABA-content of the leaf is usually correlated with lower stomatal conductance. Acclimation has mostly been found in plants exposed to slowly developing water deficits in the soil. It is not clear to what extent this could occur in well-watered plants in the greenhouse, which may suffer from water stress only for repeated short periods.

An after-effect of water stress is often observed (Hsiao, 1973). When water stress is relieved, stomata start to increase opening only some time later, the more so when stress has been more severe.

2.2.3.5 Effect of water on leaf photosynthesis

Leaf photosynthesis may be affected by humidity and plant water status, but what the mechanisms are, and whether a changed photosynthesis rate is cause or consequence of a concomitant change in stomatal aperture is often not clear (Schulze, 1986).

Humidity

Decreasing humidity frequently causes a decrease of leaf photosynthesis via decreased g_s. At moderate humidity levels, causing small leaf water deficits, the response of leaf photosynthesis to C_i is usually not affected (Kaiser, 1987; Morison, 1987). As a consequence of prolonged high transpiration rates leaf water content can decrease to levels where physiological processes are hampered, and water stress arises. This can especially occur when stomata are less responsive to humidity.

Reports on the effect of decreased humidity on leaf photosynthesis show very variable effects between various species, and pertain mostly to VPD responses at levels higher than 1 kPa. For example, El-Sharkawy *et al.* (1985) reported decreases in P_n in 19 different C_3 species from 25 to 90% when leaf-air VPD was increased from 1.25 kPa to about 4 kPa. Stanghellini & Bunce (1992) found a reduction in P_n of about 10% in tomato when leaf-air VPD was increased from 1.1 to 2.2 kPa at 25 °C.

VPD's in the greenhouse in the North-West of Europe are commonly below 1.5 kPa. How much P_n will be affected in the short-term by low humidity in this range is difficult to estimate. It is estimated that reductions will usually be less than 20%. Morison & Gifford (1983) found that P_n in growth-chamber grown grasses was almost unaffected by a leaf-air VPD increase from 0.5 to 1.4 kPa.

Water stress

Decreased leaf water content could directly affect the photosynthetic capacity in the mesophyll. However, this is not likely to occur often in greenhouse crops as photosynthetic capacity of the mesophyll appears to be rather insensitive to short-term dehydration, i.e., up to a leaf relative water content (RWC) of 50–70% (Bradford & Hsiao, 1982; Kaiser, 1987). These effects of low leaf water content on photosynthetic reactions appear to be related more to RWC than to water potential (Kaiser, 1987).

In greenhouse crops low leaf water contents are almost inevitably connected with high light intensity. Under this condition leaf photosynthesis can be reduced by photoinhibition of the light reactions. Photoinhibition at high light intensities can be an important mechanism in the reduction of photosynthetic capacity at reduced water contents (Kaiser, 1987). In general, the light level that is necessary to induce photoinhibition decreases when stress by water deficit or temperature increases (Chaves, 1991). When the RWC of the leaf does not go below 30% (this is well below the wilting point of leaves) the photosynthetic capacity will probably be restored rapidly (Kaiser, 1987). However, with high light intensities excessive (leaf) temperature increase is possible, especially when a low g_b significantly reduces exchange of latent and sensible heat. Temperatures above 35 °C could irreversibly damage the photosynthetic machinery.

With field-grown tomato, Bunce (1988[a]) found that P_n increased 60% when leaf-air VPD was decreased from 3 to 1 kPa. Leaf temperature varied between 30 and 35 °C. As g_s did not change, photosynthetic capacity was likely to be directly inhibited.

The optimal behaviour of stomata

The regulation of stomatal conductance may be understood by assuming a strategy where the plant is maximizing the daily carbon gain with respect to daily water loss (Cowan & Farquhar, 1977). These authors calculated that a minimal ratio of water loss to a given carbon gain would be reached when stomatal conductance varies in a manner which keeps the ratio of partial derivatives constant

$$\frac{dE_l}{dg_s} \bigg/ \frac{dP_n}{dg_s} = \frac{dE_l}{dP_n} = \text{Constant} \qquad \text{(Eq. 2.2.34)}$$

where dE_l/dP_n indicates the marginal water cost of carbon assimilation. The theory predicts that stomata will close in response to decreasing humidity. It is assumed that neither past leaf photosynthesis nor transpiration affect the rate of these processes in future. Thus, after-effects of water stress or increasing negative feedback of accumulated assimilates are not considered. Up till now only a few investigations have shown such optimal behaviour.

Perhaps other criteria are also to be met by optimizing stomata. During prolonged periods of low radiation intensities combined with high humidities, uptake of nutrients along with the transpiration stream could be too low to meet requirements for growth. Enhanced stomatal opening could enhance nutrient uptake, and have little effect on photosynthesis in the short-term. At another extreme, i.e., at periods of high radiation levels, high leaf temperatures could be supra-optimal for photosynthesis, or could even cause irreversible damage to tissue. Stomata could then, as suggested by Schulze & Hall (1982), increase opening to increase transpiration, reduce leaf temperature (closer to the photosynthetic optimum), or prevent tissue necrosis. The boundary layer could affect the optimization behaviour of stomata. A high boundary layer resistance tends to dominate total (boundary layer + stomatal) resistance when stomatal opening becomes large. In that case variation of stomatal conductance has little effect on transpiration.

2.2.3.6 Effect of the boundary layer

Greenhouse crops have a low boundary layer conductance as compared with field crops. A low g_b could have significant effects on gas exchange of leaves and stomatal response. The presence of a thick boundary layer places a significant resistance in the diffusion pathways of H_2O and CO_2; this also causes conditions at the leaf surface different from those of the greenhouse bulk air. As a result of a low g_b, the air at the leaf surface may be humidified. Consequently D_s may be decreased compared with D_a. It restricts latent and sensible heat transfer, thereby affecting leaf temperature. However, as the saturated water vapour pressure of the air increases exponentially with temperature, D_s could also be increased compared with D_a, for example at high radiation and high air temperature. Collatz *et al.* (1991) calculated that, under these conditions, stomatal closure could result from low values of g_b. Additionally, the leaf boundary layer may decrease C_i if stomata do not fully restore the original C_i value by opening, in response to decreased C_i and increased humidity at the leaf surface.

Very few reports were found on responses of stomata to decreased g_b. Stomatal conductance increased in response to a decrease in g_b from 0.03 to 0.015 m s^{-1} (Bunce, 1985). However, in another experiment this author found that stomatal conductance was not altered with decreasing g_b (Bunce, 1988[b]). g_b decreased from 0.035 to 0.01 m s^{-1} when air speed was lowered from 4 to 0.4 m s^{-1}. At high light intensities and leaf-air VPD at 1.7 kPa, C_i decreased by 50 µmol mol^{-1}, which resulted in a 20% reduction in CO_2 assimilation.

2.2.3.7 Interaction at the crop level

In the previous sections the interactions between water loss and CO_2 uptake at the scale of a single leaf have been discussed. At the crop level, the rate of water uptake by the root system and plant internal conductance for water transport play an important role in determining the water content of plants. They could affect water loss and CO_2 uptake via stomatal response to plant water content. Water loss and CO_2 uptake is now made up of the combined rates of transpiration and photosynthesis of all leaves in the canopy. A gradual decline exists in the canopy from top to bottom in rates of leaf photosynthesis and leaf transpiration that is much dependent on the decreasing level of (average) radiation intensity.

Still very little is known about stomatal behaviour, photosynthetic characteristics and diurnal water contents of the major greenhouse crops. Only a global picture can be drawn here of the interactions between crop photosynthesis and crop transpiration and water status.

Plants should control internal water status so as to prevent levels of water contents occurring that would cause damage to physiological processes. In the short-term, most active regulation takes place via stomatal opening; in the long-term, acclimation may take place during the growing season.

Short-term behaviour

Plant water content varies as a consequence of diurnally varying transpiration rates and plant internal resistances to water flow (section 2.2.2.4). Water content could decrease as stomata normally close not so much as to prevent lowering of the leaf water content. This may result in a decrease of plant water potentials, so that (passive) water uptake by the roots increases. As greenhouse plants are usually well-watered, it may be expected that under most conditions water uptake by the roots will not pose significant limitations to stomatal opening (section 2.2.2.5).

As the time course of environmental conditions is unknown beforehand, stomata must perform some feedforward control of transpiration, to reduce the risk of desiccation and water stress. A response to increasing transpirational demand by decreased stomatal opening limits photosynthesis to a larger extent. Thus plants should optimize between posing stomatal limitations to photosynthesis and damaging the photosynthetic capacity at extreme levels of climatic conditions. This short-term stomatal behaviour is adapted to the capacity of the root system for water uptake and to hydraulic conductances within the plant (Meinzer & Grantz, 1991).

At low to moderate radiation levels, plants can allow stomatal conductance to increase with increasing radiation in response to increased demand for CO_2 by photosynthesis, without causing water content to become too low. Humidity in the greenhouse would be increased by this response, enabling high conductances. Thus, commonly little negative effects of stomatal limitation or water stress on crop photosynthesis are likely to occur.

At high radiation the chance that transpiration and plant water content interfere with photosynthesis is greatly increased. The concomitant air humidity will greatly determine what will happen. Low air humidity could strongly increase transpiration. Depending on the crop response, stomatal opening could be reduced, or very low plant water contents could occur. High air humidity would reduce transpirational cooling and could, at high air temperature, lead to leaf temperatures which would be supra-optimal for photosynthesis. Different crop species could differ markedly in their response to these extreme conditions.

Long-term acclimation

During the growing season the plants may acclimate to changing average climatic conditions through morphological and anatomical changes. For example, photosynthetic capacity of leaves will probably increase with the increase in average light intensity from early spring to summer. As P_{nmax} and g_{smax} are often correlated, maximal transpiration rates would also increase. Higher stomatal conductances and higher photosynthetic rates are also promoted by the presence of actively growing organs (i.e. sinks for assimilates), which stimulate higher photosynthetic rates. Hall & Brady (1977) found that both P_{nmax} and g_{smax} were markedly reduced when flowering sweet pepper plants were prevented to develop fruits. When transpirational demand increases from spring to mid-summer, the size of the root system relative to canopy size may increase, osmotic adjustment may take place allowing lower water contents to be tolerated, or smaller leaves may be developed, aimed at diminishing the total transpiring leaf area. Also the short-term behaviour of stomata may change, so that a different balance between CO_2 uptake and water loss may be maintained. A different ratio of g_{smax} to P_{nmax} could reflect this change.

Strong acclimation to climate conditions in periods with low transpirational demand could result in a reduced potential for water uptake or stomata that are less sensitive to low air humidities. This increases the chances of desiccation or inadequate transpirational cooling when a sudden transition occurs to a period with increased transpirational demand, for example with the change from dull to bright weather.

In general, fast growing crops such as fruit vegetable crops, are likely to have higher transpiration rates than slowly growing crops, such as many ornamentals. Note that a high radiation load would be more harmful for crops that have genotypically fixed low stomatal conductances and that are less able to cool the leaves by transpiration.

2.2.3.8 Summary

Stomatal opening is correlated with the rate of leaf photosynthesis. Thus, stomatal responses to varying climatic conditions to a large extent reflect the effects of these conditions on leaf photosynthesis. A higher CO_2 concentration reduces stomatal conductance, but at high air humidity this reduction appears to be small. A decreasing humidity decreases stomatal conductance and generally increases transpiration, and increases stomatal limitation to photosynthesis. The capacity for photosynthesis is not affected by low or moderate humidities, at least in the short-term.

In greenhouse crops, stomatal conductance will normally be high as humidities are commonly high and water stress in the root environment is absent. Little limitation to photosynthesis is likely to occur. However, this freely transpiring behaviour could lead to very high rates of water loss in periods of high radiation and lower air humidities. Then crop photosynthesis is likely to become limited, either by increased stomatal limitation of CO_2 diffusion, or by water stress.

2.3 Long-term crop responses

2.3.1 Crop growth and development

H. CHALLA, E. HEUVELINK AND U. VAN MEETEREN

2.3.1.1 Introduction

In this chapter, dealing with the relation between environmental factors and crop growth, the focus in the first sections has been on photosynthesis and water relations. Though they have a great impact on crop growth and yield, they should be studied within the framework of the integrated response of crops to environmental factors, where the short-term response is superimposed on long-term reactions.

Long-term reactions are defined here as reactions that become manifest only after a period of, at least, one to several days, as opposed to those with a response time in the order of seconds, minutes, or hours, such as photosynthesis or the water status of the crop. Clear examples of long-term reactions are the formation of leaves, the flowering response, fruit set, or adaptations of the morphology of the plant.

For a proper discussion of the phenomena involved it is useful to distinguish growth and development. This is certainly necessary because different definitions are in use that are not quite compatible. Crop growth is defined here as the increase of biomass, or dimensions of a plant (quantitative aspects). Crop development is defined here according to Bidwell (1974) as "ordered change or progress often (but not always) towards a higher, more ordered, or more complex state". In this definition development is a phenomenon that is distinct from growth (in some definitions growth is considered as an aspect of development). Examples are not only phase transitions (e.g. juvenile to adult), but also formation and development of new organs, ageing, etcetera.

Both processes may proceed to a certain extent independently, for example after planting a freesia corm will initially lose weight, while development proceeds, and leaves are formed at the expense of the mother corm.

Growth, as defined before, is closely connected with carbon fixation and the carbon balance. The relation between the instantaneous rate of crop photosynthesis and climatic factors has been described in section 2.2.1, but obviously this relation depends on the light interception by the crop, which in turn depends on the leaf area index (LAI, the ratio of (single sided) leaf area to greenhouse area) and hence on previous crop growth. This is an example of feedback, where the transformation of photosynthate to leaf area also depends on environmental conditions. Crop growth thus has a delayed response to climatic factors (essentially with respect to leaf area growth) in addition to the immediate response (crop photosynthesis). Likewise, maintenance respiration, according to the present theory, responds immediately to (air) temperature, but since it is also linked to the amount of biomass, the rate of maintenance respiration at a given moment also depends on the integrated (and hence delayed) effect of environmental factors on accumulated dry matter.

The growth of crops, in general, follows a sigmoidal pattern (Figure 2.3.1). Initially, when the crop essentially exists of individually growing, young plants, the limiting factor is light interception (essentially LAI), and growth proceeds approximately exponentially. When LAI increases, light interception becomes less sensitive to LAI and growth proceeds approximately linearly. This stage of establishment and maturation is characterized by formation and growth of the harvestable products.

2.3.1.2 Crop growth

Growth of young plants

After emergence or transplanting plants usually grow as individuals, in the sense that competition for radiation among plants is still negligible. In that stage, which is characterized by a low LAI, the rate of crop photosynthesis is almost proportional to LAI (Figure 2.3.2). Crop photosynthesis in turn provides the necessary building blocks and energy for growth and hence for increase in dry weight per unit greenhouse area (W) and in LAI. Because increase in LAI is positively correlated with daily crop photosynthesis and the rate of crop photosynthesis is proportional to LAI, crop growth in this stage represents a self amplifying process, where crop growth rate, in general, will not be constant (Figure 2.3.1). In order to compare growth of young crops, of different weight, or at different environmental conditions, the absolute growth rate is therefore not a suitable criterion.

In order to account for the (LAI mediated) effects of plant weight on growth rate a common solution is to evaluate weight increase on a logarithmic basis:

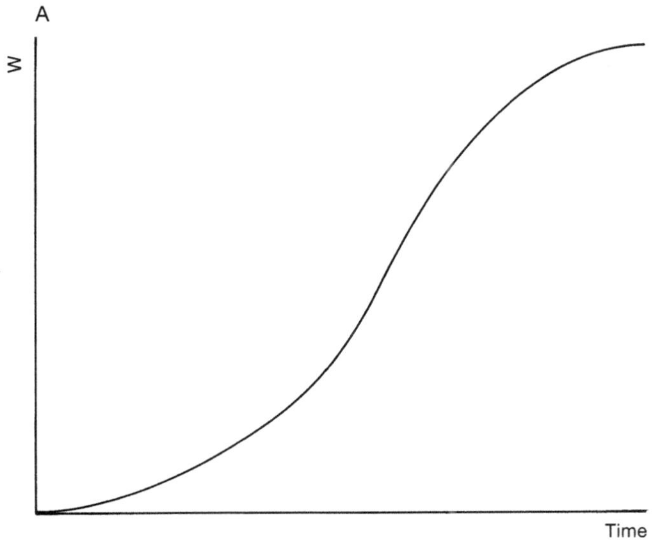

Figure 2.3.1 – Schematic, generalized representation of the time course of the weight of a crop: approximately exponential growth in the young stage, followed by approximately linear growth after canopy closure, and finally slowing down of growth due to ageing.

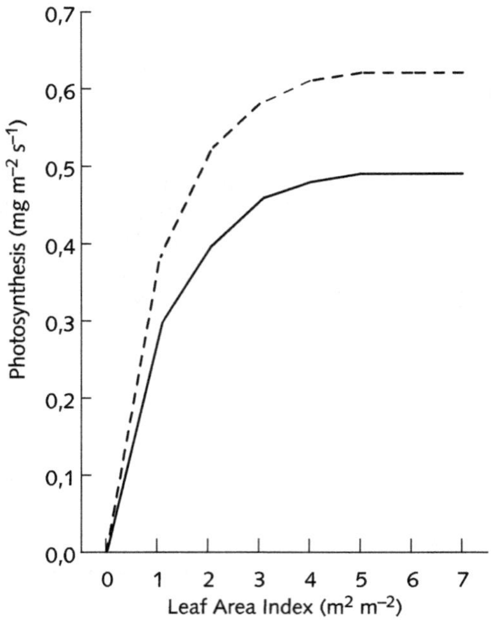

Figure 2.3.2 – Relation between LAI and crop photosynthesis at 340 µmol mol^{-1} CO_2 (solid line) and at 1000 µmol mol^{-1} CO_2 (broken line) under the following conditions: PAR = 50 W m^{-2}; fraction diffuse = 0.5; sun height = 30 °; temperature = 20 °C, spherical leaf angle distribution.

$$RGR = [\ln(W_{t2}) - \ln(W_{t1})] / (t_2 - t_1) \qquad \text{(Eq. 2.3.1)}$$

and to define RGR as the average relative growth rate over period t_1 to t_2 (g g^{-1} d^{-1}), where W_t is crop dry weight in g per unit greenhouse area at time t (compare Figure 2.3.9a). Equation (2.3.1), after rearranging gives rise to:

$$W_t = W_i \, e^{RGR \times dt} \qquad \text{(Eq. 2.3.2)}$$

where W_i is initial weight, W_t weight after t days. In other words, at constant RGR growth is exponential (Figure 2.3.1), and RGR thus may be compared with the rate of interest in compound interest computations (Hunt, 1978).

From equation (2.3.1) the instantaneous relative growth rate (r_t) can be derived when considering a time interval dt:

$$r_t = dW/(W_t \, dt) \qquad \text{(Eq. 2.3.3)}$$

which demonstrates the proportionality between the instantaneous absolute growth rate and the weight of the crop:

$$dW/dt = W_t \, r_t \qquad \text{(Eq. 2.3.4)}$$

As long as the environmental conditions remain the same and the relation between LAI and W does not change, r_t remains also the same, provided that little intra- and inter-plant shading occurs (low LAI). Obviously this is, however, rarely the case over prolonged periods of time where gradually a transition will occur to a closed canopy (next paragraph).

For further analysis of growth of young plants it is useful to introduce an important concept: the leaf area ratio (LAR), defined as the ratio of leaf area over plant dry weight:

$$\mathrm{LAR}_t = A_t / W_{p,t} \qquad \text{(Eq. 2.3.5)}$$

where A_t is leaf area per plant (m^2) at time t and $W_{p,t}$ is plant dry weight (g) at time t. LAR reflects the "effort" a crop puts in using dry weight in the formation of leaf area. Note that by expressing leaf area and crop weight on a greenhouse area basis equation (2.3.5) can be rewritten as:

$$\mathrm{LAR}_t = \mathrm{LAI}_t / W_t \qquad \text{(Eq. 2.3.6)}$$

When comparing two crops of young plants of given weight, under otherwise identical conditions, the crop with the higher LAR will have a higher LAI, and hence a higher rate of photosynthesis. This is a very important conclusion, because it shows that, apart from the rate of photosynthesis per unit leaf area (P_g), LAR also affects crop growth at this stage, though the response time is slower.

The rate of photosynthesis per unit leaf area is in a way reflected in the rate of growth per unit leaf area (g m^{-2} d^{-1}):

$$\mathrm{NAR}_t = \mathrm{d}W_{p,t} / A_t \, \mathrm{d}t \qquad \text{(Eq. 2.3.7)}$$

which is the definition of the net assimilation rate (NAR). NAR reflects P_{gp}/A, but it is not the same; compare equation (2.2.1) with the following:

$$\mathrm{NAR}_t = 0.68\, C_f (P_{gp,d} - R_{mp,d}) / A_t \qquad \text{(Eq. 2.3.8)}$$

where 0.68 C_f is the conversion efficiency of CO_2 fixation to dry matter (note that the factor 0.68 makes the conversion from CO_2 units to CH_2O, or glucose units), $P_{gp,d}$ the rate of gross photosynthesis and $R_{mp,d}$ the rate of maintenance respiration, both expressed per plant. If $R_{mp,d}$ is ignored, which is acceptable if it is small compared to $P_{gp,d}$, NAR is proportional to (average daily) $P_{gp,d}/A_t$.

From the definitions of NAR and LAR follows that the instantaneous relative growth rate:

$$r_t = \mathrm{NAR}_t \times \mathrm{LAR}_t \qquad \text{(Eq. 2.3.9)}$$

or

$$\mathrm{d}W / (W_t \mathrm{d}t) = (\mathrm{d}W_p / A_t \mathrm{d}t) \times A_t / W_{p,t} \qquad \text{(Eq. 2.3.10)}$$

Though strictly spoken this relation does not hold for the *average* values of these components of the growth analysis over a given period of time, there is nevertheless a close relation between RGR and the product of average NAR and LAR. In other words, the relative growth rate of a crop of individually growing young plants can be explained by the average net rate of photosynthesis per unit leaf area and by the efficiency of the plant in producing leaf area per unit dry weight formed (Hunt, 1978).

Though growth analysis is a valuable tool in understanding and analysing growth of young plants, an important complication in predicting RGR is the mutual dependence of NAR and LAR: these components often show a negative correlation (Thornley & Hurd, 1974; Bruggink & Heuvelink,

1987; Heuvelink, 1989; Bruggink, 1992). This problem will be discussed later in more detail.

A further analysis of LAR is possible by distinguishing its two components: the leaf weight ratio (LWR) expressing the dry matter distribution over leaves and other organs (g g^{-1}), and specific leaf area (SLA), representing the leaf area per unit dry weight in the leaves (m^2 g^{-1}):

$$LAR_t = SLA_t \times LWR_t \qquad \text{(Eq. 2.3.11)}$$

where:

$$LWR_t = W_{l,t} / W_{p,t} \qquad \text{(Eq. 2.3.12)}$$

$$SLA_t = A_t / W_{l,t} \qquad \text{(Eq. 2.3.13)}$$

where $W_{l,t}$ is dry weight of the leaves per plant at time t.

In general SLA is more sensitive to environmental change and more prone to ontogenetic drift than LWR (Hunt, 1978), so variations in LAR are usually primarily caused by variations in SLA. Note that 1/SLA, the leaf dry weight per unit leaf area, can be interpreted as a measure of (dry) leaf thickness. For further details and in depth discussions on growth analysis see Hunt (1978, 1982).

Growth of a closed canopy

The growth analysis described above applies for analysis of the growth of young, widely spaced plants. In closed canopies, leaf area expansion is of minor importance, because most light will be intercepted at higher LAI anyway and further increase of LAI has only a marginal effect on crop photosynthesis (Figure 2.3.2). Therefore it is more meaningful here to consider absolute crop growth rate (GR), defined as $\Delta W/\Delta t$ (g m^{-2} d^{-1}), which in analogy with equation (2.3.8), is related to crop photosynthesis as (section 2.2.1.):

$$GR = 0.68\, C_f (P_{gc,d} - R_{mc,d}) \qquad \text{(Eq. 2.3.14)}$$

where $P_{gc,d}$ is average daily rate of crop gross photosynthesis (g CO_2 m^{-2} d^{-1}), and $R_{mc,d}$ average daily rate of crop maintenance respiration (g CO_2 m^{-2} d^{-1}). Since $R_{mc,d}$ is proportional to W (section 2.2.1) it usually cannot be ignored in closed canopies but rather may play an important role at high W and low radiation (low $P_{gc,d}$).

GR may be computed according to the elaborate procedures described in section 2.2.1, but an approximation may be obtained according to Goudriaan (1982), where radiation intercepted by the crop is multiplied by an average crop light use efficiency (α_c):

$$P_{gc,d} = \alpha_c I_\Sigma (1 - \rho)(1 - e^{-K \times LAI}) \qquad \text{(Eq. 2.3.15)}$$

where α_c is average crop light use efficiency for CO_2 uptake (g MJ^{-1}), I_Σ daily integral of photosynthetic active radiation (PAR) at the top of the canopy (MJ m^{-2} d^{-1}), ρ reflection coefficient of the canopy, K extinction coefficient of PAR in the canopy and LAI is the leaf area index. Numbers obtained for field crops are (Goudriaan, 1982): $\alpha_c = 6.6$ g MJ^{-1} (at ambient CO_2), $\rho = 0.1$, $k = 0.7$. At an LAI of 3 ("closed" canopy) $P_{gc,d} = 6.6 \times 0.79\, I_\Sigma$ (g CO_2 m^{-2} d^{-1}). If we adopt a value of 0.7 for C_f (equation (2.3.14)), it can be concluded that $\Delta GR \approx 2.5 \times \Delta I_\Sigma$ (g m^{-2} d^{-1}) under atmospheric CO_2 concentration (the relation has been given as a difference equation to eliminate maintenance respiration). Note that I_Σ refers to PAR at crop level, which is appreciably lower than radiation outside the greenhouse.

Another point that should be observed here is that the relation between radiation and crop photosynthesis, though less pronounced than for individual leaves (section 2.2.1) is not linear. The average light use efficiency of a crop therefore, also because of variations in radiation interception by the crop will depend on the radiation conditions. On the other hand, more detailed calculations based on the theory of section 2.2.1, covering potential greenhouse production in a wide range of climates and seasons, provided evidence for a linear, but reasonably accurate relation between (predicted) potential daily crop growth and the daily integrals of direct and diffuse global radiation (Challa & Bakker, 1995). Based on this relation a value of 2.7 (30% diffuse radiation), and 4.8 g MJ^{-1} (100% diffuse radiation) can be derived for $\Delta GR/\Delta I_{\Sigma}$ at ambient (340 µmol mol^{-1}) CO_2, and respectively 3.6 and 6.0 g MJ^{-1} at 1000 µmol mol^{-1} CO_2, assuming an average transmission percentage of 64% of the greenhouse cover and 47% PAR in global radiation (Gijzen, 1992).

Influence of environmental factors on crop growth

Introduction

After the previous descriptions of the mechanism of dry matter production, in this paragraph we focus on the (long term) response of crop growth to environmental factors. A distinction should be made between extreme conditions, where plants are experiencing stress, manifested by hampered growth and development and the occurrence of damage, and the range of conditions where stress does not play a role. In this treatise we will mainly consider conditions without stress, though prevention of stress or damage is an important aspect of climate control. The problem in dealing with stress phenomena is the complex physiological background, the resulting lack of understanding and the lack of systematic studies on individual crops. Existing knowledge is mostly empirical, often quite crop specific and not easily accessible (Challa, 1990).

Effects of environmental factors on processes with a short term response, such as photosynthesis and respiration have already been discussed before (section 2.2.1). The relation between photosynthesis and structural dry matter production in the long run (where short-term storage of assimilates may be ignored) is given by equation (2.3.14).

As has been pointed out the quantitative basis for growth and its analysis of young, isolated growing plants and crops growing in a more or less closed canopy differs. This distinction will be reflected in the following treatise. Before dealing with the various external growth factors, however, it is important to notice that apart from the role of environmental factors, large genotypic (between species and within species) differences may exist with respect to the level and the kind of response to these factors. For young plants such differences are particularly reported for SLA and LWR, and hence for LAR (e.g. Nieuwhof *et al.*, 1991; Vlahos *et al.*, 1991; Bruggink, 1992). In closed canopies photosynthetic properties of leaves and light interception and distribution within the canopy may differ to a certain extent among species and varieties, but the most important genetically determined variations in yield should be ascribed to assimilate (re-)distribution (Evans, 1990; section 2.3.2).

Although the major environmental factors will be discussed separately, it should be noticed that (strong) interactions may exist. In section 2.2 the interaction between climatic factors and short-term responses of crops, as well as the interactions between the processes involved have been considered. Such interactions, of course, are also expressed in crop growth. Often, however, interactions between factors are poorly understood, such as e.g. the effect of CO_2 enrichment at different humidity levels (Mortensen, 1987), or the harmful effect of high humidity (calcium deficiency in tomato leaves) at high solar irradiance (Aikman & Houter, 1990).

Irradiance

Only quantitative effects of irradiance will be considered here; qualitative effects of day length and light quality are discussed in section 2.3.3. The major effect of irradiance on growth is through photosynthesis. For closed canopies this photosynthetic effect was already related to crop growth, but the response of young isolated plants is more complex, due to the already mentioned interaction of NAR and LAR, and the direct exposure of all leaves to radiation, resulting in a much stronger non-linear relation between photosynthesis and radiation.

A rectangular hyperbolic equation adequately describes the relation between daily radiation integral (I_Σ) and NAR (Harssema, 1977; Bruggink & Heuvelink, 1987; Bruggink, 1992):

$$NAR = NAR_{max} \times I_\Sigma / (I_\Sigma + I_{1/2s}) \qquad \text{(Eq. 2.3.16)}$$

where NAR_{max} is the light-saturated net assimilation rate and $I_{1/2s}$ the irradiance level at half the light-saturated rate.

This response resembles the photosynthesis–light response curve of leaves, but it differs from it because daily light integrals are used as an input. Therefore the parameters will be more sensitive to the radiation conditions of the experiments. This might explain the higher values of NAR for young tomato plants observed in growth chamber experiments compared to greenhouse experiments, the latter showing day-to-day fluctuation in radiation sum as well as fluctuation in radiation intensity during the day (Bruggink, 1992). For the same reason, a lower light intensity at the same daily light integral, thus resulting in a longer photoperiod, should give rise to a higher NAR. Vlahos *et al.* (1991) observed a higher NAR and a higher RGR (LAR remained unchanged), at lower light intensity at the same daily light integral with three morphologically different *Achimenes* cultivars. Craker *et al.* (1983) observed with radish that RGR was approximately 10% higher when the same light integral was given during 16 h d^{-1} instead of 8 h d^{-1}.

LAR, in general, is negatively correlated with light, which is mainly caused by the response of SLA, whereas LWR is not much affected (Nilwik, 1981[b]; Bruggink, 1992). At high radiation, in young tomato, cucumber and sweet pepper plants the response of NAR to changes in radiation was partly compensated for by adaptations in LAR (Bruggink, 1987). An increase in NAR of 10% resulted in summer in a decrease in LAR of 4% and therefore RGR increased by only 6%. The relation between NAR and LAR can be described by a hyperbolic equation (Thornley & Hurd, 1974; Bruggink & Heuvelink, 1987; Bruggink, 1992):

$$1/LAR = A + B \times NAR \qquad \text{(Eq. 2.3.17)}$$

Combining equations (2.3.16) and (2.3.17) and taking into account that, by approximation, RGR = NAR × LAR, it follows that RGR, like NAR (equation (2.3.16); Bruggink & Heuvelink, 1987) is related to irradiance according to a rectangular hyperbola. Such a response has been reported for tomato, cucumber, sweet pepper and carnation plants (Nilwik, 1981[a]; Bruggink & Heuvelink, 1987; Bruggink, 1992).

The close link between NAR and photosynthesis suggests that differences in RGR among species are most likely caused by differences in LAR, since photosynthesis responses of many species resemble each other. Bruggink (1992) observed for young tomato and carnation plants, a fast and a slowly growing plant species respectively, that the response of NAR to the mean daily light integral was not much different. The higher RGR of tomato was caused by a higher LAR, which resulted from a higher SLA and a higher LWR in tomato. For both species the reciprocal of LAR was linearly related to NAR (equation (2.3.17)). The sensitivity of LAR to NAR increased with increasing NAR and was considerably higher for tomato than for carnation. Thus RGR of carnation is much more sensitive to radiation than that of tomato (Bruggink, 1992; Figure 2.3.3).

In the previous paragraph it was pointed out that growth rate of a closed canopy is closely related to crop photosynthesis and hence to radiation. Theoretically growth rate should be approximately linearly related to the daily light integral intercepted by a crop, according to equation (2.3.15), but

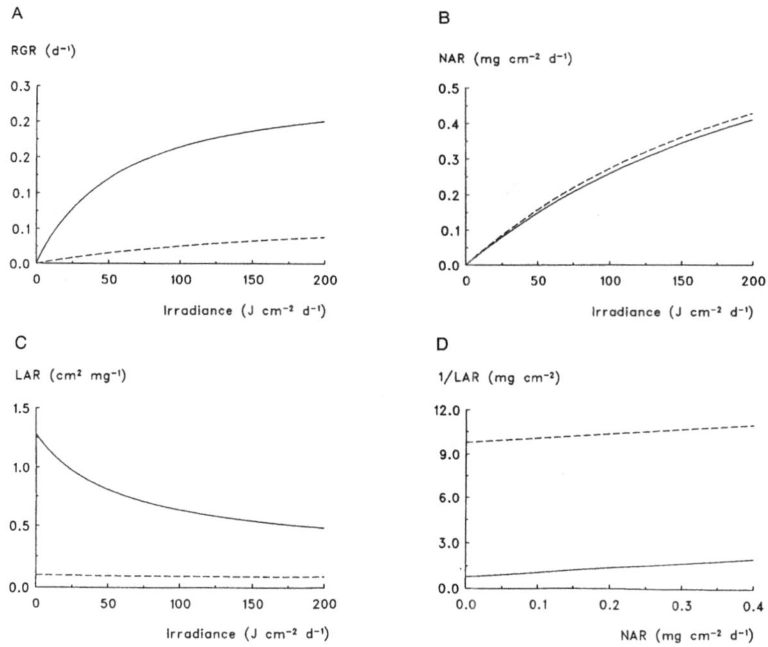

Figure 2.3.3 – The relation between (A) RGR and irradiance; (B) NAR and irradiance; (C) LAR and irradiance; and (D) 1/LAR and NAR for tomato (———) and carnation (- - - -). Irradiance is mean daily light integral (PAR). After Bruggink (1992).

experimental proof is still scarce. De Koning (1993) observed a hyperbolic relationship between PAR intercepted by a tomato crop (range 2–7 MJ m^{-2} d^{-1}) and crop growth rate (g m^{-2} d^{-1}), with an average light use efficiency of about 2.5 g MJ^{-1}, which compares well with the values derived theoretically. The observed non-linearity may be attributed to the coincidence of high radiation and low CO_2, the negative correlation between radiation level and the fraction of diffuse radiation (reduced light use efficiency with more direct radiation), and the low LAI values prevailing in summer.

More results are reported on the relationship between yield and radiation. Yield is related to crop growth through the harvest index (dry matter distribution) and the dry matter content of the harvestable product (Challa & Schapendonk, 1986). For several crops a linear relationship between cumulative crop growth and cumulative dry weight of harvestable product was observed (Challa & Heuvelink, 1993). Dry matter content of the harvestable product, however, may vary considerably during the season, e.g. De Koning & De Ruiter (1992) reported variations in dry matter content in tomato fruits between 5.1% in spring and 6.4% in summer. Such differences may seem unimportant, but they give rise to great differences in (fresh) yield at equal dry weight production. Therefore the response of yield to irradiance, may not be proportional to that of dry weight production.

Cockshull (1988) reported a linear relationship between cumulative yield and cumulative solar radiation at crop height within a greenhouse for tomato. Drews *et al.* (1980) and De Visser & Vesseur (1982) showed a linear relationship between the amount of radiation received and the weight of fruit produced in cucumber. These authors concluded that 1% less light would reduce production of cucumber by about 1%, at least in the early part of the year. Cockshull (1988) discussed the problems associated with comparing cumulative yields with cumulative light integrals. If the linear relationship passes through the origin, 1% less light will always give 1% less yield, but other ratios may be obtained if this is not the case. Often the effect of radiation reduction is evaluated by comparing cumulative yield at a predetermined moment during cultivation. It has been shown that, theoretically, this procedure, especially when the moment of evaluation is chosen early in the production phase, is highly sensitive to the developmental stage of the crop (Challa & Schapendonk, 1984), and therefore basically incorrect (Figure 2.3.4).

De Koning (1989[b]) described a linear relationship between cumulative (fresh) yield and total PAR received by a tomato crop from first anthesis. The slope of this relationship (41.5 g MJ^{-1}) reflects the average light-utilisation efficiency from the start of harvest. Cockshull (1988) reported a value of 39.6 g MJ^{-1} for this parameter, whereas Bailey & Hunter (1988) obtained a value of 42.6 g MJ^{-1} in an experiment with 6 glasshouses covered with different materials. In a shading experiment (6.4% and 23.4% reduction of solar radiation incident on tomato plants) Cockshull *et al.* (1992) observed over the first 14 weeks of harvest (February to May), regardless of treatment, 2.01 kg fresh weight of fruit harvested for every 100 MJ of global radiation incident on the crops from the onset of harvest. This corresponds to 43 g MJ^{-1} PAR, if PAR is 47% of global radiation (Gijzen, 1992). If we consider 40 g MJ^{-1} PAR as a good average for the light use efficiency of tomato yield and assume a fraction of assimilates diverted to the fruits of 0.72 (De Koning & De Ruiter, 1992) and a fruit dry matter content of 0.055, the corresponding average light use efficiency for dry matter production would be about 3.1 g MJ^{-1} PAR, which compares well with the values indicated before.

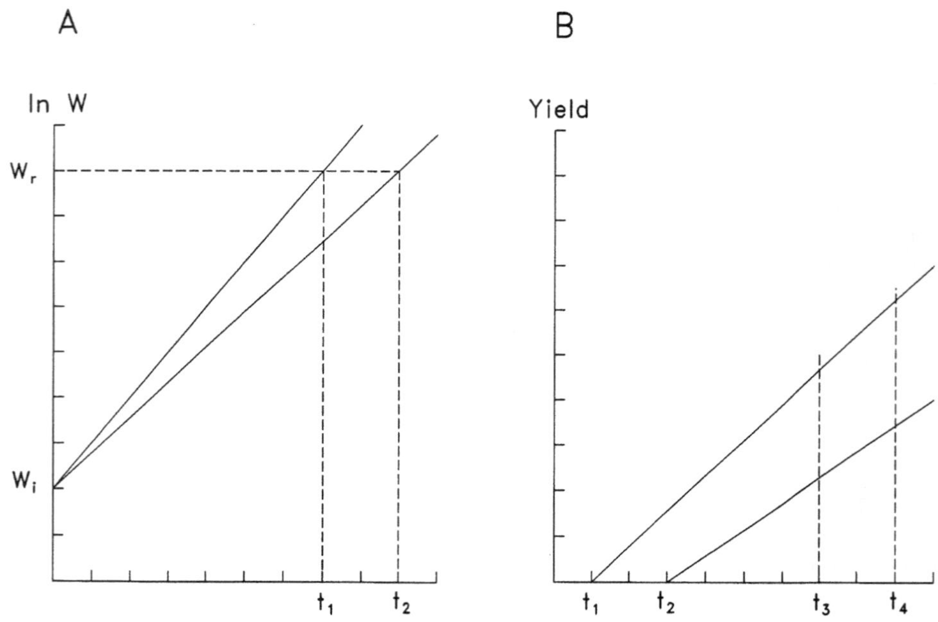

Figure 2.3.4 – Schematic representation of the problem of evaluating effects of PAR on yield by comparing cumulative yield at a given reference moment t3 or t4. This results in quite different assessments: the delay in growth (a) results in a later (t2-t1) start of the production phase (b), and a relative effect on yield strongly dependent on the choice of the reference time. (after Challa & Schapendonk, 1984).

Shading is widely used to reduce greenhouse temperature during the summer months, but it frequently results in yield reduction, for example in eggplant (Wolff and Coltman, 1990), roses (Chandler & Watson, 1954; Coker & Hanan, 1988) and tomato (Cockshull *et al.*, 1992). In The Netherlands tomato yield was reduced by 10%, even when a mobile sun screen was only closed at greenhouse air temperature > 25 °C and at outside global radiation level > 650 W m^{-2} (Van Holsteijn, 1990). In the cultivation of shade plants (many pot plants), however, shading may be necessary in summer, as they are not able to withstand high radiation levels. Without shading they may show reduced growth or even necrosis, e.g. cyclamen, *Kalanchoë* and *Saintpaulia ionantha* (Larson, 1992).

Supplementary lighting (section 4.7) is increasingly used in practice, mainly at higher latitudes when radiation conditions are limiting. The main objectives of its application are: to prevent failure with certain, light demanding crops, to increase yield and to improve product quality. In addition improved control of the production process and levelling out of labour requirements throughout the year may play a role. Supplementary lighting is most commonly applied in the cultivation of ornamentals, but it is also used with vegetables (Andersen & Hansen, 1989; McAvoy & Janes, 1988; Mortensen & Grimstad, 1990; Dorais *et al.*, 1991). Hendriks (1992) estimated the greenhouse area in Europe equipped with supplementary lighting to be over 1600 ha, with the largest area in The Netherlands (approximately 800 ha) and the highest percentage in Denmark (about 50% of the area of ornamentals).

For rose, vegetative growth and flower production are reported to correlate with intensity (< 250 µmol m^{-2} s^{-1} PAR) and duration of supplementary lighting (reviewed by Zieslin & Mor, 1990). In chrysanthemum the level and the duration of artificial lighting had significant effects on growth and time up to harvest (Andersson, 1990). For many years supplementary lighting of stock plants (e.g. *Pelargonium*) has been used to improve production and quality of cuttings during the winter season (Hendriks, 1992).

The light use efficiency of artificial light may differ from that of natural light, due to the different spectrum and the wavelength dependency of quantum efficiency (McCree, 1972). Sagar *et al.* (1982) concluded, based on a theoretical evaluation, that the relative photosynthetic yield of high pressure sodium lamps, the most common light source in horticulture, should be 34% higher than that of natural light. Several authors observed differences in growth with different lamp types, at the same photon flux density (Andersen, 1986; Mortensen & Strømme, 1987), but these differences could also be attributed to morphogenetic effects (next paragraph).

Negative effects of supplementary light on some crops have been reported, for example leaf necrosis in certain rose varieties, leaf yellowing, necrotic spots (Hendriks, 1992), chlorosis and reduced growth in tomato following several days of uninterrupted light (Bradley & Janes, 1985).

Temperature

Over a wide range temperature has only a minor effect on photosynthesis (section 2.1.1). In particular, at the level of a whole crop the response is even less than that observed for individual leaves (Challa, 1990). Its effect becomes mainly manifest at high radiation and CO_2 levels through enhancement of the intrinsic photosynthetic capacity, and at low CO_2 levels through its role in the competition between CO_2 and O_2 for Ribulose-1,5-*bis*-phosphate-carboxylase-oxygenase (photorespiration). The main effect of temperature on crop growth should be considered in relation to the stage of the crop: in young crops it plays a role in leaf expansion and hence in the interception of radiation, in clos-ed canopies its main effect is through maintenance respiration.

Growth analyses confirm that in young plants of many crops the main effect of temperature on RGR may be attributed to its effect on LAR whereas there is only a minor effect on NAR. This has been found for cucumber (Challa & Brouwer, 1985), sweet pepper (Nilwik, 1981[b]), tomato (Heuvelink, 1989), *Ficus benjamina* and *Schefflera* (Vogelezang, 1993). Only at temperatures below 18 °C did a negative effect of temperature on NAR become manifest in cucumber, which was accompanied by a lack of chlorophyll in the leaves (Kleinendorst & Veen, 1983). Brouwer observed for oat (1973), and maize (1974) an interaction with irradiance: at low irradiance the temperature effect on LAR was most important, whereas at higher irradiance a rise in temperature affected RGR mainly by stimulating NAR.

LAR is positively correlated with temperature, mainly due to the response of SLA, whereas LWR tends to be independent of temperature (Harssema, 1977; Nilwik, 1981[b]; Heuvelink, 1989). LAR and SLA, just as stem length, respond not only to the average diurnal temperature, but also to DIF, the difference between day and night temperature (section 2.3.1.3). In tomato SLA was positively affected by DIF (Heuvelink, 1989). The response of LAR and SLA to varying temperature may be very complex: in cucumber alternating high (25 °C) and low (15 °C) temperatures during the night for 2 or 4 h resulted in a higher LAR than for alternating nights at 25 °C and 15 °C (Challa & Brouwer, 1985). These authors also observed that LAR was more affected by night temperature following a bright day than by night temperature following a dull day, suggesting that the effect of temperature on leaf expansion may depend on availability of assimilates and hence on plant growth rate.

The effect of temperature on the growth of closed canopies, as pointed out before, is mainly through maintenance respiration. The rate of maintenance respiration $R_{mc,d}$ has been described as an exponential function of temperature (Penning de Vries & Van Laar, 1982), with a Q_{10} of about 2 (meaning that its rate is doubled with a rise in temperature of 10 °C). The (relative) sensitivity of crop growth for $R_{mc,d}$, according to equation (2.3.14), depends on the daily rate of crop gross photosynthesis, $P_{gc,d}$. Under poor light conditions, therefore, temperature effects on growth will be greater than with ample light (Figure 2.3.5). As maintenance respiration is proportional to crop dry weight (section 2.2.1), the effect of temperature on crop growth will be greater the higher the weight per unit greenhouse area, W (Figure 2.3.5).

Though air temperature is, in general, more important for crop growth than root zone temperature (e.g. Harssema, 1977; Kleinendorst & Veen, 1983; Vogelezang, 1993), its effect cannot be fully ignored, since modern greenhouse technology enables independent control and it has some distinct effects on certain crops. The effect of root temperature on plant growth was reviewed by Cooper (1973) and more recently by Vogelezang (1993). Vogelezang (1993) concluded that, in general, root-zone heating had a positive effect on crop production, but interactions may exist with above ground factors, such as air temperature, radiation and day length. Root-zone temperature was most critical

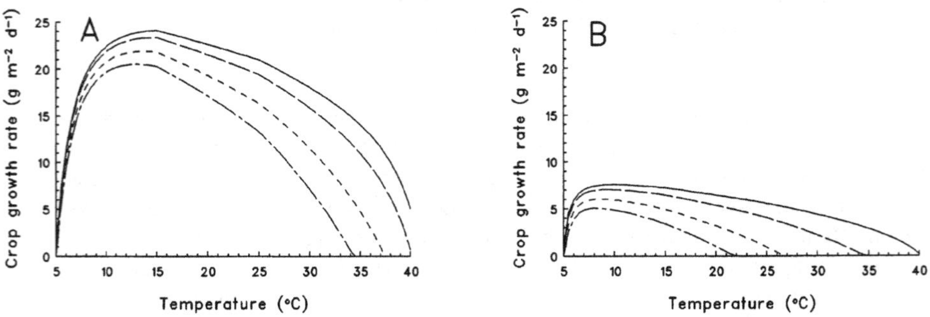

Figure 2.3.5 – Simulated crop growth rate at 340 mol mol^{-1} CO_2 on May 30 (52° North) as a function of temperature at crop dry weights of 100 (——), 200 (– – – –), 400 (- - - - -), and 600 g m^{-2}(– · – · –), at a daily global radiation of (A) 20 and (B) 5 MJ m^{-2} d^{-1}. LAI was 3 and the ratio leaves : stem : fruits : roots was 4 : 2 : 4 : 1.

for the developmental processes shoot formation and flowering (Vogelezang, 1993). Root-zone temperature, however, also had a positive effect on LAR in *Saintpaulia ionantha* (Vogelezang, 1988), *Poinsettia* and tomato (Janes *et al.*, 1981). Vogelezang (1993) ascribed this positive effect to a reduction in root resistance to water flow and hence to an improved water balance of the crop.

Carbon dioxide

A positive effect of CO_2 enrichment on growth rate, irrespective of light conditions, is now well established (Mortensen, 1987), with effects on yield of mature C_3 crops in the order of 27% (Kimball, 1986). This positive effect is mainly attributed to enhanced photosynthesis (Enoch, 1990), due to an increased rate of CO_2 fixation by the photosynthetic enzyme Ribulose-1,5-*bis*-phosphate-carboxylase-oxygenase (Rubisco) and concomitant suppression of photorespiration (Hand, 1990; section 2.2.1). CO_2 concentrations higher than 1000 μmol mol^{-1} may cause growth reduction and leaf injury (Madsen, 1968; Auge *et al.*, 1984; Ehret & Jolliffe, 1985), but it is also possible that contaminants, such as nitric oxide, nitrogen dioxide, and unburned hydrocarbons such as ethylene and propylene play a role, when flue gases are used as a source (Hand, 1990). The influence of CO_2 concentration in the air on the rate of crop photosynthesis and its interactions with temperature and radiation have been discussed above (section 2.2.1). Idso *et al.* (1987), for example, demonstrated this interaction in several crops (e.g. carrot and radish), where above 19 °C CO_2 enrichment (640 μmol mol^{-1}) increased biomass, whereas control plants (ambient CO_2) were heavier below 19 °C. A consequence of the non-linear response of crop photosynthesis to CO_2 and the interactions with radiation and temperature is that it is difficult, and not justified, to try to quantify the effect of CO_2 on crop growth other than in a very approximate way: its effect strongly depends on the diurnal control strategy followed, in combination with the dynamics of all other relevant factors.

Leaves of CO_2 enriched plants are usually thicker (Madsen, 1968; Enoch, 1990), resulting in a lower SLA and LAR. Thornley and Hurd (1974) concluded that with young tomato plants the relation between LAR and NAR (equation (2.3.17)) holds for different levels of radiation and CO_2. The effect of CO_2 concentration in the air on RGR of young plants can thus be largely compared with that of radiation, since both factors affect photosynthesis and hence NAR.

After prolonged exposure to increased CO_2 concentration, the positive effect of CO_2 enrichment may decline, due to acclimation of the crop (Bruggink, 1984) and associated deterioration of the photosynthetic properties. Although the occurrence of acclimation has been doubted (Picken *et al.*, 1986), recent literature provides strong evidence of the phenomenon, though it does not prove that it always occurs. Hicklenton & Jolliffe (1980), Yelle *et al.* (1990) and Besford *et al.* (1990) all found that the positive effect of CO_2 on the rate of photosynthesis in tomato decreased after prolonged exposure to CO_2 enrichment. They concluded that the main cause of acclimation was a decline in activated Rubisco.

Humidity

The relation between air humidity or water vapour pressure of the air and crop growth is rather complex. Grange and Hand (1987) concluded in their literature review that humidities between 0.2 and 1.0 kPa vapour pressure deficit (VPD) had little effect on the physiology and development of horticultural crops. Picken (1984) mentioned the same range in his review of humidity effects on pollination in tomato. Hoffman (1979) reviewed the effects of humidity on 26 crops and reported that growth was adversely affected when the VPD was above 1 kPa, but that there was little or no effect between 0.3 and 1.0 kPa. A comprehensive analysis has been made by Bakker (1991[a]). The main effect is on leaf expansion which is favoured by high humidity (through an improved water balance), but may be counteracted by a negative effect, in some crops, caused by calcium deficiency in the leaves, through a reduction of transpiration (Bakker, 1990; Holder & Cockshull, 1990). These authors also indicated that the occurrence of calcium deficiency depends on the period (day or night) when VPD is reduced. According to Bakker (1991[a]) the positive effect of air humidity on stomatal conductance results in only a minor effect on crop photosynthesis.

Increased air humidity (reduced VPD) enhanced RGR of young cucumber (Van de Sanden & Veen, 1992; Figure 2.3.6) and of young tomato plants (Klapwijk, 1975), the latter only at high irradiance. However, the causes reported in the literature are not consistent. Van de Sanden & Veen (1992) reported that at high VPD, in the range of 0.8–1.4 kPa, the raise in RGR was attributed to an increase in NAR, caused by an increased stomatal conductance. At low VPD (0.2–0.8 kPa) RGR was increased by a higher SLA. Also Burrage (1988) observed an increased SLA for tomato plants at high humidity, but growth increase at higher relative humidity (low VPD) observed by Klapwijk (1975) resulted from a higher NAR. Also Bakker (1991) observed a small increase in NAR of young tomato plants, grown at high humidity, without any significant effect on SLA or LAR. It is likely that, at least in young plants, other factors interact with the effect of VPD on growth.

Humidity also plays a role in the occurrence of diseases and their epidemics. Spores of many fungal diseases require liquid water for germination. High air humidity inside the greenhouse promotes condensation on the crop (Hand, 1988).

Relation between instantaneous crop growth and yield

In this chapter the effects of environmental factors on crop growth have been discussed, but in order to apply this knowledge in relation to climate control, it is important to consider the effect of instantaneous variations in growth (ΔW) on yield and on economic output. Various solutions for this problem have been suggested (Van Henten & Bontsema, 1991; Challa & Heuvelink, 1993; Seginer, 1993). Seginer (1993) and Van Henten & Bontsema (1991) derived an optimal path for the crop state over a complete crop cycle by means of a Hamiltonian function, where the relation between the greenhouse environment and crop performance is evaluated by means of a crop growth model. In their approach the so-called co-state variable provides the economic value of a change in crop dry weight as a function of time. This method, however, does not provide insight into the physiological backgrounds, which are implicitly represented in the model. Therefore, we will follow here an analysis based on the crop physiological and horticultural principles discussed in the previous paragraphs (Challa & Heuvelink, 1993).

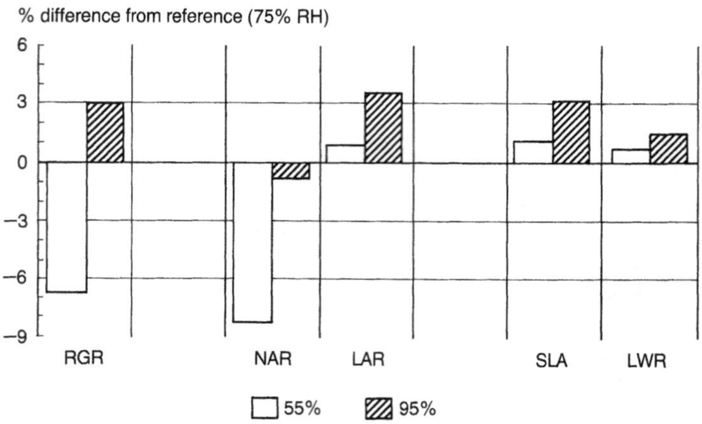

Figure 2.3.6 – Main effects of relative humidity on growth parameters of cucumber (after Van de Sanden & Veen, 1992).

In the young plant phase (phase of crop establishment) the primary objective is to reach a closed canopy or, if LAR does not vary much, the attainment of a target crop weight (W_{ref}). W_{ref} is attained at time t_p, which may, for example, be the time of transition from the young to the production phase. The relation between a change in weight ΔW brought about in the young phase and the change in time required to attain W_{ref}, Δt_p can be quantified according to equation (2.3.18) (Figure 2.3.7a):

$$\Delta t_p = \Delta \ln[W_t] / RGR_t = (\ln[W_t + W] - \ln[W_t]) / RGR_t \qquad \text{(Eq. 2.3.18)}$$

or

$$\Delta t_p = \ln[1+\Delta W/W_t] / RGR_t \qquad \text{(Eq. 2.3.19)}$$

where RGR_t is relative growth rate at time t.

From equation (2.3.19) follows that the effect of ΔW on t_p depends on RGR_t and W_t. At a low RGR, the same ΔW will give rise to a greater Δt_p, because the crop needs more time to attain W+ΔW. At a low W the same ΔW has a much greater effect on t_p than at a high W, because of a much greater impact on the instantaneous relative growth rate r_t (equation (2.3.3)).

A change in the start of the production phase (Δt_p), which arises in the young phase, will give rise to a change in dry weight in the production phase of ΔW (Figure 2.3.7b):

$$\Delta W = \Delta t_p \, GR(t_p) \qquad \text{(Eq. 2.3.20)}$$

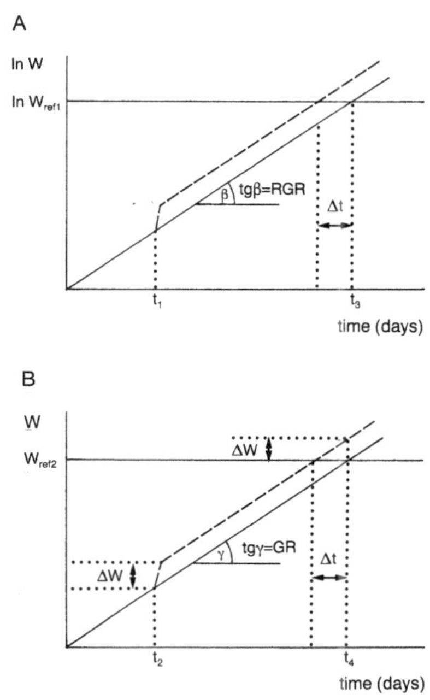

Figure 2.3.7 – Evaluation of possible effects of a change in daily growth rate of crops in the young phase (logarithmic ordinate, A) and when growing in a closed canopy (linear ordinate, B). For further details see text.

where $GR(t_p)$ is growth rate of the crop at the onset of the production phase (t_p).

When the fraction of dry weight diverted to harvestable product, F_{hp}, is constant during the production phase, as observed for several crops (Challa & Heuvelink, 1993; Figure 2.3.8), a change in dry weight will give rise to a change in dry weight of the harvestable product ΔW_{hp}:

$$\Delta W_{hp} = \Delta W F_{hp} \qquad \text{(Eq. 2.3.21)}$$

The economic value of ΔW_{hp} depends strongly on the type of crop and the way the culture is managed. A first distinction can be made between products sold by weight or by piece. For products that are sold by weight ΔW_{hp} will give rise to a change in economic yield ΔYield (NLG m^{-2}; NLG = Dutch guilders):

$$\Delta \text{Yield} = \Delta W F_{hp} \, \text{PRICE}(t_h) / \text{DMC} \qquad \text{(Eq. 2.3.22)}$$

where PRICE(t_h) is expected price per unit fresh weight at harvest time t_h (NLG g^{-1}) and DMC is dry matter content (g g^{-1}) of the harvestable product. If it is assumed that PRICE(t_h) and DMC are not affected by changes in Δt_p, ΔYield will be proportional to Δt_p.

If products are sold by piece (e.g. cucumber, cut flowers) equation (2.3.22) may still be used if the average dry weight per piece is not affected, because W will give rise to a corresponding change in the number produced. Elsewhere the relation between the weight per piece and the price per piece still has to be established.

An example of the economic value of a change in crop weight ΔW is given for a spring and an autumn tomato crop in Figure 2.3.9. Similar responses for the co-state variable for lettuce have been reported by Van Henten (1994).

It can be shown, that the evaluation of ΔW in the young phase is consistent with that in the production phase. Therefore the choice of the transition between young phase and production phase (W_{ref} attained at t_p) is not critical and may be chosen according to the crop concerned. Further details of the principles involved in the assessment of the economic value of ΔW are beyond the scope of the present outline and can be found elsewhere (Challa & Heuvelink, 1993).

2.3.1.3 Developmental processes

Introduction

During cultivation, depending on the crop, the plant usually passes through different phases. The starting material may be seeds, bulbs, or tubers, representing quiescent material requiring moisture and temperature (in the case of seeds sometimes also light) for germination or sprouting. More commonly, however, rooted cuttings or young plants, that have already passed through this initial phase, are used as starting material, which may require some adaptation before continuation of normal growth, after transplanting. After the start of the cultivation, following germination, sprouting, or transplanting, a certain period of growth is usually required to attain sufficient size and leaf area (photosynthetic "machinery"), before actually entering the production phase. During the production phase, depending on the crop, leaves (leafy vegetables, green pot plants), flowers (cut flowers, flowering pot plants), fruits, storage organs (tubers, bulbs), etc. are formed.

The control of phase transitions is an important issue in the cultivation of many greenhouse crops, in order to obtain a balanced crop, to meet product requirements, and for proper timing of harvest. Within each phase development of the plant proceeds by formation and development of organs. The rate of formation and the quantitative and qualitative characteristics of the organs formed have to be controlled properly in order to realize the objectives of the grower.

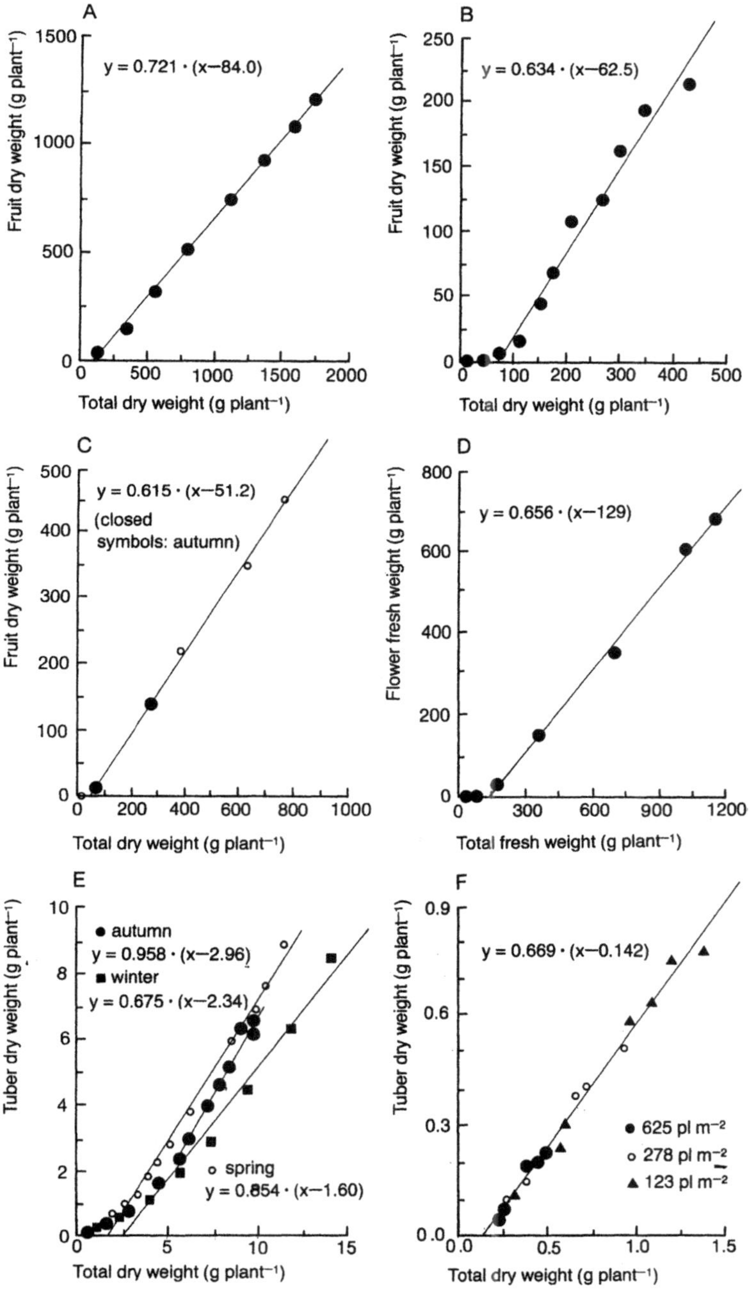

Figure 2.3.8 – Relation between total plant dry weight and dry weight in the harvestable product for six crop species. All dry weights include harvested plant parts. (A) tomato (De Koning, 1991), (B) cucumber (Liebig, 1978), (C) sweet pepper (Vegter, 1989), (D) rose (De Vries & Dubois, 1994), (E) kohlrabi (Liebig, unpublished), (F) radish (Heuvelink, unpublished) (after Challa & Heuvelink, 1993)

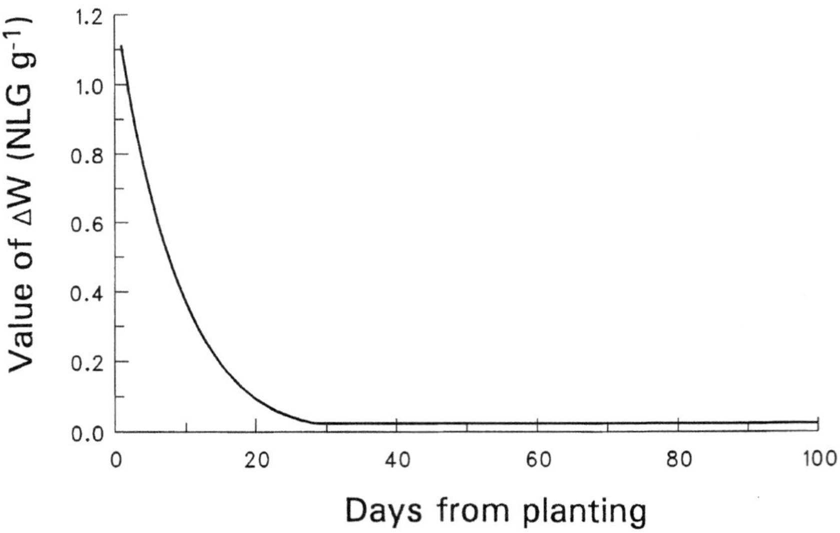

Figure 2.3.9 – Economic value of a change in crop dry weight ΔW in relation to crop development for a tomato crop planted in May (after Challa & Heuvelink, 1993).

The development of a crop can be considered as a basic pattern, reflecting the genetic properties of the crop, that can be modified, but not changed by the environment. In contrast to growth, where processes dominate with characteristics common to most crops, the relation between environmental factors and development reflects the genetic adaptation of crops to the original habitat, and to selection in breeding programmes. This relation, therefore, may be quite specific for a crop, or even a cultivar. This makes it difficult to deal with developmental processes in general, where it may be easier to find the exception than the rule.

In this section no detailed account can be given of the physiological background of developmental processes of greenhouse crops. Instead, common physiological knowledge will be presented in order to provide a general understanding and overview of the role of greenhouse climate in the control of development of greenhouse crops.

Rate of development

During their development (according to the definition in section 2.3.1.1), the plant as a whole, as well as the individual organs, are subjected to irreversible qualitative changes over time. Often it is possible to identify markers in the sequence of events related to development. Examples are the formation of subsequent leaves, anthesis, fruit set, and fruit ripening. When markers are defined it is possible to quantify development in terms of a stage that is attained.

For the whole plant the developmental stage is commonly indicated by the number of leaves. The time between the formation of two successive leaves is called the plastochron and hence the developmental stage of a plant is defined by its plastochron age (PA), or plastochron index (Erickson & Michelini, 1957). Interpolation in intermediate stages is possible by evaluation of the size of the youngest leaves. This concept can be extended to individual leaves by defining a reference size of that leaf, e.g. the moment of reaching a size of 5 mm, where its plastochron age is 0 (plastochron age of an individual leaf will be negative when its size < reference size) (Maksymowich, 1973). The concept of plastochron age has proved to be useful for comparison of leaves, where it is a much more suitable indicator than the chronological age.

The definition of a developmental stage besides the chronological age gives rise to the concept of rate of development (DVR):

$$DVR = \Delta DVS/\Delta t \qquad \text{(Eq. 2.3.23)}$$

where DVS is developmental stage, and Δt the amount of time required to attain a change in developmental stage ΔDVS. In the case of the whole plant DVS can be defined by the plastochron age PA (the number of leaves formed on that plant) and hence equation (2.3.23) transforms to:

$$DVR = \Delta PA/\Delta t \qquad \text{(Eq. 2.3.24)}$$

In the case where there are only two markers, as in the example of a fruit, it is common to define the start (anthesis) as DVS=0 and the harvesting stage as DVS=1. In that case the (average) rate of development is the inverse of the growth duration (time from anthesis to harvest) of that fruit.

Having defined the rate of development, it is now possible to discuss the relation with environmental factors. The dominant factor affecting the rate of development is temperature (Cockshull, 1992): in general, within the "normal" range, it is observed that the higher the temperature, the higher the rate of development. This rule is quantified by the well-known concept of heat units (HU) (Hesketh *et al.*, 1980; Vos *et al.*, 1982):

$$DVS_t = \Sigma (T_i - T_{min})/HU, \text{ for } T_i \geq T_{min}, \text{ else } T_i - T_{min} = 0 \qquad \text{(Eq. 2.3.25)}$$

where DVS_t is the developmental stage after t days, T_i average temperature at day i, and T_{min} is minimum temperature, where development stops, HU heat units required to attain a given developmental stage, expressed in degree-days. DVS_t as defined here provides a way to interpolate between the start (DVS_0=0) and end (DVS_f=1) of the developmental process under consideration. It should be stressed that DVR represents not one universal criterion within the plant, but that it is linked to the particular process considered. For example the DVR of a tomato fruit responds differently to temperature than that of the whole plant (De Koning, 1994). In many crops DVR depends mainly on temperature, but radiation may also have an effect, such as on the plastochron of cucumber (Newton, 1963) and on development of flower buds of chrysanthemum (Andersson, 1990).

Note that equation (2.3.25) implicitly assumes a linear relationship between DVR and T, which is not always the case, as has been demonstrated for maturation of tomato fruits (De Koning, 1994). Many authors, however, did observe a linear relationship between temperature and various developmental processes over a wide temperature range (Roberts & Summerfield, 1987), for example the response of leaf unfolding rate in *Hibiscus rosa-sinensis* (Karlsson *et al.*, 1991) and in Easter lily (Karlsson *et al.*, 1988) and the rate of flowering in tomato (De Koning, 1994).

In those cases where developmental processes respond linearly to temperature, it may be expected that they respond to average temperature, within certain limits, independent of the temperature regime (day/night temperature). This has been confirmed by many authors, for example for flower development in chrysanthemum within the range of 10–20 °C (Cockshull *et al.*, 1981), and for flower induction, initiation and development of chrysanthemum (Lepage *et al.*, 1984). Apart from the influence of mean temperature, an influence of temperature regime on the time to anthesis has been observed in *Pelargonium zonale* (Pytlinski & Krug, 1989) and in *Campanula isophylla* (Moe *et al.*, 1991).

Several authors (Krug & Liebig, 1980; Slack & Hand, 1983; Van den Berg, 1987; De Koning, 1988) observed that in closed crops (producing crops) growth, development and production is related to average 24h-temperature, within certain limits, independent of the temperature regime. The ability of crops to buffer fluctuations in temperature over time is, however, not necessarily confined to periods of 24 h, as has been demonstrated with kohlrabi (Liebig, 1988) and tomato (De Koning, 1990). The reason that such compensation is observed with closed crops and not with young plants is the importance of LAR for growth of young plants (Heuvelink, 1989) and possibly also the higher "physical capacity of assimilate buffering" in producing crops (De Koning, 1990). The ability of crops to buffer fluctuations in temperature over time is an important characteristic in relation to climate control, since tolerance for short term deviations of the average temperature requirements may create opportunities to satisfy other requirements, or to use resources more efficiently (Aikman & Picken, 1989).

Phase transitions

Phase transitions, as pointed out before, represent an important event in the cultivation of many crops. Accurate control of these transitions enables the grower to:

- Reduce yield loss (e.g. reduction of the number of pot plants without flowers, prevention, or delay of flowering in relation to bolting of vegetables, production of bulbs, or corms);
- Control and synchronize harvesting time (for example important in relation to occasions such as Christmas, Easter, Mother's Day, out-of-season production, and in relation to labour efficiency: year-round production);
- Control the number of leaves before flowering or formation of storage organs (important for quality and for the balance between vegetative and generative growth);
- Control the number of flowers per plant (quality, yield of flowers or fruits).

Because climatic factors play an important role in the occurrence and the timing of phase transitions a short review will be given of the general principles.

The phase transitions that play a role in greenhouse crops are:

- Germination of seeds, sprouting of bulbs and corms;
- Flowering;
- Formation of storage organs.

A general observation with respect to the role of environmental factors in these transitions is that, apart from a direct effect on key processes, there may be also indirect effects that are related to prerequisites of the transitions, such as developmental stage, or assimilate requirements. This complicates the understanding of the role of environmental factors greatly. In tomato transplants, for example, flowers are already initiated in an early stage (Dieleman & Heuvelink, 1992), and here, obviously, the role of temperature in earliness of tomato has nothing to do with a promoting effect of temperature on flowering, but should be explained by the role of temperature on the rate of leaf formation (developmental stage) and on fruit development.

Germination, sprouting of bulbs and corms

For the majority of greenhouse crops that are propagated by seed cultivation starts with a transplant. However, there are a number of crops where sowing *in situ* is common practice, such as radish, spinach, and carrot. Provided that a proper pre-treatment for dormancy break, when needed, has been given, the major requirements for germination are moisture and correct temperature. Some seeds also require light (e.g. purslane). The response of germination to temperature occurs according to the heat-unit concept discussed above (equation (2.3.25)) (Bierhuizen & Wagenvoort, 1974).

At emergence, the seedling goes through another transition, which, however, requires only exposure to light. During this transition the leaves will unfold, expand and turn green. Stem elongation is reduced, compared to the initial phase after germination. The optimal temperature in the seedling stage is lower than during germination, but higher than in the phase of crop establishment, because of the importance of leaf area formation and the limited role of maintenance respiration in that stage (section 2.3.1.2).

Bulbs and corms, before planting, usually have received a pre-treatment in order to break dormancy and/or to prepare them for the particular cultivation. After planting, the availability of water and a suitable temperature suffices for further development. They are planted either with existing roots (lilies), or, in most cases, with root primordia only. Growth of shoot and roots proceeds more or less independently, quite contrary to the concept of the functional balance discussed in section 2.3.2. A major concern in this phase is therefore to obtain a proper balance between root and shoot. Control of this balance relies on a proper choice of air and soil temperature. In order to restrict sprouting and to stimulate root growth and activity for sufficient supply of water and nutrients to the shoot, air temperature in this stage is kept low compared to that of the roots.

Flowering

Flowering is an extremely complex process and therefore only some general points concerning its control will be discussed here. For a full treatise refer for example to Bernier *et al.* (1981). Regardless of the growing conditions most plants remain vegetative for some time after emergence. In this "juvenile" phase the plant is insensitive to conditions that later promote flowering. A pronounced juvenile phase, often of many years, is common in woody perennials, but may also be present in some herbaceous annuals and biennials, where its duration may vary from a few days to several months. Its existence may be expressed by an increasing sensitivity to day length with increasing age of the plants. It is therefore common practice to call plants juvenile during their early period of growth, when they exhibit a poor photoperiodic response. Woody perennials and evergreen shrubs which are used as ornamentals (potted plants) have a juvenile phase lasting from between some months to as much as eight years (e.g. *Camellia japonica*: Rünger & Cockshull, 1985). In bulbous plants this phase may last from less than one (freesia) to six years (tulip, narcissus) (De Hertogh & Le Nard, 1993). Often crops are already in the adult stage when planted and may even already possess flowers or flower primordia, such as tulip, tomato, cucumber. In crops with a distinct juvenile phase the role of environmental factors in the transition to the adult phase is probably mainly through the effect on plant size (Hackett, 1985): optimum conditions for growth, or conditions that prevent the development of dormant periods, in general shorten the juvenile phase (e.g. Bromeliads: De Greef *et al.*, 1989).

In literature on the physiology of flowering, different phases are distinguished: induction of the floral stimulus, evocation (events at the shoot apex which commit the meristem to formation of flower primordia), initiation (organogenesis), growth of the floral parts and flower opening (anthesis) (Bernier *et al.*, 1981). During each of these phases, depending on the crop, requirements for normal development may differ.

Flower induction is not required in some crops, where flowers are initiated immediately after attaining the adult phase (endogenous control). In many crops, however, flower induction depends on exogenous signals. Inducing environmental factors may be:

1. Photoperiodic signals;
2. Temperature, also in combination with 1;
3. Radiation, also in combination with 1.

Day length, or better the length of the dark period (Vince-Prue, 1975) controls flower induction in obligate (qualitative) long day plants (LDP), or short day plants (SDP) and advances or delays flower induction in facultative (quantitative) LDP's or SDP's. Length of the night is primarily sensed by the photoreceptor phytochrome in the leaves, in interaction with an internal circadian rhythm (Hart, 1988). In protected cultivation day length can be controlled by means of darkening screens and with supplementary light and its use is widely spread, for example in year round production of Chrysanthemum (obligate SDP). In the initial phase after planting, flowering is undesirable, because first sufficient LAI and stem length have to be formed. Under natural short days flowering can be postponed by increasing the day length with (low level) artificial light. The same effect may also be obtained by interruption of the night by a short period of light, but here the timing of the moment of interruption may play a role in connection with the internal circadian rhythm of the plant (Hart, 1988). In practice this phenomenon has resulted in systems of cyclic lighting, where the crop is exposed to short periods of light followed by (periods of) darkness during the night.

There may be an interaction between day length and temperature with respect to flower induction. In some LDP's the critical day length may increase with increasing temperature, and there are examples of obligate SDP's where the day length is only important within a certain temperature range (some strawberry and Begonia cultivars). Vogelezang (1990) demonstrated that in Begonia × hiemalis "Toran" also root temperature may affect flower induction. In some LDP's the day length requirement may be replaced by a low, and in some by a high (> 30 °C) temperature treatment. The interaction between day length and temperature in relation to flower induction is complex and not well understood, and, moreover, highly crop and cultivar specific. It should be noted here that the concept of heat units as discussed in the previous paragraphs may not apply at all in relation to the flowering response and flower development, where also no or an adverse effect of temperature may be observed (Roberts & Summerfield, 1987).

A well known example of a *direct* role of temperature in flower induction is the low temperature requirement, called vernalization (Wiebe, 1989). The phenomenon is common in perennials and reflects their adaptation to the seasonal pattern, where flowering should be postponed until after the winter. In greenhouse cultivation vernalization mainly plays a role in the pre-treatment of starting material, but in some cases there is a low temperature requirement for floral initiation during cultivation (freesia). Vernalization may be reversed (de-vernalization) by exposure to high temperature (for example, temperature > 18 °C for 4 weeks will prevent flowering of vernalized Alstroemeria plants, according to Healy and Wilkins (1985)).

An example of undesired flower induction by low temperature, though less important in greenhouse cultivation than outdoors, is the phenomenon of bolting, occurring for example in kohlrabi (Habegger, 1985) and chinese cabbage (Elers & Wiebe, 1984). In some crops there is a minimum temperature requirement for flowering. A night temperature of 18 °C is considered the minimum for uniform flower bud initiation for azalea.

It should be noted here that, although flower induction and initiation are important phenomena, further development up to anthesis and fruiting may play an equally important role in the development of crops. The importance of radiation conditions for flower development has long been recognized. It is clearly established that, besides the radiation level, day length may also influence flower development (Kinet et al., 1985). Similar to the response of flower induction, the response to day length may be absolute (qualitative), or facultative (quantitative), depending on the species or the cultivar. In some species the optimal day length may change with the developmental stage of the reproductive organs and may differ from that for flower initiation. The same temperature interactions described in relation to day length sensitivity of induction may be observed for flower development.

In many crops abortion of flower primordia, flowers or fruits, or related disturbances in their development, may occur due to an unfavourable balance between assimilate supply and requirement, when radiation is limiting or the competition among sinks is too high (Kinet *et al.*, 1985). Apart from radiation, temperature is a major factor involved, because it may strongly influence the formation of organs and the balance between vegetative and generative sinks (e.g. leaves versus flower competition in freesia, too many flowers in rose, too many growing fruits in tomato or cucumber). This point will be further discussed in relation to dry matter distribution in section 2.3.2.

In some species a low temperature may induce flower abortion, as in rose, easter lily (root temperature), and *Gladiolus* (night temperature) (Kinet *et al.*, Chapter 5, 1985). Disturbances in flower development have also been observed, following marginal, interrupted, or otherwise perturbed flower induction, which become manifest as malformations within inflorescences or flowers (appearance of vegetative structures, inflorescence branching, foliaceous bracts, proliferation of flowers) (Kinet *et al.*, 1985).

Formation of storage organs

Salisbury & Marinos (1985) in their treatise on the ecological role of plant growth substances provide a short overview of the general principles of the control of initiation of storage organs. These principles in many respects resemble those of flower induction: growth regulators appear to play a role and the process may be controlled by daylength and temperature, including their interactions. Formation of storage organs plays a minor role in most greenhouse crops: apart from a few greenhouse vegetables, storage organs mainly play a role in the cultivation of bulbous flowers, where their formation is often considered in relation to competition with production of flowers. In these crops the control of initiation and growth of storage organs is an important element in the design of optimal regimes for supplementary lighting and temperature.

Control of shape and size

Shape and size are important characteristics of the crop and of the product. Essential characteristics are length and diameter of the internodes and hence of the stem; number, size and shape of leaves, flowers, fruits, and of storage organs, and number and quality of branches. The control of shape and size of greenhouse crops may be of importance for a number of objectives:

1. Increased, or advanced production. With many crops, before the product is harvested, a given crop architecture has to be built in order to obtain optimum light interception (crop photosynthesis) and formation of sufficient and/or early sinks for production of harvestable products;
2. Synchronisation of the crop (also in relation to 3.) in order to obtain a uniform crop, for increased efficiency (labour, greenhouse space utilization), and optimal crop management;
3. Obtaining the desired quality of the product, in terms of optimum size, shape and appearance, and uniformity of a batch (pot plants) (see section 2.3.3).

Apart from crop management by training, pruning, grafting, bending and use of growth regulators, greenhouse climate also plays a role in the control of shape and size of greenhouse crops. The following discussion should be considered as an extension of the previous more specific treatise on flowering and is closely related to the problem of dry matter partitioning that will be considered later (section 2.3.2).

The level, the duration and the spectral quality of radiation, have a profound effect on plant morphogenesis (Kendrick & Kronenberg, 1986). These factors affect the elongation and diameter of internodes, branching, as well as the size and shape of leaves, flowers and fruits.

Effects of radiation on size of plants and plant organs may be attributed to a large extent to availability of assimilates, as has been demonstrated for example with tomato, where fruit size decreased following reduction of radiation by shading, while the number of fruits remained the same (Cockshull, 1992). In crisp lettuce, however, the relation between size (weight) of the head and radiation is complex and likely to be mediated by morphogenetic reactions of the leaves (Wurr & Fellows, 1991).

The possibilities for controlling the radiation level are, however, quite limited. Screens can be used to reduce radiation, but their application is mainly in protecting the crop from excessive radiation and reducing temperature (section 4.5). Supplementary radiation is used mainly in the cultivation of ornamentals under poor light conditions, in winter at higher latitudes, where the contribution to the daily radiation integral may be substantial. In addition to this quantitative effect, the spectral quality of the lamp type and the timing of radiation play a role in the control of plant size and shape. Light sources with a high red to far-red ratio stimulated lateral branching (Moe *et al.*, 1991;

Hendriks, 1992), whereas blue light has a pronounced effect on stem elongation (Mortensen & Strømme, 1987).

Temperature is a prominent factor in controlling plant (and product) shape and size. In general, a higher temperature gives rise to longer internodes and less branching. Moreover, the higher rate of development, as discussed before, will modify the dry matter distribution pattern in the plant, by creating more sinks, competing for the same supply of assimilates.

An example of the effect of temperature on the shape of leaves is the problem of poor head formation (open base) in lettuce under low light conditions, which can be explained by the different sensitivity of expansion of the leaf blade and elongation of the midrib to radiation and temperature (Bensink, 1971).

Recently the discovery of the DIF-concept has led to new opportunities to control growth in height of, in particular, pot plants and bedding plants (Moe & Heins, 1990; Moe *et al.*, 1992). The DIF-concept states that plants react not only to the average temperature, but that there is a distinct response in internode length on the difference between day- and night temperature (=DIF). At high positive DIF (day temperature > night temperature) the length increases, whereas a negative DIF gives rise to compact plants. The importance of the DIF concept is that it represents a more or less independent control of plant size, because, as discussed before, (most) developmental processes respond primarily to the average temperature, and not to DIF. There are indications that the mechanism of DIF relates to the action of phytochrome (Moe & Heins, 1990).

Increased CO_2 concentration in the greenhouse air, in general, has a positive effect on dry weight, plant height, number of leaves and lateral branching (Mortensen, 1987). The main effect of elevated CO_2 seems to result from enhanced photosynthesis, but there are also morphogenetic changes, which may have a different background. At elevated CO_2 higher root to shoot ratios are generally observed (Goudriaan & De Ruiter, 1983; Pearcy & Björkman, 1983), leaves are usually thicker (one or two extra palisade cell layers; Enoch, 1990), and the development of lateral shoots, side branches, tillers and basal shoots is promoted (Enoch, 1990), which may be interpreted as a suppression of apical dominance (Enoch & Zieslin, 1988).

Humidity plays a role in relation to plant height (some potplants) and flower induction, but here the dominant factor is water supply, rather than air humidity. In some fruit vegetables an effect of air humidity has been reported on leaf size. A high humidity promotes the formation of large leaves in cucumber (Bakker, 1991[a]) and this effect is probably common to many crops, but in tomato this positive effect is overruled by a negative effect on calcium nutrition, which gives rise to smaller leaves (Bakker, 1991[a]). Humidity also may affect branching of some crops (McIntyre, 1977), but the physiological background is still unclear.

2.3.2 Biomass partitioning in plants

L.F.M. Marcelis and A.N.M. de Koning

2.3.2.1 Introduction

The final yield of a crop is determined by the accumulation of biomass (fresh and dry weight) of the harvestable organs and their quality (section 2.3.3). An increase in the biomass partitioned into these organs proportionately enhances the yield, provided that the total plant growth rate is not altered. The harvestable organs can be for instance the above-ground plant parts (e.g. lettuce), vegetative storage organs (e.g. radish, kohlrabi), flowers with stem and leaves (cut flowers) or fruits (e.g. tomato, cucumber, sweet pepper, eggplant). Often not only the total weight of the harvestable organs is important, but also their number, the weight per piece, as well as form and quality.

Climatic factors such as light, temperature, CO_2 and air humidity may strongly affect the biomass partitioning among the different plant parts. Therefore, climate control can be used as a tool to manipulate the biomass partitioning. In young plants the climate control should be aimed at a rapid leaf area formation in order to increase light interception. Then, the climate control should be aimed at an optimal earliness, as discussed in section 2.3.1. In plants which grow determinately, after an initial period the harvestable organs are initiated and then the distribution to these organs increases gradually until the organs are all harvested in a single harvest (e.g. radish, kohlrabi, chrysanthemum). In these types of plants, after the initial period, the climate control should aim at a maximum proportion of assimilates diverted to the harvestable organs. In plants which show an indeterminate growth, after an initial period without growth of the harvestable organs, new harvestable organs are continuously formed and harvested over an extended period, while growth of other plant parts also continues (e.g. fruit vegetables, rose, carnation). In these types of plants the biomass partitioning between the harvestable organs and the rest of the plant may change cyclically. A sufficient amount of assimilates should continuously be diverted to the non-harvestable plant parts to maintain a high production capacity. An optimal balance between growth of the harvestable organs and the rest of the plant should be aimed for. In addition, in both types of plants the control of biomass partitioning should aim at an optimal size and quality of the harvestable organs.

In this section, first some general principles of the regulation of biomass partitioning are discussed, then the main factors determining the distribution of biomass to the different plant organs are described. Finally, tools to manipulate the biomass allocation in fruit vegetables are discussed.

2.3.2.2 General principles of the regulation of biomass partitioning

Assimilates are produced by photosynthesis in the leaves. The assimilates are transported to the sink organs or stored in the storage pools. When the rate of assimilate supply increases, the growth rates of the organs or storage of assimilates may increase. In the long term the number of sink organs can also increase, because of an increase in initiation rate and/or decrease in abortion rate. Sometimes a negative feedback occurs resulting in a reduction of the photosynthetic rate.

Wardlaw (1990) and Wareing & Patrick (1975) proposed that the transport path (phloem) can affect dry matter distribution, but they also stated that these effects diminish readily when sink demand increases. Evans (1975), Murakami *et al.* (1982) and Webb & Gorham (1964) found the transport path to be of only minor importance in regulating translocation of assimilates. Several authors (Evans, 1975; Farrar, 1988; Ho, 1988) have found indications that the biomass allocation is primarily regulated by the sink strengths of the individual organs. Sink strength can be defined as the capacity of a sink to accumulate assimilates (e.g. Marcelis *et al.*, 1989; Wolswinkel, 1985). It is generally assumed that this capacity can be quantified by the potential growth rate of a sink, i.e. the growth rate under conditions of non-limiting assimilate supply (Marcelis *et al.*, 1989). In this view the biomass allocation to an organ is determined by its sink strength relative to the total sink strength of all plant organs together. The pattern of biomass partitioning can change during crop development because the sink strengths of the individual organs or the number of sink organs may change. Climatic factors influence the biomass partitioning in the short term as a result of a difference in response of the sink strengths of the individual organs and in the long term also through effects on the number of organs (initiation, abortion, harvest or senescence).

Although the concept of sink strengths could be used to describe the dry matter distribution between any plant parts, the dry matter distribution between root and shoot is often described by a functional equilibrium between root activity (water or nutrient uptake) and shoot activity (photosynthesis); i.e. the ratio of root to shoot mass is proportional to the ratio of shoot to root specific activity (Brouwer, 1963). However, when plants become reproductive, problems with the approach of a functional equilibrium arise.

2.3.2.3 Biomass partitioning among plant organs

Roots

The roots often comprise only a small fraction of the total plant dry weight in greenhouse crops (Table 2.3.1). De Willigen & Van Noordwijk (1987) proposed that in greenhouse culture where plants are grown in artificial substrates and supply of water and nutrients is optimal, maximal plant growth can be achieved with a small root system.

The distribution of dry matter towards the roots generally decreases with age or size of young herbaceous plants (see for example Cooper & Thornley, 1976; Wilson, 1988). For fruit vegetables at the onset of reproduction root growth is strongly reduced while even some root death has been observed (Hurd *et al.*, 1979; Van der Vlugt, 1987). In perennial plants the biomass distribution to the roots often shows periodicity (Klepper, 1991). Sometimes the partitioning to the roots is low in spring and high in summer, but fluctuations have also been observed over a period of 10 days (Klepper, 1991). When parts of the shoot are removed, the dry matter distribution to the roots usually decreases (Brouwer, 1963; Wilson, 1988). In cut-flowers such as rose, the total root weight even decreases after harvesting flowering branches (Fuchs, 1986).

The effects of environmental conditions are generally in accordance with a functional equilibrium between root activity (water or nutrient uptake) and shoot activity (photosynthesis) (Brouwer, 1963). Factors which reduce the specific activity of the roots, such as a decrease in supply of water or major nutrients (especially nitrogen), a decrease in water potential or temperatures above or below the optimum temperature for root functioning, increase the dry matter distribution towards the roots (e.g. Marcelis, 1993[d]; Wilson, 1988). The dry matter distribution to the roots also increases by factors which stimulate the specific activity of the shoot, such as an increase in CO_2-concentration, light intensity or length of photoperiod. However, for photoperiod, temperature, and CO_2 the literature is ambiguous.

Vegetative storage organs

The ratio of the vegetative storage organ (root, stem or hypocotyl) to the total plant dry weight of crops, such as carrot, kohlrabi and radish increases with time/size (Nieuwhof, 1976; Hole *et al.*, 1984; Pell *et al.*, 1990; Challa & Heuvelink, 1993). For radish Nieuwhof (1976) showed that the optimum temperature for hypocotyl growth was lower than for shoot growth, resulting in a decreasing distribution of dry matter to the hypocotyl with increasing temperature. The dry matter distribution to the hypocotyl of radish decreases with decreasing light intensity (Starck, 1973) and increasing plant density (Hole *et al.*, 1984). However, Challa & Heuvelink (1993) observed no effect of plant density on the ratio between hypocotyl and total plant weight. The dry matter distribution to the hypocotyl of radish is also greatly dependent on the cultivar (Nieuwhof, 1976). Although the distribution to the

Table 2.3.1 – Cumulative dry weight of the roots as a proportion of the total plant dry weight (including harvested plant parts) for plants in the vegetative and generative stage.

Species	Vegetative stage (%)	Generative stage (%)	References
Cucumber	8–15	3–7	Challa, 1976; Marcelis, 1994[a].
Eggplant	29	8	Claussen, 1976.
Sweet pepper	17–22	5	Hall, 1977; Nielsen & Veierskov, 1988.
Tomato	17–20	1–10	Ward, 1964; Yoshioka & Takahashi, 1979; Richards, 1981; Ehret & Ho, 1986[b].
Pot chrysanthemum	25	10	Cockshull & Hughes, 1968.

storage organ can be affected by several factors, an almost linear relationship is often found between the logarithmic weights (Hole *et al.*, 1984) or, after an initial period, between the absolute weights (Figure 2.3.8E) of the storage organ and the shoot.

Stem and leaves

The ratio between stem and leaf weight is an important quality aspect of ornamental plants and of young vegetable plants. In young plants the stem to leaf ratio generally increases (sometimes after an initial decrease) with age/size (Harssema, 1977; Horie *et al.*, 1979; Nilwik, 1981[a]). In older cucumber plants the dry matter distribution between leaves and stem seems to be constant with age/size (Schapendonk & Brouwer, 1984). In general, temperature has no great effect on the distribution between stem and leaves (Harssema, 1977; Vogelezang, 1990; Marcelis, 1994[a]), but sometimes the stem to leaf ratio is reported to increase with increasing temperature (Nilwik, 1981[a]; Kleinendorst & Veen, 1983). Moreover, an increase in stem to leaf ratio with increasing temperature might be observed, when the ontogenetic effects of temperature are not taken into account, which probably at least partly explains the increase in stem to leaf ratio of cucumber as reported by Kleinendorst & Veen (1983).

Smith (1981[a]) stated that many crop plants tend to avoid shade and hence the stem to leaf ratio generally increases with decreasing light intensity or increasing plant density. Indeed, an increase in the stem to leaf ratio with decreasing light intensity was observed in tomato, sweet pepper and cucumber (Harssema, 1977; Nilwik, 1981; Marcelis, 1994[a]). However, Acock *et al.* (1979) reported that for chrysanthemum the effect of light intensity on dry matter distribution between stem and leaves was variable and probably dependent on the cultivar. In contrast to shade avoiding plants, shade tolerant plants usually show an increasing stem to leaf ratio with increasing light intensity (Smith, 1981). This might hold for many pot plants, which in their natural habitat often grow under shade conditions (Meeuwissen, 1985).

Fruits

Crops of fruit vegetables such as tomato, cucumber, sweet pepper and eggplant are characterized by indeterminate growth. After a short initial phase of only vegetative growth, fruits are initiated and harvested continuously (Figure 2.3.10). The fruits compete strongly with each other and with the vegetative parts for the assimilates available. Although the cumulative weights of all fruits (including harvested fruits) appeared to be linearly related to the cumulative total plant weight (including harvested plant parts) (Challa & Heuvelink, 1993; Figure 2.3.11a), the instantaneous distribution of the dry matter increment between fruits and vegetative parts may change cyclically (Hall, 1977; De Koning, 1989[a]; Figure 2.3.11b). At the end of a growing season the total cumulative fruit weight constitutes a large fraction of the total cumulative plant dry weight (Table 2.3.2).

At any moment the dry matter distribution between fruits and vegetative parts and among individual fruits is to a large extent determined by their sink strengths. The sink strength of individual organs is reflected in the potential growth rate, that is the growth rate under conditions of non-limiting assimilate supply. The (potential) growth rate of a fruit depends on its developmental stage (Marcelis, 1992[b]). The potential growth rate increases for cucumber but not for tomato, while for both species the growing period decreases with increasing temperature (Marcelis, 1993[c]; Figure 2.3.12). Conditions during initiation of the fruit such as light intensity or position of the fruit on the plant also affect the sink strength (De Koning, 1994). In seeded fruits the number of seeds set stimulate the sink strength of a fruit (Picken, 1984). During fruit growth, factors such as light intensity or CO_2-concentration are believed to affect only the availability of assimilates, but not the sink strength of the individual organs.

The partitioning of dry matter into fruits can be controlled by manipulating the sink strength of the individual organs or the number of organs. Since the total sink strength of all fruits together

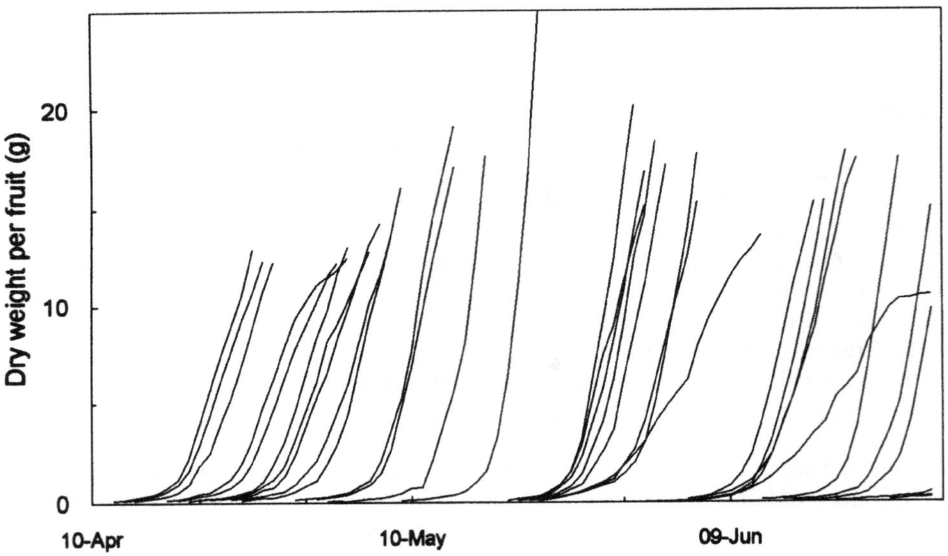

Figure 2.3.10 – Time course of dry weights of individual fruits of one cucumber plant. The dry weight of each fruit is shown between time of flowering and harvest of that fruit.

increases with the number of fruits on a plant, a close positive correlation is often found between the dry matter distribution to the fruits and the fruit load (Nielsen & Veierskov, 1988; Marcelis, 1992[a], 1993[a]).

After an extended period of a high light level the average number of fruits on a cucumber plant increases and as a result the dry matter distribution to the fruits (Marcelis, 1993[b]), while Widders & Price (1989) observed a reduction in the dry matter distribution to the fruits with increasing plant density. These effects of plant density can probably be ascribed to a diminishing light interception per plant. However, Cockshull *et al.* (1992) did not find any effect of a 23% decrease in solar radiation on the dry matter distribution to fruits in tomato.

With the same fruit load the dry matter distribution to the fruits in cucumber increases with increasing temperature (Marcelis, 1993[a]). However, when cucumber plants are grown over an extended period at a raised temperature the fruit load decreases and the dry matter distribution to the fruits is not markedly affected by the temperature on the long term (Marcelis, 1993[a]). Neither in pepper, nor in were clear effects of temperature on the ratio between the final dry weights of all fruits together and total plant were found (Bhatt & Srinivasa Rao, 1989; De Koning, 1989[a]). However, earliness of production is enhanced by increasing temperature.

The initiation rate of new fruits increases with increasing temperature. The initiation rate also often increases with increasing light intensity or with decreasing competition by other sink organs (Marcelis, 1994[b]).

A large fraction of the initiated fruits may abort in cucumber (up to 60 or 70%; Marcelis, 1992[a]), eggplant (up to 85%; Bakker, 1991[a]) and sweet pepper (up to 85%; Bakker, 1991[a]), which indicates that in these crops the initiation rate of flowers or fruits does not limit the number of fruits growing on a plant. Fruit abortion often increases with decreasing air humidity by day (Bakker, 1991[a]) or increasing temperature (Picken, 1984). Abortion of a fruit occurs shortly after anthesis. Whether or not

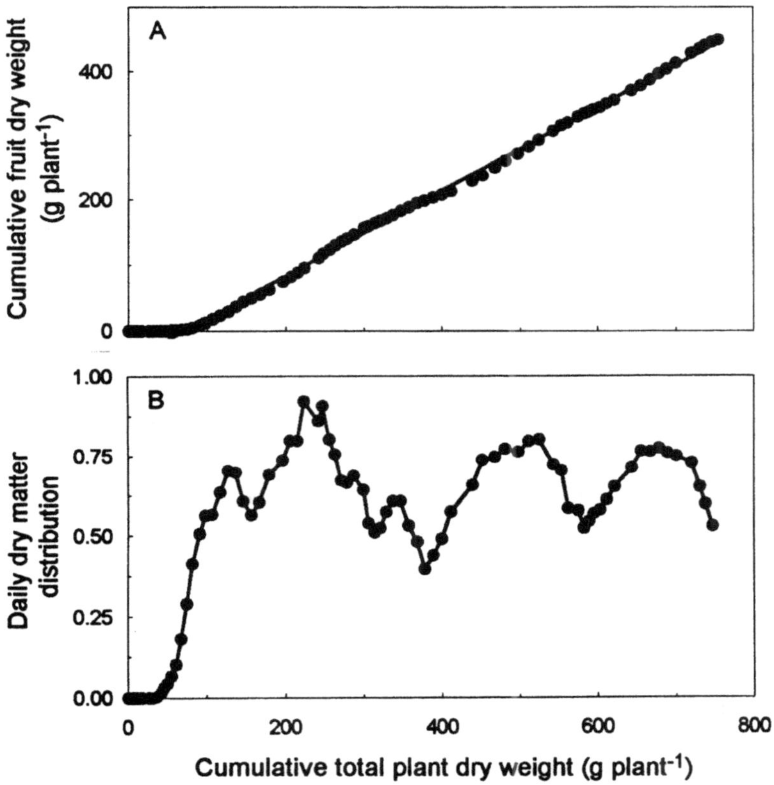

Figure 2.3.11 – The relationship between growth of the fruits and the total plant in cucumber. (A) Cumulative weight of the fruits against the cumulative weight of the total plant (including harvested plant parts). (B) The daily dry matter distribution to the fruits (fruit growth rate divided by total plant growth rate) against the cumulative weight of the total plant (including harvested plant parts).

Table 2.3.2 – Cumulative dry and fresh weight of the fruits as a proportion of the total cumulative plant weight (including harvested plant parts) after an extended growth period for several greenhouse crop species.

Species	Dry weight (%)	Fresh weight (%)	References
Cucumber	50–64	70–85	Marcelis, 1992[a].
Eggplant	50	-	Claussen, 1976.
Sweet pepper	45–60	-	Hall, 1977; Nielsen & Veierskov, 1988.
Tomato	52–72	77–84	Ward, 1964; Hurd *et al.*, 1979; Ehret & Ho, 1986[b]; De Koning, 1993.

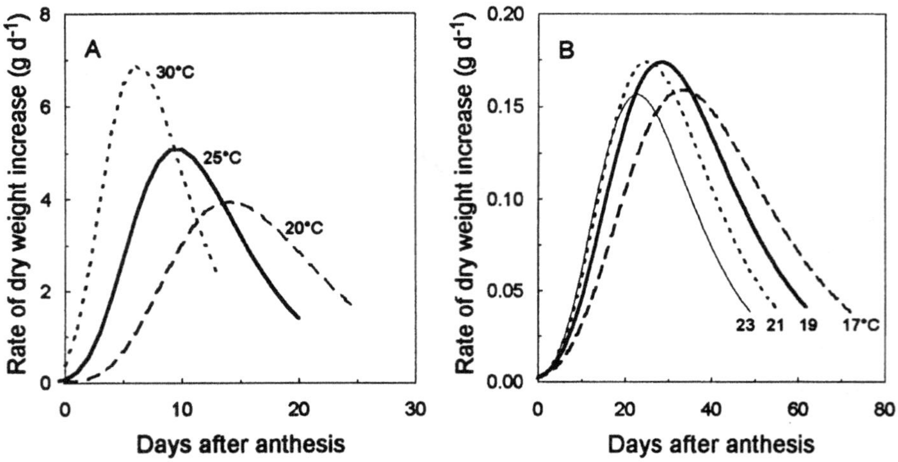

Figure 2.3.12 – Effect of temperature on the potential growth rate of individual cucumber (A) and tomato fruits (B).

a fruit aborts seems to be mainly determined by the availability of assimilates during a short period before and after anthesis (Kinet, 1977; Schapendonk & Brouwer, 1984; Marcelis, 1992[a]). With increasing temperature the availability of assimilates decreases due to an increase in total assimilate demand, which may explain the effects of temperature on fruit abortion. Schapendonk & Brouwer (1984) found indications that fruit abortion in cucumber was not solely due to assimilate shortage, but also to the dominance of competing fruits. Moreover, fruit abortion is negatively correlated with the number of seeds set (Picken, 1984); this correlation may explain the effects of air humidity on fruit abortion.

The number of newly developed fruits growing on a plant strongly depends on the sink/source ratio, i.e. the ratio between the sum of sink strengths of all organs and the photosynthetic rate. With increasing fruit load or with decreasing light intensity or CO_2 the sink/source ratio increases. When the sink/source ratio is low, a sufficient amount of assimilates is available for many young fruits to start growing and as a result the dry matter distribution to the fruits increases. Subsequently, the sink/source ratio increases and many young fruits abort. After some fruits are harvested, the sink/source ratio, the fruit load and the dry matter distribution to the fruits will be low. This cyclic process of fruit load and dry matter distribution can then start again.

Flowers and flowering branches

The flowers often are important sinks for assimilates in flowering pot plants and cut-flowers (Cockshull & Hughes, 1967; Harris & Jeffcoat, 1972; Halevy, 1986). The dry matter partitioning into the flowers generally increases with increasing assimilate supply (Halevy, 1986; Wardlaw, 1990). However, Cockshull & Hughes (1967) observed in pot chrysanthemum no effect of light intensity or CO_2-concentration on the dry weight ratio of the flowers to the total plant. They found a close positive relationship between this ratio and the stage of flower development.

In some indeterminately growing cut-flowers, such as rose and carnation, new branches are formed continuously and are harvested when flowers have been formed. These flowering branches have features in common with fruits of indeterminate greenhouse vegetable crops. Although branches, due to the presence of leaves, have a greater photosynthetic capacity than fruits, there is also a noticeable competition for assimilates among branches and between branches and the rest of the plant (Harris & Jeffcoat, 1972; Mor & Halevy, 1979). Like fruits, the cumulative weight of all flowering branches seems to be linearly related to the cumulative total plant weight (Challa & Heuvelink, 1993), while branch growth shows a strong cyclic pattern in roses (Van den Berg, 1987). Comparable to the cyclic process of fruit load and dry matter distribution in fruit vegetables, a cyclic process of branch load and dry matter distribution often occurs in indeterminately growing cut-flowers.

2.3.2.4 Tools to control biomass partitioning in fruit vegetables

In order to establish a fast increase in leaf area, young crops of fruit vegetables are grown at relatively high temperature. High temperature also increases abortion of flowers and young fruits but as long as those plants have a low source capacity and cannot support growth of good quality fruits, abortion may be even advantageous. Although in sweet pepper and eggplant flowers or young fruits drop easily, fruit pruning is sometimes necessary to prevent a too early fruit load. In winter-grown tomato removal of the first truss or some distal fruits of the first trusses and in cucumber pruning of the first flowers are common practice. Despite an increase in the abortion of flowers and young fruits with increasing temperature, the earliness of fruit production can be enhanced by high temperature because of a faster plant development.

When plants are large enough to bear fruits, in winter the fruit set in tomato is improved by subjecting the crop to restricted water supply (De Koning & Hurd, 1983), high salinity and enhanced transpiration by low air humidity. However, Ehret & Ho (1986[b]) and Bakker (1991[a]) did not observe clear effects of salinity and air humidity on dry matter partitioning. In sweet pepper low night temperature is frequently used to improve fruit set. From the moment fruits are retained, keeping the sink and source strength of a plant in balance is of utmost importance. A too high sink/source ratio will cause sub-optimal vegetative growth, small fruits and abortion of young fruits. A too low sink/source ratio as a result of a too low fruit load directly reduces yield by leaving an insufficient number of fruits growing. Moreover, a low sink/source ratio may decrease leaf area growth and potential yield as observed with tomato at high radiation (Nederhoff *et al.*, 1992) or even decrease the net assimilation rate (Hall, 1977). Ideally, vegetative growth should be just sufficient to renew the vegetative parts such that sufficient growth potential is maintained in future, while the remaining photosynthates should be partitioned into the fruits, as these are the marketable part of crop production. Daily light sum and therefore potential photosynthesis varies significantly during a year, causing a five-fold difference in crop growth rate between winter and summer (De Koning, 1993). It is obvious that the sink/source ratio and hence fruit load should be adapted to those differences, i.e. a high fruit load in periods with high irradiance.

In commercial practice the sink/source ratio is mainly controlled by three measures. Firstly, the potential fruit load per ground area can be adjusted to the seasonal variation in potential photosynthesis by varying the plant density or number of shoots retained per plant. In most crops, optimal plant density is low in winter and high in summer. In an early tomato crop side shoots can be retained in spring, in order to increase the fruit load per ground area. In cucumber more axillary shoots are retained in summer than in spring. The second level of controlling the sink/source ratio is pruning of fruits. Pruning of flowers or young fruits is carried out if the sink/source ratio is expected to become too high. In cucumber a too heavy fruit load can be decreased by picking the fruits in an early but marketable developmental stage. Plant density and fruit pruning are effective measures to adjust the fruit load per ground area to the seasonal variation. However, the daily or weekly light sum can be very different from "normal". Therefore, for the short term control (third level of sink/source control) a directly responding measure is needed. Here, in commercial practice, temperature seems to be the most suitable variable as the sink strength of individual fruits is immediately affected by temperature and in a glasshouse temperature can be altered quickly. Owing to a buffering capacity of plants, for tomato, adjustment of temperature to the prevailing light conditions may be spread over several days without markedly affecting the crop (De Koning, 1990). The maximum period for such temperature compensation will be largely dependent on the maturation period of individual fruits. Although the sink/source ratio is reduced by CO_2 enrichment, in commercial practice the CO_2-concentration is only based on the effect on the source (Chapter 5) without considering the effect on the sink/source ratio.

Summarising, of all environmental factors, mainly temperature is used to control biomass distribution in glasshouse crops, as temperature has a direct effect on the sink strength of the individual organs. Temperature also affects biomass partitioning because high temperature enhances development and increases the initiation of flowers, buds and fruits, as well as their abortion due to increasing total assimilate demand. No clear effects of humidity on dry matter partitioning have been observed. Light and CO_2-concentration primarily affect the total availability of assimilates and have no immediate effect on the biomass distribution.

If crop growth models include the effects of plant density, fruit pruning and temperature on the sink/source ratio, they might help the grower considerably in controlling the crop. Suitable models will be available in the near future (e.g. Bertin & Gary, 1993; De Koning, 1994; Marcelis, 1994[b]).

2.3.3 Product quality

C. Vonk Noordegraaf and G.W.H. Welles

2.3.3.1 Introduction

As mentioned in the previous section, besides (fresh)weight of the harvestable products, their quality largely determines the final yield in glasshouse yield. Production can easily be quantified in terms of weight or number. Quality, however, is harder to quantify as it is a combination of various aspects, some of them subjective.

Quality can be defined as how the product and the way of production fits the demand for handling, trading, retailing and the expectations of consumers. So the criteria for quality differ for each link in the chain from producer to consumer. For example, a long shelf life can be important for the retailer but may not favour the consumer because of poor taste.

Some of the quality aspects are visible: external quality, such as shape and colour, but others are not: internal quality, such as taste, shelf life, ornamental value. The visible ones can influence the price of the product directly, the invisible ones will not influence the price in the short-term but may influence the "image" of the product in the long run. Several aspects of quality can be measured (analytical quality) while others are subjective (emotional quality) (Cramwinckel, 1989). This emotional quality is often related to aspects of the production process. For example, the use of integrated pest control will be appreciated by most consumers, so products cultivated using these techniques will be deemed to be of better quality. To ensure the market position in the short- and long-term, external quality, internal quality and emotional quality all need to be given high priority.

At harvest products are graded on the basis of visible quality characteristics. Some of them can be measured in terms of number of flowers, length, size. Other external quality aspects are more subjectively graded such as defects, form and damage. For many crops a clear description is available of the external product characteristic required for first or second quality classes. For the main crops information is available on the climatic influences on these quality parameters and production. From this information the grower can deduce a compromise for the environmental strategy, based on whether he chooses for quality or quantity, when these are in competition with each other.

However, a high external quality does not always imply a high internal quality which may show during transport or shelf life (e.g. flower and leaf dropping in pot plants, yellowing of cucumber).

These features are caused by differences in internal quality and up until now they can not be detected at harvest.

Some can be tested using a described method as shelf life, vase life of cut-flowers, nitrate content or taste. Others are more difficult to define (changes in colour, hardiness of plants, sensitivity to diseases, hidden defects).

As climate consists of a complex of interacting factors (light, temperature, humidity and carbon dioxide are discussed in Chapter 3) it is hard to ascribe the various quality aspects directly to certain climate factors. For example, many (external) quality aspects are related to the transpiration of the plants and the uptake of nutrients (section 2.2.2). A wide range of products is grown in glasshouses: cut-flowers, foliage plants, flowering pot plants, vegetables consisting of fruits (tomatoes) or leaves (lettuce), each with its own criteria for quality. For practical reasons, in this section the description of environmental effects on quality is therefore restricted to some examples of the effects of light, temperature, humidity and CO_2 on the main crops of pot plants, cut-flowers and on vegetables (Welles *et al.*, 1992).

2.3.3.2 Effects of light on product quality

Effects of light intensity and quantity

Pot plants

Plants grown at high light intensity will have smaller and thicker leaves, a smaller shoot/root ratio and be more branched than plants grown under lower light conditions. Density, height and structure of the crop influence the distribution of light within the crop and thus the form of the plants. Spacing plants at the right time can improve the form as this will allow for a better development of lateral branches. So not only the light intensity at the top of the plants is important, but also the distribution within the crop, specially when large, branched foliage plants such as Ficus are produced (Uitermark & Benninga, 1991).

For some foliage plants such as Codiaeum the marketing value will increase when they have more yellow/red variegated leaves, as these are deemed more attractive. This can be achieved by increasing light quantity and temperature which promote the formation of anthocyan (Preissel *et al.*, 1980). Placing plants grown at high light quantities in a dark indoor position can cause defoliation as has been shown in Ficus and Codiaeum (Conover & Poole, 1975; Van Spronsen, 1981). Growing some pot plants at lower light intensities or giving them an adaptation period, can be desirable for a good keeping quality under indoor conditions (Conover & Poole, 1990).

Light quantity can influence size and colour of flowers, quality of inflorescences and flower abortion. Also, flower abscission in response to a dark period is strongly influenced by light conditions during the previous growing period (Fjeld, 1992; Moe *et al.*, 1992). More light means larger and sometimes more flowers in different flowering pot plants and a longer keeping quality.

Cut-flowers

For standard carnations flower size and weight are promoted by light quantity (Harris & Harris, 1962; Bunt, 1978), and for spray types the number of flowers can be increased by improved development of the flower bearing lateral shoots. Crops which are sensitive to flower abortion such as Iris, Gladiolus and Lilium, will show less abortion and abscission (Kamerbeek & Durieux, 1971) and roses will form less blind shoots when the light quantity is high (Moe & Kristoffersen, 1969).

For rose production high intensity lighting (5 to 6 Wm^{-2}) with high pressure sodium lamps (SON) is widely used. Both production and external quality can be promoted in this way. An after effect of lighting on the keeping quality of roses is caused by the adaptation of the stomatal behaviour due to the altered light/dark cycle. The stomata behaviour, which has been programmed by the lighting period, continues after harvesting. The stomata tend to remain open even under dark conditions which increases transpiration of the cut-roses (Slootweg & Van Meeteren, 1991). When the water uptake is hampered by blockage of the vessels due to air embolism or bacteria the vase life will be shortened.

Vegetables

A high rate of photosynthesis affects the production of sugars and acids, both important components in determining the flavour of fruit vegetables. More light will favour the sugar content, and reduce the acid content (Janse, 1984). A high light intensity also lowers nitrate accumulation in leaf vegetables (Blom-Zandstra, 1990).

At low light levels the synthesis of chloroplasts in the skin of the fruits will be low, which means less chlorophyll and more chromoplasts. In cucumber these fruits will be lighter green at harvest and can easily turn yellow during shelf life. In this crop a poor fruit colour is related to a poor keeping quality (or shelf life).

Besides taste and shelf life the external quality of vegetables is also affected by the light level. In Table 2.3.3 the major effects of light on quality of some vegetables are summarized.

Effects of photoperiod and spectral distribution

Quality can be influenced also by photoperiod and the spectral distribution of light, especially through its effects on flowering and photomorphogenesis. These aspects are of primary importance in flowering plants and pot plants, but less so for vegetables. Examples mentioned here are therefore restricted to floriculture plants. The flowering process of many floricultural crops is decided (qualitatively) or influenced (quantitatively) by daylength (section 2.3.1.3). The same daylength may be beneficial to the whole process from flower induction to flowering, but it may also be possible that induction, initiation and development as well size, shape and abnormalities of the inflorescence are influenced by different daylengths. In this way daylength can be used as a managing agent for different quality processes.

When light quantity is low in the winter season, the quality of Chrysanthemum can be improved by providing long days when flower induction in the top meristem and those of the uppermost leaf axils has been taken place. In this way vegetative growth will be stimulated instead of generative development, resulting in a heavier and stronger stem, larger top leaves and more petals in the flowers. In Begonia x hiemalis a period of induction brought on by short days is followed by a period of long days. This will achieve vegetative growth by preventing some the meristems from initiating flower buds.

With Alstroemeria the number of branches per umbel is influenced by daylength and thickness of the main stem. As daylength increases flower induction increases but the initiation of flowering stalks within the umbel stops earlier when daylength is longer, which results in less flower stalks

Table 2.3.3 – Influence of radiation (400 – 4000 $Jcm^{-2}dag^{-1}$)on the quality of some vegetables (Janse, 1984, 1985). – = reduction or less, + = better or more.

Tomato	Cucumber	Lettuce	Sweet pepper
colour –		glassiness –	cracking –
		compactness +	
shelf life + / –	shelf life +	shelf life +	
taste +	fruit rot +	nitrate –	
		(less nitrate)	

within the umbel (Vonk Noordegraaf, 1981).

Where flower formation is influenced by daylength, shoot formation or branching can be retarded by adopting daylengths which are favourable for flowering, especially when light quantity is low. This form of competition has been described for Alstroemeria (Vonk Noordegraaf, 1981), carnation (Heins & Wilkins, 1977) and Gerbera (Leffring, 1981).

The morphogenesis of plants can also be affected by the light quality (section 2.3.1.3). A lack of blue light as well as a ratio of red to far-red of less than about 1.2 may result in an increase in stem elongation in many species (Maas, 1992). Rose shoots grown in artificial light were 43% longer when all the blue wavelengths were filtered out. However, the practical application for glasshouse production is until now very limited. The research in the field of growth regulation and photomorphogenesis by specific use of artificial light (levels and wavelengths) will probably lead to improved plant production systems without the need for growth regulators combined with controllable plant quality.

2.3.3.3 Effects of temperature on product quality

Most processes of growth and development are influenced by temperature (section 2.3.1) so there is a clear influence on different aspects of quality. When temperature is low in relation to the light conditions during the growing period the texture of the plants will be strong and the firmness of the stems high. This may be positive for the external quality of many pot plants and cut-flowers (Cyclamen, carnation). In many crops both the internode length and the stem length increase at lower temperatures (Vonk Noordegraaf, 1973). Height is an important quality aspect for many pot and bedding plants. It has been shown for different crops that plant height can be reduced by high night and low day temperatures (section 2.3.1.3). The difference between day and night temperature (DIF) can be used to control plant height as far as this difference can be realized, dependent on time of the year and location. The reversed day/night temperature is partly practised as an alternative to growth retardants which are also commonly used for the height control of many pot and bedding plants (Moe & Heins, 1990).

Flower initiation of many crops is influenced by temperature (section 2.3.1.3). Some crops (Calceolaria, Pelargonium grandiflorum, Brunfelsia) need a period of low temperature before flower initiation starts (vernalization) or to break the dormancy of the flower buds (Anthurium scherzerianum) (Vonk Noordegraaf, 1973). In many flowering plants temperature can have a beneficial influence on flower formation within certain limits. Daylength sensitivity is also influenced by temperature and in some crops daylength and temperature are interchangeable (Runger, 1976).

Changing the temperature during the process of flower initiation can lead to irregular flower formation. This can be observed in many bulbous and tuberous crops (Hartsema, 1961). The normal floral spike of Freesia bears flowers at the same distance. If the temperature rises above 18 °C during flower initiation, the next flower will develop some distance above the flower already initiated. So the first flower(s) is positioned too low, which is known as "thumbing" (De Lint, 1969).

Lowering the mean daily temperature decreases growth rate and development (section 2.3.1.2). Roses have less lateral breaks and lateral buds break more slowly. The stems are thicker and flower buds longer and broader (Van den Berg, 1987). Rose flowers initiated at lower temperatures develop more petals in comparison with those grown at higher temperatures, which can cause malformed flowers (bull-heads) (Moe & Kristoffersen, 1969).

Carnations show the same response with respect to the formation of petals. When night temperature is low (5 °C) for a longer period extra whorls of petals will arise, causing the calyx to split afterwards.

In vegetables many other processes are disturbed by low temperatures. For example fruit set of tomatoes is hampered resulting in irregularly shaped fruits, bad fruit colouring and slow ripening. Low night temperature will increase the percentage of malformed sweet pepper fruits and fruits with a detached style (Rylski & Spigelman, 1982). In winter the compactness of iceberg lettuce will decrease at lower temperatures. Table 2.3.4 presents a review of some effects of temperature on various quality aspects of vegetables.

Temperature may have different effects on quality aspects of one crop. A compromise in temperature level is necessary (Table 2.3.4) where taste and shelf life are concerned.

At high irradiance, greenhouse and leaf temperatures may increase excesively. The temperature of sunlit plant parts (leaves, fruits, flowers) may be up to 10 °C higher than the surrounding air (Van Holsteijn, 1988). Besides damage to the photosynthesis machinery this may lead to secondary injuries (heat stress injuries or drought stress, associated with the high transpiration rates) such as e.g. sunscald (Rylski & Spigelman, 1986) and other detrimental effects on quality such as lower keeping quality (Janse, 1988), uneven ripening of fruits or necrosis (Larson, 1992). These various detrimental effects of high tissue temperatures may impose measures to reduce the light under extreme conditions.

2.3.3.4 Effects of humidity on product quality

Transpiration of plants, which is important for cooling and the transport of nutrient elements, is highly affected by air humidity. When the vapour pressure deficit is low, transpiration will be low, and plant cells can easily maintain a high turgor and reach their maximum size (section 2.2.2). Plants grown at high air humidity tend to make larger leaves, while under extreme situations some plants (Ficus) show uncontrolled cell growth at the main nerves. To maximize the growth large cells are desirable but in many cases restricted growth and stimulated transpiration result in a better quality.

In floriculture plants quality aspects differ largely between species and varieties. Generally the plants must be firm and be able to adapt to low humidity conditions compared with the humidity levels normally maintained during the cultivation period in the glasshouse. Irregular transpiration must be avoided during cultivation as well as during the post harvest period. Sudden changes in transpiration, caused either by changing humidity or other environmental conditions, may cause disturbances in the balance between water uptake and transpiration, resulting in large variations in the water content of the crop (section 2.2.2). If cell enlargement becomes excessive, it may cause deformation of cell walls, which in turn may lead to leaf burning (lettuce, lily, Alstroemeria), burst or broken stems (Freesia). The latter problems are probably also related to the lower calcium content of the various tissues of plants grown under high humidity (and low transpiration). Stimulation of transpiration by reducing humidity improves the calcium transport to the transpiring organs which reduces the risks of these particular problems (Bakker, 1991[a]).

Besides quality aspects directly related to humidity (Table 2.3.5), this environmental factor also influences infection by some pathogens, such as Botrytis. Infection of leaves or flowers by spores of different fungal diseases during the production period may reduce the market value of these products, as these spores will germinate during the post harvest period when the air humidity will be high.

Table 2.3.4 – Influences of temperature (for cucumber and sweet pepper 15 – 35 ° C, for lettuce 5 – 20° C) on quality of some vegetables. 0 = unaffected, – = less, + = more or better.

Tomato	Cucumber	Lettuce	Sweet pepper
colour 0	shelf life –	glassiness + / –	cracking –
shelf life –	fruit rot 0	firmness –	taste +
taste +		shelf life –	shelf life –

Table 2.3.5 – Influence of air humidity (range 0.2 – 1.0 kPa VPD) on some vegetables. 0 = unaffected, – = less, + = more or better (Janse, 1987, 1988).

Tomato	Cucumber	Lettuce	Sweet pepper
colour 0	shelf life –	glassiness –	russeting –
shelf life 0	fruit rot –	compactness 0	shelf life –
taste 0		shelf life –	blossom red rot –

2.3.3.5 Effects of CO_2 on product quality

As a high carbon dioxide concentration enhances photosynthesis (section 2.2.2), it generally improves both production and quality. Through the higher level of dry matter production (more reserves) the external quality of many crops is improved: firmer stems and leaves, improved colour and size of flowers, especially under the poor light conditions of the winter period. When the CO_2 concentration is high in relation to the temperature the texture of some crops can be too fleshy (Saintpaulia) which means that stems and leaves break easily (Enoch, 1990).

Information on the effects of CO_2 on internal quality aspects such as taste, shelf life and longevity is extremely limited but from the little information available it can be concluded that effects are minor.

2.2.3.6 Concluding remarks

Besides being affected by climatic conditions, quality is strongly influenced by nutrition, electrical conductivity and water supply. All these factors interact with the genetic characteristics of the plant. Thus the choice of variety is a decisive factor determining potential quality. The extent to which this potential quality will be realized depends on the growing conditions during the production period. In the post harvest period the present quality will come to expression and the success rate will be decided by the post harvest conditions.

2.4 Synthesis

H. CHALLA AND J.C. BAKKER

In this chapter the complex relations between environmental factors and crop response have been highlighted in some detail. In this section an attempt is made to use this knowledge to formulate requirements and rules in relation to the design of advanced, scientifically founded climate control strategies. Theoretically it may seem desirable to develop models that describe the response of greenhouse crops to the environment perfectly and completely and to use such models in advanced control algorithms. In practice this approach will not work for a number of reasons:

- Perfect and complete crop models are not and will not become available;
- Uncertainty is an element which is inherent in the crop production system;
- Decisions with respect to climate control have to be considered within the framework of operational management of the greenhouse and should therefore be subjected to the judgement of the grower (Chapter 6).

For these reasons present and future climate control systems in greenhouses will have to rely, at least partly, on information and judgements provided by the grower. A strategy for future climate control systems should therefore take this fact into account.

A second point that needs to be stipulated is that greenhouse cultivation is an economic activity, where climate requirements have to be considered within the framework of the management of the nursery, and where besides the performance of the crop other factors such as relations with pests and diseases, biological control, pollination by bees or bumble bees, labour conditions, and required economic inputs are of interest. Climate requirements should therefore be considered in relation to the objective of achieving a high and good quality production at the right time, at reasonable cost and acceptable risk.

The analysis of the production process presented clearly demonstrates that crop production is the final result of a complex of processes. The individual processes that contribute to production may exhibit quite contrasting responses to the environmental factors, contrasting time constants, contrasting dynamics and moreover there are many interactions between the processes. Since various processes occur within the same crop, within the same greenhouse, they are essentially subjected to the same environmental conditions. The requirements of the crop as a whole therefore reflect the potentially conflicting requirements of individual processes.

A very important characteristic of crop production is the distinction between energy fixation in the photosynthetic process resulting in the formation of assimilates, and the use of these assimilates for growth and production of harvestable product. This distinction is important because, in contrast to assimilates that can be stored for long periods of time, radiative energy, trapped within the thylakoids of the chloroplasts, has to be utilized efficiently and immediately. Sub-optimal conditions for photosynthesis will give rise to irrecoverable losses in production potential. In section 2.3.1 the close relation between intercepted radiation and dry matter production and yield was discussed. This relation supports the hypothesis that, in general, the utilization of radiation is the dominant rate-limiting process for production: under normal conditions all assimilates formed by photosynthesis are used in the production process. As a rule short-term discrepancies between photosynthesis and assimilate consumption are matched through storage (surplus) and substrate concentration mediated feed-back (shortage). Although accumulation of assimilates may result in a reduction of photosynthesis we believe that in greenhouse cultivation with a high degree of environmental control this is not a normal situation and that tuning of processing and formation of assimilates is not an important issue for (short-term) climate control, though it may play an important role in the long-term production strategy.

The dominant environmental factor determining the efficiency of light utilization is CO_2 concentration (section 2.2.1), which has only marginal side effects on other aspects of crop growth and development (section 2.3.1). Conversely, temperature and water vapour pressure of the air have only a minor effect on crop photosynthesis (section 2.2.1 and 2.2.3). Only indirectly do the control of temperature and humidity interfere with CO_2 availability because of the link between heating and CO_2 generation and the loss of CO_2 in relation to ventilation. CO_2 has only minor effects on utilisation of assimilates for growth, but water vapour pressure and particularly temperature have pronounced effects (section 2.3.1 and 2.3.2). Because assimilates can be stored, the response to these factors is in general less time-critical and can often be described in relation to average diurnal values. An exception, however, has to be made with respect to stress phenomena, where immediate damage may occur, such as in the case of water shortage, chilling or heat damage.

The main effect of water vapour pressure deficit is on crop transpiration, which influences the water status of the crop. Transpiration also plays a role in the transport of nutrients and other substances within the plant, for example in relation to local Ca deficiencies. In addition water vapour pressure plays a role in the occurrence of condensation on the crop or organs of the crop (e.g. fruits). The water status of the crop (more specifically the turgor in growing cells) is important for cell extension and hence a factor affecting morphogenesis and crop quality (section 2.2.2: physiogenic disorders, water stress). It is not easy to formulate general rules for the control of water vapour pressure on the basis of the literature and it is likely that more specific research is required to come to a better founded judgement. Based on knowledge presented in this chapter and additional personal judgement we come to the following preliminary evaluation.

The major processes influenced by water vapour pressure (deficit) are condensation and transpiration. Some condensation may possibly be acceptable for some time, depending on the situation, but limits could be formulated by the grower (e.g. with respect to the time of the day and the duration). Transpiration requirements are determined by its transport function and its role with respect to the water balance. To meet the transport requirements formulation (by the grower) of a minimum (average?) amount of transpiration per 24 h or per day and per night is probably the best choice. The relation between transpiration and water status of the crop is complex and it is even more difficult to establish a quantitative relation with crop growth, yield and product quality. Taking these uncertainties into consideration, together with the expense involved in controlling transpiration, it seems plausible to aim for a target transpiration rate, allowing a certain range around this value and placing a limit on the amplitude. The parameters should be judged by the grower, based on experience and evaluation of the crop. The control of transpiration may become a time-critical factor in the case of desiccation, but in other cases this is probably not so.

Temperature has primarily an impact on crop development and morphogenesis, though it may affect the diurnal carbon budget more in winter, at low radiation and a high biomass per unit greenhouse area, through its effect on maintenance respiration. Moreover, there is a delayed effect on photosynthesis (section 2.3.1), mediated through formation of leaf area and related light interception, particularly at low LAI. The relations between temperature and development and morphogenesis, though highly complex and insufficiently understood, are (within certain limits) in many cases approximately linear, and consequently the average diurnal temperature or even the average temperature over longer periods of time is more important than the actual time course. The concept of DIF ($T_{day} - T_{night}$) has been discussed in relation to morphogenesis and may lead to additional requirements with respect to the distribution of temperature (integral) over day and night. In conclusion, temperature requirements may be characterized by the average diurnal temperature, or average day and night temperature, and the acceptable minimum and maximum day and nighttime temperatures. This combination should allow room to deal with tolerance to damage, as well as with, for example, flowering or quality requirements. Given the complex relations governing the choice and the importance of feed back from observation by the grower, the choice should be left to the grower, possibly supported by models dealing with part of the relevant processes. Avoidance of temperature extremes is time-critical, but for other aspects there seems to be room for averaging over periods longer than a day.

It was stated before that the climate requirements of the crop result from conflicting requirements at process level. Reconsidering our preliminary conclusions on how climate requirements should be formulated, this problem of conflicting requirements has to be solved or reduced to a certain extent. The problem can be reduced by distinguishing quantitative and qualitative or threshold response types. Quantitative responses become manifest over a wide range of conditions, whereas threshold responses are only observed when a given threshold value is surpassed. In principle only quantitative responses have to be considered, within the range of conditions allowed by the threshold responses. A complication, however, is the problem that threshold values may not be constants but vary with crop state and, worse, may depend on other climatic factors. An example is the minimum temperature, which depends on the radiation level. An additional problem is that with an increasing number of processes described by threshold values the chance of conflicting requirements increases and that the greenhouse climate becomes fully dictated by this type of responses. This problem can only be solved by evaluating in one way or another the relative importance of various phenomena and creating some kind of hierarchy, probably in close consultation with the grower (Bakker, 1995).

Under normal conditions light periods alternate with periods of darkness. This alternation offers another interesting opportunity to solve the problem of contrasting requirements for different processes, by taking the response time into consideration. The occurrence of processes with a short response time beside those with a slow response time offer the opportunity to optimize both, to a certain extent independently, under non-steady-state conditions. Over longer periods of time a good compromise could be achieved by formulating requirements of processes with a slow response as average day/night, diurnal, or even week temperature. If, for example, during the light period priority is given to photosynthesis requirements, the night may be used to "correct" the average values with respect to the long term requirements.

An important conclusion of this review is that the present greenhouse climate computers, controlling different climate factors according to pre-defined set-points, do not adequately meet the requirements that follow from the present analysis, and certainly do not provide the grower with the right choices for control of the production process.

Finally it should be noted that it is principally impossible to develop sensors that could be used to evaluate crop performance directly, an approach known as the "speaking plant" approach (Udink ten Cate *et al.*, 1978). Only when a process is evaluated within the framework of the production process as a whole may its role be interpreted correctly. A different situation, however, prevails when sensors are used in relation to stress conditions, an approach sometimes called the "squeaking plant" approach, but here the purpose is not to measure crop performance, but to detect stress conditions in the crop.

References

Acevedo, E., E. Fereres, Th.C. Hsiao & D.W. Henderson, 1979. Diurnal growth trends, water potential, and osmotic adjustment of maize and sorghum leaves in the field. Plant Physiology 64: 476–480.

Acock, B., D.A. Charles-Edwards & S. Sawyer, 1979. Growth response of a *Chrysanthemum* crop to the environment. III. Effects of radiation and temperature on dry matter partitioning and photosynthesis. Annals of Botany 44: 289–300.

Acock, B., D.W. Hand, J.H.M. Thornley & J. Warren Wilson, 1975. Photosynthesis in stands of green peppers. An application of empirical and mechanistic models to controlled-environment data. Annals of Botany 40: 1293–1307.

Acock, B., D.A. Charles-Edwards, D.J. Fitter, D.W. Hand, L.J. Ludwig, J. Warren Wilson & A.C. Withers, 1978. The contribution of leaves from different levels within a tomato crop to canopy net photosynthesis: an experimental examination of two canopy models. Journal of Experimental Botany 29: 815–827.

Adams, P., 1991. Effect of diurnal fluctuations in humidity on the accumulation of nutrients in the leaves of tomato. Journal of Horticultural Science 66: 545–550.

Adams, P. & L.C. Ho, 1989. Effects of constant and fluctuating salinity on the yield, quality and calcium status of tomatoes. Journal of Horticultural Science 64: 725-732.

Aikman, D.P. & G. Houter, 1990. Influence of radiation and humidity on transpiration: implications for calcium levels in tomato leaves. Journal of Horticultural Science 65: 245–253.

Aikman, D.P. & A.J.F. Picken, 1989. Wind-related temperature setting in glasshouses. Journal of Horticultural Science 64: 649–654.

Andersen, A., 1986. Comparison of fluorescent lamps as an energy source for production of tomato plants in a controlled environment. Scientia Horticulturae 28: 11–18.

Andersen, H. & M. Hansen, 1989. Nutrient content of various pot plant species as influenced by supplementary lighting. Acta Horticulturae 138: 109–112.

Andersson, N.E., 1990. Effects of level and duration of supplementary light on development of chrysanthemum. Scientia Horticulturae 44: 163–169.

Aphalo, P.J. & P.G. Jarvis, 1991. Do stomata respond to relative humidity? Plant, Cell and Environment 14: 127–132.

Appleby, R.F. & W.J. Davies, 1983. The structure and orientation of guard cells in plants showing stomatal responses to changing vapour pressure differences. Annals of Botany 52: 459–468.

Aubinet, M., J. Deltour & D. de Halleux, 1989. Stomatal regulation in greenhouse crops: analysis and simulation. Agricultural and Forest Meteorology 48: 21–44.

Auge, R., H. Vidalie, M. Laffaire & L. Guerin, 1984. Influence du gaz carbonique sur la croissance et le développement des plantes en pots. Revue Horticole, 244: 29–35 (in French).

Bailey, B.J. & A. Hunter, 1988. Plant response and energy use in five high thermal resistance greenhouses. Acta Horticulturae 229: 165–171.

Baker, N.R., W.D. Davies & C. Ong (Eds), 1985. Control of leaf growth. SEB Seminar series 27. Cambridge University Press, Cambridge, 350 pp.

Bakker, J.C., 1990. Effects of day and night humidity on yield and fruit quality of glasshouse tomatoes (*Lycopersicon esculentum* Mill.). Journal of Horticultural Science 65: 323–331.

Bakker, J.C., 1991[a]. Analysis of humidity effects on growth and production of glasshouse fruit vegetables. Dissertation, Wageningen Agricultural University, Wageningen, 155 pp.

Bakker, J.C., 1991[b]. Leaf conductance of four glasshouse vegetable crops as affected by air humidity. Agricultural and Forest Meteorology 55: 23–36.

Bakker, J.C., 1995. Greenhouse climate control: constraints and limitations. Acta Horticulturae, ISHS, Kyoto, Japan, 1994. (in press).

Bakker, J.C., & C. Sonneveld, 1988. Calcium deficiency of glasshouse cucumber as affected by environmental humidity and mineral nutrition. Journal of Horticultural Science 63: 241-246.

Ball, J.T., 1987. Calculations related to gas exchange. In: E. Zeiger, G.D. Farquhar & I.R. Cowan (Eds), Stomatal function. Stanford University Press, Stanford, pp. 445–476.

Ball, J.T. & J.A. Berry, 1981. The C_i/C_s ratio: a basis for predicting stomatal control of photosynthesis. Carnegie Institution Yearbook 81: 88–92.

Ball, J.T., I.E. Woodrow & J.A. Berry, 1987. A model predicting stomatal conductance and its contibution to the control of photosynthesis under different environmental conditions. In: J. Bigges (Ed.), Progress in photosynthesis research, Vol. IV.5. Martinus Nijhoff Publishers, Dordrecht, pp. 221–224.

Bangerth, F., 1979. Calcium-related physiological disorders of plants. Annual Review of Phytopathology 17: 97-122.

Barlow, E.W.R., 1982. Water relations of the mature leaf. In: J.E. Dale & F.L. Milthorpe (Eds), The growth and functioning of leaves. Cambridge University Press, pp. 315–345.

Barrs, H.D., 1973. Controlled environment studies of the effects of variable atmospheric water stress on photosynthesis, transpiration and water status of *Zea mays* L. and other species. In: R.O. Slatyer (Ed.), Plant response to climatic factors. Proceedings Uppsala Symposium, UNESCO, 1970, pp. 249–258.

Barrs, H.D. & P.E. Weatherley, 1962. A re-examination of the relative turgidity technique for estimating water deficits in leaves. Australian Journal of Biological Sciences 15: 413–428.

Behboudian, M.H., 1977[a]. Water relations of cucumber, tomato, and sweet pepper. Mededelingen Landbouwhogeschool Wageningen 77-6, Wageningen, 84 pp.

Behboudian, M.H., 1977[b]. Responses of eggplant to drought. I. Plant water balance. Scientia Horticulturae 7: 303–310.

Behboudian, M.H. & H.M.C. Van Holsteijn, 1977. Water relations of lettuce. I. Internal physical aspects for two cultivars. Scientia Horticulturae 7: 9–17.

Bengtsson, B., 1982. Uptake and translocation of calcium in cucumber. Physiologia Plantarum 54: 107-111.

Bensink, J., 1971. On morphogenesis of lettuce leaves in relation to light and temperature. Mededelingen Landbouwhogeschool Wageningen 71–15, Wageningen, 93 pp.

Bernier, G., J.-M. Kinet & R.M. Sachs, 1981. The physiology of flowering. Vol. I and Vol. II. CRC Press Inc., Florida, 149 and 231 pp. resp.

Berry, J. & O. Bjorkman, 1980. Photosynthetic response and adaptation to temperature in higher plants. Annual Review of Plant Physiology 31: 491–543.

Bertin, N. & C. Gary, 1993. Tomato fruit-set: a case study for validation of the model TOMGRO. Acta Horticulturae 328: 185–193.

Besford, R.T., L.J. Ludwig & A.C. Withers, 1990. The greenhouse effect: acclimation of tomato plants growing in high CO_2, photosynthesis and ribulose-1,5- Bis phosphate carboxylase protein. Journal of Experimental Botany 41: 925–931.

Bhatt, R.M. & N.K. Srinivasa Rao, 1989. Photosynthesis and dry matter partitioning in 3 cultivars of *Capsicum annuum* L. grown at 2 temperature conditions. Photosynthetica 23: 21–27.

Bidwell, R.G.S., 1974. Plant physiology. Macmillan Publishing Co., Inc. New York; Collier Macmillan Publishers, London, 643 pp.

Bierhuizen, J.F. & W.A. Wagenvoort, 1974. Some aspects of seed germination in vegetables. 1. The determination and application of heat sums and minimum temperature for germination. Scientia Horticulturae 2: 213–219.

Björkman, O., 1981. Responses to different quantum flux densities. Physiological Plant Ecology I. Encyclopedia of Plant Physiology, Vol. 12A. Springer Verlag, Berlin, pp. 57–107.

Björkman, O. & B. Demmig, 1987. Photon yield of O2 evolution and chlorophyll fluorescence characteristics at 77K among vascular plants of diverse origins. Planta 170: 489–504.

Blom-Zandstra, M. 1990. Some physiological aspects of nitrate accumulation in lettuce (*Lactuca sativa*). PhD-thesis State University, Utrecht, 81 pp.

Boyer, J.S., 1974. Water transport in plants: mechanism of apparent changes in resistance during absorption. Planta 117: 187–207.

Boyer, J.S., 1985. Water transport. Annual Review Plant Physiology 36: 473–516.

Boyer, J.S., 1989. Water potential and plant metabolism: comments on Dr P.J. Kramer's article 'Changing concepts regarding plant water relations', Volume 11, Number 7, pp. 565–568, and Dr J.B. Passioura's Response, pp. 569–571. Plant, Cell and Environment 12: 213–216.

Bradfield, E.G. & C.G. Guttridge, 1979. The dependence of calcium transport and leaf tipburn in strawberry on relative humidity and nutrient solution concentration. Annals of Botany 43: 363-372.

Bradfield, E.G. & C.G. Guttridge, 1984. Effects of night-time humidity and nutrient solution concentration on the calcium content of tomato fruit. Scientia Horticulturae 22: 207–217.

Bradford, K.J. & T.C. Hsiao, 1982. Physiological responses to moderate water stress. In: O.L. Lange, P.S. Nobel, C.B. Osmond & H. Ziegler (Eds), Physiological Plant Ecology II. Water relations and carbon assimilation. Encyclopedia of Plant Physiology, Vol. 12B, Springer Verlag, Berlin, pp. 264–324.

Bradford, K.J., T.D. Sharkey & G.D. Farquhar, 1983. Gas exchange, stomatal behaviour, and d13C values of the *flacca* tomato mutant to abscisic acid. Plant Physiology 72: 245–250.

Bradley, F.M. & Janes, H.W., 1985. Carbon partitioning in tomato leaves exposed to continuous light. Acta Horticulturae 174: 293–302.

Brouwer, R., 1963. Some aspects of the equilibrium between overground and underground plant parts. Mededeling Instituut voor Biologisch en Scheikundig onderzoek van Landbouwgewassen, Wageningen, 213: 31–39.

Brouwer, R., 1973. Dynamics of plant performance. Acta Horticulturae 32: 31–54.

Brouwer, R., 1974. A comparison of the effect of drought and low root temperatures on leaf elongation and photosynthesis in maize. Acta Horticulturae 39: 141–145.

Brouwer, R., 1983. Functional equilibrium: sense or nonsense? Netherlands Journal of Agricultural Science 31: 335–348.
Bruggink, G.T., 1984. Effects of CO_2 concentration on growth and photosynthesis of young tomato and carnation plants. Acta Horticulturae 162: 279.
Bruggink, G.T., 1987. Influence of light on the growth of young tomato, cucumber and sweet pepper plants in the greenhouse: calculating the effect of differences in the light integral. Scientia Horticulturae 31: 175–183.
Bruggink, G.T., 1992. A comparative analysis of the influence of light on growth of young tomato and carnation plants. Scientia Horticulturae 51: 71–81.
Bruggink, G.T. & E. Heuvelink, 1987. Influence of light on the growth of young tomato, cucumber and sweet pepper plants in the greenhouse: effects on relative growth rate, net assimilation rate and leaf area ratio. Scientia Horticulturae 31: 161–174.
Bruggink, G.T., H.E. Schouwink & Th.H. Gieling, 1988. Modelling of water potential and water uptake rate of tomato plants in the greenhouse: preliminary results. Acta Horticulturae 229: 177–185.
Bunce, J.A., 1981. Comparative responses of leaf conductance to humidity in single attached leaves. Journal of Experimental Botany 32: 629–634.
Bunce, J.A., 1984. Effects of humidity on photosynthesis. Journal of Experimental Botany, 35: 1245–1251.
Bunce, J.A., 1988[a]. Non-stomatal inhibition of photosynthesis by water stress. Reduction in photosynthesis at high transpiration rate without stomatal closure in field-grown tomato. Photosynthesis Research 18: 357–362.
Bunce, J.A., 1988[b]. Effects of boundary layer conductance on substomatal pressures of carbon dioxide. Plant, Cell and Environment 11: 205–208.
Bunt, A.C., 1978. Effect of season on the carnation (*Dianthus caryophyllus* L.). III Flower quality. Journal of Horticultural Science 53: 75-84.
Burrage, S.W., 1988. Growth and ion uptake in tomatoes grown in high and low humidities. In: K.E. Cockshull (Ed.), The effects of high humidity on plant growth in energy-saving greenhouses. Report EUR 11261, Office of Official Publications of the European Communities, Luxembourg, pp. 9–18.
Burrows, F.J. & F.L. Milthorpe, 1976. Stomatal conductance in gas exchange control. In: T.T. Kozlowski (Ed.), Water deficits and plant growth, IV. Academic Press, New York, pp. 103–153.
Charbonneau, J., A. Gosselin & M.-J. Trudel, 1988. Influence de la conductivité électrique de la solution nutritive sur la croissance et le développment de la tomate de serre cultivée avec ou sans éclairage d'appoint. Canadian Journal of Plant Science 68: 267-276. (in French).
Challa, H., 1976. An analysis of the diurnal course of growth, carbon dioxide exchange and carbohydrate reserve content of cucumber. Agricultural Research Reports 861: 1–88.
Challa, H., 1990. Crop growth models for greenhouse climate control. In: R. Rabbinge, J. Goudriaan, H. Van Keulen, F.W.T. Penning de Vries & H.H. Van Laar (Eds), Theoretical production ecology: reflections and prospects. Simulation Monographs 34. Pudoc, Wageningen, pp. 125–145.
Challa, H. & M.J. Bakker, 1995. Potential production within the greenhouse environment. In: H.Z. Enoch & G. Stanhill (Eds), The greenhouse ecosystem. (Elsevier's Series: D.W. Goodall (Editor in chief) Ecosystems of the world). Elsevier, Amsterdam (in press).
Challa, H. & P. Brouwer, 1985. Growth of young cucumber plants under different diurnal temperature patterns. Acta Horticulturae 174: 211–217.
Challa, H. & E. Heuvelink, 1993. Economic evaluation of crop photosynthesis. Acta Horticulturae 328: 219–228.
Challa, H. & A.H.C.M. Schapendonk, 1984. Quantification of effects of light reduction in greenhouses on yield. Acta Horticulturae 148: 501–510.
Challa, H. & A.H.C.M. Schapendonk, 1986. Dynamic optimalization of CO_2 concentration in relation to climate control in greenhouses. In: H.Z. Enoch & B.A. Kimball (Eds), Carbon dioxide enrich-

ment of greenhouse crops. Volume I, Status and CO_2 resources. CRC Press, Inc., Boca Raton, Florida, pp. 147–160.

Chandler, E.L. & O.P. Watson, 1954. Contributions of various light intensities to the growth and yield of greenhouse roses. Proceedings American Society for Horticultural Science 64: 441–447.

Charles-Edwards, D.A. & L.J. Ludwig, 1974. A model for leaf photosynthesis by C_3 plant species. Annals of Botany 38: 921–930.

Chaves, M.M., 1991. Effects of water deficits on carbon assimilation. Journal of Experimental Botany 42: 1–16.

Claussen, W., 1976. Einfluss der Frucht auf die Trockensubstanzverteilung in der Aubergine (*Solanum melongena* L.). Gartenbauwissenschaft 41: 236–239. (in German).

Cockshull, K.E., 1988. The integration of plant physiology with physical changes in the greenhouse climate. Acta Horticulturae 229: 113–132.

Cockshull, K.E., 1992. Crop environments. Acta Horticulturae 312: 7 – 85.

Cockshull, K.E. & A.P. Hughes, 1967. Distribution of dry matter to flowers in *Chrysanthemum morifolium*. Nature 215: 780–781.

Cockshull, K.E. & A.P. Hughes, 1968. Accumulation of dry matter by *Chrysanthemum morifolium* after flower removal. Nature 217: 979–980

Cockshull, K.E., D.W. Hand & F.A. Langton, 1981. The effects of day and night temperature on flower initiation and development in *Chrysanthemum*. Acta Horticulturae 125: 101–110.

Cockshull, K.E., C.J. Graves & C.R.J. Cave, 1992. The influence of shading on yield of glasshouse tomatoes. Journal of Horticultural Science 67: 11–24.

Coker, F.A. & J.J. Hanan, 1988. The effect of shading on 'Samantha' roses. Colorado State Research Bulletin 455: 1–5.

Collatz, G.J., J.T. Ball, C. Grivet & J.A. Berry, 1991. Physiological and environmental regulation of stomatal conductance, photosynthesis and transpiration: a model that includes a laminar boundary layer. Agricultural and Forest Meteorology 54: 107–136.

Conover, C.A. & R.T. Poole, 1975. Acclimatization of tropical trees for interior use. HortScience 10: 600-601.

Conover, C.A. & R.T. Poole, 1990. Acclimatization revisited. Foliage Digest, 13: 6–7.

Cooper, A.J., 1973. Root temperature and plant growth. Research review no. 4. Commonwealth Bureau Hortic. Plant. Crops, East Malling, UK, 73 pp.

Cooper, A.J. & J.H.M. Thornley, 1976. Response of dry matter partitioning, growth and carbon and nitrogen levels in the tomato plant to changes in root temperature: experiment and theory. Annals of Botany 40: 1139–1152.

Cosgrove, D., 1986. Biophysical control of plant cell growth. Annual Review of Plant Physiology 37: 377–405.

Cowan, I.R., 1972. Oscillations in stomatal conductance and plant functioning associated with stomatal conductance: observations and a model. Planta 106: 185–219.

Cowan, I.R. & G.D. Farquhar, 1977. Stomatal function in relation to leaf metabolism and environment. In: Integration of activity in the higher plant. Soc. Exp. Biol. Symp., 31. Cambridge University Press, Cambridge, pp. 471–505.

Craker, L.E., M. Seiber & J.T. Clifford, 1983. Growth and development of radish (*Raphanus sativus*, L.) under selected light environments. Annals of Botany 51: 59–64.

Cramwinckel, A.B., 1989. What is quality? The meaning of 'analytical' and 'emotional' quality. Trade Journal of Technology of Foods in Holland and Belgium, 6: 17-21.

Dale, J.E., 1988. The control of leaf expansion. Annual Review of Plant Physiology and Plant Molecular Biology 39: 267–295.

Dale, J.E., & F.L. Milthorpe (Eds), 1982. The growth and functioning of leaves. Cambridge University Press, Cambridge, 540 pp.

Darrall, N.M., 1989. The effect of air pollutants on physiological processes in plants. Plant, Cell and Environment 12: 1–30.

Davies, W.J., & J. Zhang, 1991. Root signals and the regulation of growth and development of plants in drying soil. Annual Review of Plant Physiology and Plant Molecular Biology 42: 55–76.

Davies, W.J., T.A. Mansfield & A.M. Hetherington, 1991. Sensing of soil water status and the regulation of plant growth and development. Plant, Cell and Environment 13: 709–919.

De Graaf, R. & J. Van den Ende, 1981. Transpiration and evapotranspiration of the glasshouse crops. Acta Horticulturae 119: 147–158.

De Greef, J.A., M.P. De Proft, O. Mekers, R. Van Dijck, L. Jacobs & L. Philippe, 1989. Floral induction of Bromeliads by ethylene. In: H. Clijsters, M. De Proft, R. Marcelle & M. Van Poucke (Eds), Biochemical and physiological aspects of ethylene production in lower and higher plants. Kluwer Academic Publishers, Dordrecht/Boston/London, pp. 313–322.

De Hertogh, A. & M. Le Nard, 1993. The physiology of flower bulbs. Elsevier, Amsterdam, 811 pp.

De Koning, A.N.M., 1988. The effect of different day/night temperature regimes on growth, development and yield of glasshouse tomatoes. Journal of Horticultural Science 63: 465–471.

De Koning, A.N.M., 1989[a]. The effect of temperature on fruit growth and fruit load of tomato. Acta Horticulturae 248: 329–336.

De Koning, A.N.M., 1989[b]. Development and growth of a commercially grown tomato crop. Acta Horticulturae 260: 267–273.

De Koning, A.N.M., 1990. Long-term temperature integration of tomato. Growth and development under alternating temperature regimes. Scientia Horticulturae 45: 117–127.

De Koning, A.N.M., 1991. Gewaswaarnemingen praktijkbedrijven tomaat 1990. Plantslachtingen bij: W. Van Schie en P. Grootscholten. Intern Verslag 9b, Proefstation voor Tuinbouw onder Glas, Naaldwijk, 39 pp. (in Dutch).

De Koning, A.N.M., 1993. Growth of a tomato crop: measurements for model validation. Acta Horticulturae 328: 141–146.

De Koning, A.N.M., 1994. Development and dry matter distribution in glasshouse tomato: a quantative approach. Dissertation, Wageningen Agricultural University, Wageningen, 240 pp.

De Koning, A.N.M. & H.W. De Ruiter, 1992. Growth and yield measurements at commercial nurseries. Annual Report 1991 Glasshouse Crops Research Station, Naaldwijk, pp. 30–31.

De Koning, A. & R.G. Hurd, 1983. A comparison of winter-sown tomato plants grown with restricted and unlimited water supply. Journal of Horticultural Science 58: 575–581.

De Lint, P.J.A.L., 1969. Flowering in freesia: temperature and corms. Acta Horticulturae 14: 125-131.

De Stigter, H.C.M. & A.G.M. Broekhuysen, 1984. Weight-change patterns in cut and intact rose shoots. II. Floral and foliar contributions as related to competition for water. Journal of Plant Physiology 115: 319–329.

De Visser, A. & W.P. Vesseur, 1982. Daglicht, een van de vele factoren die de produktie bepalen (*Daylight, one of the production influencing factors*). Tuinderij 62(9): 38–39. (in Dutch).

De Vries, D.P. & L.A.M. Dubois, 1994. The root and shoot growth of harvested or non-harvested 'Sonia' rose bushes, bench grafted onto *R. canina* 'Inermis'. Gartenbauwissenschaft (submitted).

De Willigen, P. & M. Van Noordwijk, 1987. Roots, plant production and nutrient use. Dissertation, Wageningen Agricultural University, Wageningen, 282 pp.

De Wit, C.T., 1965. Photosynthesis of leaf canopies. Agricultural Research Reports. Centre for Agricultural Publications and Documentation, Wageningen, 57 pp.

Dieleman, J.A. & E. Heuvelink, 1992. Factors affecting the number of leaves preceding the first inflorescence in the tomato. Journal of Horticultural Science 67: 1–10.

Dimond, A.E., 1966. Pressure and flow relations in vascular bundles of the tomato plant. Plant Physiology 41: 119–131.

Dorais, M., A. Gosselin & M.J. Trudel, 1991. Annual greenhouse tomato production under a sequential intercropping system using supplemental light. Scientia Horticulturae 45: 225–234.

Drew, M.C., 1987. Function of root tissues in nutrient and water transport. In: P.J. Gregory, J.V. Lake & D.A. Rose (Eds), Root development and function. SEB Seminar Series no. 30. Cambridge University Press, Cambridge, pp. 71–101.

Drews, M., A. Heissner & P. Augustin, 1980. Die Ertragsbildung der Gewächshausgurke beim Frühanbau in Abhängigkeit von der Temperatur und Bestrahlungsstärke. Archiev für Gartenbau 28: 17–30. (in German).

Duniway, 1971. Water relations of Fusarium wilt in tomato. Physiological Plant Pathology 1: 537–546.

Düring, H., & F. Oggionni, 1986. Transpiration und Mineralstoffeinlagerung der Weinbeere. Vitis 25: 59-66. (in German).

Ehleringer, J. & R.W. Pearcy, 1983. Variation in quantum yield for CO_2 uptake among C_3 and C_4 plants. Plant Physiology 73: 555–559.

Ehret, D.L., & L.C. Ho, 1986[a]. Translocation of calcium in relation to tomato fruit growth. Annals of Botany 58: 679–688.

Ehret, D.L. & L.C. Ho, 1986[b]. The effects of salinity on dry matter partitioning and fruit growth in tomatoes grown in nutrient film culture. Journal of Horticultural Science 61: 361–367.

Ehret, D.L. & P.A. Jolliffe, 1985. Leaf injury to bean plants grown in carbon dioxide enriched atmosphere. Canadian Journal of Botany 63: 2015–2020.

El-Sharkawy, M.A., J.H. Cock & A. Del Pilar Hernandez, 1986. Stomatal response to air humidity and its relation to stomatal density in a wide range of warm climate species. Photosynthesis Research: 137–149.

Elers, B., & H.-J. Wiebe, 1984. Flower formation of chinese cabbage. Scientia Horticulturae 22: 219–231, 327–332.

Enoch, H.Z., 1990. Crop responses to aerial carbon dioxide. Acta Horticulturae 268: 17-32.

Enoch, H.Z. & N. Zieslin, 1988. Growth and development of plants in response to carbon dioxide concentrations. Applied Agricultural Research 3: 248–256.

Erickson, R.O. & F.J. Michelini, 1957. The plastochron index. American Journal of Botany 44: 297–305.

Evans, L.T., 1975. Beyond photosynthesis – the role of respiration, translocation and growth potential in determining productivity. In: J.P. Cooper (Ed.), Photosynthesis and productivity in different environments. Cambridge University Press, Cambridge, pp. 501–507.

Evans, L.T., 1987. The dependence of quantum yield on wavelength and growth irradiance. Australian Journal of Plant Physiology 14: 69–79.

Evans, L.T., 1990. Assimilation, allocation, explanation, extrapolation. In: R. Rabbinge, J. Goudriaan, H. Van Keulen, F.W.T. Penning de Vries & H.H. Van Laar (Eds), Theoretical production ecology: reflections and prospects. Simulation Monographs 34. Pudoc, Wageningen, pp. 77–87.

Farquhar, G.D. & S. Von Caemmerer, 1982. Modelling of photosynthetic response to environmental conditions. In: O.L. Lange, P.S. Nobel, C.B. Osmond & H. Ziegler (Eds), Physiological plant ecology II. Water relations and carbon assimilation. Encyclopedia of Plant Physiology, Vol. 12B. Springer Verlag, Berlin, pp. 549–587.

Farquhar, G.D. & S.C. Wong, 1984. An empirical model of stomatal conductance. Australian Journal of Plant Physiology 11: 191–210.

Farquhar, G.D., S. Von Caemmerer & J.A. Berry, 1980. A biochemical model of photosynthetic CO_2 assimilation in leaves of C_3 species. Planta 149: 78–90.

Farquhar, G.D., S.C. Wong, J.R. Evans & K.T. Hubick, 1989. Photosynthesis and gas exchange. In: H.G. Jones, T.J. Flowers & M.B. Jones (Eds), Plants under stress. SEB Seminar series 39. Cambridge University Press, Cambridge, pp. 47–69.

Farrar, J.F., 1988. Temperature and the partitioning and translocation of carbon. In: F.P. Long & F.I. Woodward (Eds), Plants and temperature. Symposia of the Society for Experimental Biology 42.

Company of Biologists Ltd., Cambridge. pp. 203–235.

Ferrario, S., I. Agius & A. Morisot, 1992. Daily variations of the mineral composition of xylemic exudates in tomato. Journal of Plant Nutrition 15: 85-98.

Fiscus, E.L., 1975. The interaction between osmotic- and pressure-induced water flow in plant roots. Plant Physiology 55: 917–922.

Fiscus, E.L., A. Klute & M.R. Kaufmann, 1983. An interpretation of some whole plant transport phenomena. Plant Physiology 71: 810–817.

Fjeld, T. 1992. Effects of temperature and irradiance level on carbohydrate content and keeping quality of Christmas begonia (Begonia × cheimantha Everett.). Scientia Horticulturae 50: 219-228.

Fredrick, F., R. Lemeur & R. Zhang, 1992. Test of a Penman-Monteith evapotranspiration model for a plant dependent control of water supply to greenhouse crops. Landbouwtijdschrift 45: 41–57. (in Dutch).

Fuchs, H.W.M., 1986. Harvesting, pruning and root reactions of roses. Acta Horticulturae 189: 109–115.

Geiger, D.R., 1976. Effects of assimilate translocation on photosynthesis. Canadian Journal of Botany 54: 2337–2345.

Geiger, D.R. & J.C. Servaites, 1991. Carbon allocation and responses to stress. In: H.A. Mooney, W.E. Winner & E.J. Pell (Eds), Responses of plants to multiple stresses. Academic Press, London, pp. 103–127.

Gijzen, H., 1992. Simulation of photosynthesis and dry matter production of greenhouse crops. Simulation Report CABO-TT no. 28, Wageningen, 69 pp.

Gijzen, H, & J. Goudriaan, 1989. A flexible and explanatory model of light distribution and photosynthesis in row crops. Agricultural and Forest Meteorology 48: 1–20.

Gislerød, H.R., & P.V. Nelson, 1989. The interaction of relative air humidity and carbon dioxide enrichment in the growth of Chrysanthemum × morifolium Ramat. Scientia Horticulturae 38: 305–313.

Gislerød, H.R., A.R. Selmer-Olsen & L.M. Mortensen, 1987. The effect of air humidity on nutrient uptake of some greenhouse plants. Plant and Soil 102: 193-196.

Goudriaan, J., 1977. Crop micrometeorology: a simulation study. Simulation Monographs. Pudoc, Wageningen, 249 pp.

Goudriaan, J., 1982. Some techniques in dynamic simulation. In: F.W.T. Penning De Vries & H.H. Van Laar (Eds), Simulation of plant growth and crop production. Pudoc, Wageningen, pp. 66–84.

Goudriaan, J. & H.H. Van Laar, 1978. Relations between leaf resistance, CO_2 concentration and CO_2 assimilation in maize, beans, lalang grass and sunflower. Photosynthetica 12: 241–249.

Goudriaan, J. & H.E. De Ruiter, 1983. Plant growth in response to CO_2 enrichment at two levels of nitrogen and phosphorus supply. 1. Dry matter, leaf area and development. Netherlands Journal of Agricultural Science 31: 157–169.

Goudriaan, J., H.H. Van Laar, H. Van Keulen & W. Louwerse, 1985. Photosynthesis, CO_2 and plant production. In: W. Day & R.K. Atkin (Eds), Wheat growth and modeling. NATO ASI Series, Series A: Life Sciences, Vol. 86, pp. 107–122.

Grange, R.I. & D.W. Hand, 1987. A review of the effects of atmospheric humidity on the growth of horticultural crops. Journal of Horticultural Science 62: 125-134.

Grantz, D.A., 1990. Plant response to atmospheric humidity. Plant, Cell and Environment 13: 667–679.

Habegger, R., 1985. Vernalisations- und Devernalisationsreaktionen von Kohlrabi (*Brassica oleracea* convar. *acephala* var. *gongylodes* L.). Gartenbauwissenschaft 51(4): 184–189 (in German).

Hackett, W.P., 1985. Juvenility, maturation, and rejuvenation in woody plants. Horticultural Reviews 7: 109–155.

Halevy, A.H., 1986. Assimilate allocation and flower development. In: J.G. Atherton (Ed.), Manipulation of flowering. Butterworth, London, pp. 363–378.

Hall, A.J., 1977. Assimilate source-sink relationships in *Capsicum annuum* L. I. The dynamics of growth in fruiting and deflorated plants. Australian Journal of Plant Physiology 4: 623–636.

Hall, A.E. & C.J. Brady, 1977. Assimilate source-sink relationships in *Capsicum annuum* L. II. Effects of fruiting and defloration on the photosynthetic capacity and senescence of the leaves. Australian Journal of Plant Physiology 4: 771–783.

Hall, A.E., E.-D. Schulze & O.L. Lange, 1976. Current perspectives of steady-state stomatal responses to environment. In: O.L. Lange, L. Kappen & E.-D. Schultze (Eds), Water and plant life. Springer-Verlag, Berlin, pp. 169–188.

Hand, D.W., 1988. Effects of atmospheric humidity on greenhouse crops. Acta Horticulturae 229: 143–158.

Hand, D.W., 1990. CO_2 enrichment in greenhouses: problems of CO_2 acclimation and gaseous air pollutants. Acta Horticulturae 268: 81–102.

Hand, D.W. & K.E. Cockshull, 1975. The effects of CO_2 concentration on the canopy photosynthesis and winter bloom production of the glasshouse rose 'Sonia' (syn. 'Sweet Promise'). Acta Horticulturae 51: 243–250.

Harley, P.C. & T.D. Sharkey, 1991. An improved model of C_3 photosynthesis at high CO_2: reversed O_2 sensitivity explained by lack of glycerate reentry into the chloroplast. Photosynthesis Research 27: 169–178.

Harris, G.P. & B. Jeffcoat, 1972. Distribution of ^{14}C-labelled assimilates in flowering carnation plants. Journal of Horticultural Science 47: 25–35.

Harris, G.P. & J.E. Harris, 1962. Effects of environment on flower initiation in carnation. Journal of Horticultural Science 37: 219-234.

Harssema, H., 1977. Root temperature and growth of young tomato plants. PhD Thesis, Wageningen Agricultural University, Wageningen, 86 pp.

Hart, J.W., 1988. Light and plant growth. Unwin Hyman Ltd., London, 204 pp.

Hartsema, A.M., 1961. Influence of temperature on flower formation and flowering of bulbous and tuberous plants. Encyclopedia of Plant Physiology, Springer Verlag. Berlin, Vol 16, pp. 123-167.

Hartung, W. & W.J. Davies, 1991. Drought–induced changes in physiology and ABA. In: W.J. Davies & H.G. Jones (Eds), Abscisic acid, physiology & biochemistry. Environmental Plant Biology Series, Bios Science Publishers, pp. 63–79.

Healy, W.E. & H.F. Wilkins, 1985. Alstroemeria. In: A.H. Halevy (Ed.), Handbook of flowering. Vol. I. CRC Press, Inc. Florida, pp. 419–424.

Heins, R.D. & H.F. Wilkins, 1977. Influence of photoperiod on 'Improved White Sim' carnation (*Dianthus caryophyllus* L.) branching and flowering. Acta Horticulturae 71: 69-74.

Hellkvist, J., G.R. Richards & P.G. Jarvis, 1974. Vertical gradients of water potential and tissue water relations in Sitka spruce trees measured with the pressure chamber. Journal of Applied Ecology 11: 637–667.

Hendriks, L., 1992. Supplementary lighting for greenhouses. Acta Horticulturae 312: 65–76.

Herold, A., 1980. Regulation of photosynthesis by sink activity: the missing link. New Phytologist 86: 131–144.

Hesketh, J.D., C.D. Elmore & J.W. Jones, 1980. Predicting flowering and subsequent leaf expansion. In: J.D. Jones & J.W. Jones (Eds), Predicting photosynthesis for ecosystem models. CRC Boca Raton, Florida, pp. 123–131.

Heuvelink, E., 1989. Influence of day and night temperature on the growth of young tomato plants. Scientia Horticulturae 38: 11–22.

Hicklenton, P.R. & P.A. Jolliffe, 1980. Alternations in the physiology of CO_2 exchange in tomato plants grown in CO_2-enriched atmospheres. Canadian Journal of Botany 58: 2181–2189.

Ho, L.C., 1988. Metabolism and compartmentation of imported sugars in sink organs in relation to sink strength. Annual Review of Plant Physiology 39: 355–378.

Ho, L.C., R.I. Grange & A.J. Picken, 1987. An analysis of the accumulation of water and dry matter in tomato fruit. Plant, Cell and Environment 10: 157–162.

Ho, L.C., 1989. Environmental effects on the diurnal accumulation of ^{45}Ca by young fruit and leaves of tomato plants. Annals of Botany 63: 281-288.

Hoffman, G.J., 1973. Humidity effects on yield and water relations of nine crops. Transactions of the American Society of Agricultural Engineers 16: 164–167.

Hoffman, G.J., 1979. Humidity. In: T.W. Tibbits and T.T. Kozlowski (Eds), Controlled environment guidelines for plant research. Academic Press, New York, pp. 141–172.

Hoffman, G.J. & S.L. Rawlins, 1971. Growth and water potential of root crops as influenced by salinity and relative humidity. Agronomy Journal 63: 877–880.

Hoffman, G.J., S.L. Rawlins, M.J. Garber & E.M. Cullen, 1971. Water relations and growth of cotton as influenced by salinity and relative humidity. Agronomy Journal 63: 822–826.

Holder, R. & K.E. Cockshull, 1990. Effects of humidity on the growth and yield of glasshouse tomatoes. Journal of Horticultural Science 65: 31–39.

Hole, C.C., T.H. Thomas, A. Barnes, P.A. Scott & W.E.F. Rankin, 1984. Dry matter distribution between shoot and storage root of carrot, parsnip, radish and red beet. Annals of Botany 53: 625–631.

Horie, T., C.T. De Wit, J. Goudriaan & J. Bensink, 1979. A formal template for the development of cucumber in its vegetative stage. Proceedings of the Koninklijke Nederlandse Akademie voor Wetenschappen, Series C., 82: 433–480.

Hsiao, T.C., 1973. Plant responses to water stress. Annual Review of Plant Physiology 24: 519–570.

Hunt, R., 1978. Plant growth analysis. Studies in Biology no. 96. Edward Arnold, London, 67 pp.

Hunt, R., 1982. Plant growth curves: the functional approach to plant growth analysis. Edward Arnold, London, 248 pp.

Hurd, R.G., A.P. Gay & A.C. Mountifield, 1979. The effect of partial flower removal on the relation between root, shoot and fruit growth in the indeterminate tomato. Annals of Applied Biology 93: 77–89.

Hüsken, D., E. Steudle & U. Zimmermann, 1978. Pressure probe technique for measuring water relations of cells in higher plants. Plant Physiology 61: 158–163.

Idso, S.B., B.A. Kimbar, H.G. Anderson, & J.R. Mauney, 1987. Effects of atmospheric CO_2 enrichment on plant growth: the interactive role of air temperature. Agriculture, Ecosystems and Environment, 20: 1–10.

Janes, H.W., R. McAvoy, M. Maletta, J. Simpkins & D.R. Mears, 1981. The effect of warm root-zone temperatures on growth of tomato and poinsettia. Acta Horticulturae 115: 245–258.

Janse, J., 1984. Invloed van licht op de kwaliteit van tomaat en komkommer. *(Effects of light on quality of tomato and cucumber).* Groenten en Fruit 40(18): 28-31 (in Dutch).

Janse, J., 1985. Kwaliteit in de zomermaanden: beschermen tegen de zon en voorzichtig behandelen. *(Quality in summertime: protect against direct sunlight and handle carefully).* Tuinderij 65(9): 28-29. (in Dutch).

Janse, J., 1987. Effects of humidity, temperature, concentration of the nutrient solution on firmness, shelf life and flavour of sweet pepper fruits. Acta Horticulturae 244: 123-129.

Janse, J., 1988. Teeltmaatregelen en kwaliteit bij paprika's: tegengestelde rakties vragen om weloverwogen instellingen. *(Cultivation techniques and quality of sweet pepper: opposite responses require deliberate settings).* Tuinderij 68(4): 22-23 (in Dutch).

Jarvis, P.G., 1975. Water transfer in plants. In: D.A. De Vries & N.H. Afgan (Eds), Heat and mass transfer in the biosphere, part 1. John Wiley & Sons, New York, pp. 369–394.

Jarvis, P.G., 1976. The interpretation of the variations in leaf water potential and stomatal conductance found in canopies in the field. Philosophical Transactions of the Royal Society, London, Series B, 273: 593–610.

Jarvis, P.G., 1981. Stomatal conductance, gaseous exchange and transpiration. In: J. Grace, E.D. Ford &

P.G. Jarvis (Eds), Plants and their atmospheric environment. Blackwell Scientific Publications, Oxford, pp. 175–203.

Jarvis, P.G. & J.I.L. Morison, 1981. The control of transpiration and photosynthesis by the stomata. In: P.G. Jarvis & T.A. Mansfield (Eds), Stomatal physiology, Cambridge University Press, Cambridge, pp. 247–279.

Jarvis, P.G., 1985. Coupling of transpiration to the atmosphere in horticultural crops: the omega factor. Acta Horticulturae 171: 187–205.

Jarvis, P.G., W.R.N. Edwards & H. Talbot, 1981. Models of plant and crop water use. In: D.A. Rose & D.A. Charles-Edwards (Eds), Mathematics and plant physiology, Academic Press, London, pp. 151–194.

Jensen, R.D. & S.A. Taylor, 1961. Effect of temperature on water transport through plants. Plant Physiology 36: 639–642.

Jensen, R.D., S.A. Taylor & H.H. Wiebe, 1961. Negative transport & resistance to water flow through plants. Plant Physiology 36: 633–638.

Jolliet, O. & B.J. Bailey, 1992. The effect of climate on tomato transpiration in greenhouses: measurements and models comparison. Agricultural and Forest Meteorology 58: 43–62.

Jones, H.G., 1978. Modelling diurnal trends of leaf water potential in transpiring wheat. Journal of Applied Ecology 15: 613–626.

Jones, H.G., 1983. Plants and microclimate. Cambridge University Press, Cambridge. 323 pp.

Jones, H.G., 1985. Physiological mechanisms involved in the control of leaf water status: implications for the estimation of tree water status. Acta Horticulturae 171: 291–296.

Jones, H.G., 1990. Physiological aspects of the control of water status in horticultural crops. HortScience 25: 19–26.

Jones, H.G., 1992. Plants and microclimate: a quantitative approach to environmental plant physiology, 2nd edition. Cambridge University Press, Cambridge, 428 pp.

Jones, H.G. & R.A. Sutherland, 1991. Stomatal control of xylem embolism. Plant, Cell and Environment 14: 607–612.

Jones, H.G., T.J. Flowers & M.B. Jones (Eds), . Plants under stress. SEB Seminar series 39. Cambridge University Press, Cambridge, 257 pp.

Jones, J.W., E. Dayan, H. Van Keulen & H. Challa, 1988. Modeling tomato growth for optimizing greenhouse temperatures and carbon dioxide concentrations. Acta Horticulturae 248: 285–294.

Kaiser, W.M., 1987. Effects of water deficits on photosynthetic capacity. Physiologia Plantarum 71: 142–149.

Kamerbeek, G.A. & A.J.B. Durieux, 1971. Influence of light on flowerbud abscission in plants of lily cultivar 'Enchantment'. Acta Horticulturae 23: 71-75.

Karlsson, M.G. & R.D. Heins, 1986. Response surface analysis of flowering in chrysanthemum 'Bright Golden Anne'. Journal of the American Society for Horticultural Science 111: 253–259.

Karlsson, M.G., R.D. Heins, J.O. Gerberick & M.E. Hackmann, 1991. Temperature driven leaf unfolding rate in *Hibiscus rosa-sinensis*. Scientia Horticulturae 45: 323–331.

Kaufmann, M.R. & E.L. Fiscus, 1985. Water transport through plants: Internal integration of edaphic and atmospheric effects. Acta Horticulturae 171: 83–93.

Kendrick, R.E. & G.H.M. Kronenberg, 1986. Photomorphogenesis in plants. Martinus Nijhoff Publishers, Dordrecht, pp. 3–14.

Kimball, B.A., 1986. Influence of elevated CO_2 on crop yield. In: H.Z. Enoch & B.A. Kimball (Eds), Carbon dioxide enrichment of greenhouse crops. Volume II. CRC Press, Inc., Florida, pp. 105–115.

Kinet, J.M., 1977. Effect of light conditions on the development of the inflorescense in tomato. Scientia Horticulturae 6: 15–26.

Kinet, J.-M., R.M. Sachs & G. Bernier, 1985. The physiology of flowering. Vol. III, CRC Press, Inc. Florida, 274 pp.

Kirschbaum, M.U.F & G.D. Farquhar, 1984. Temperature dependence of whole-leaf photosynthesis in

Eucalyptus pauciflora Sieb. ex Spreng. Australian Journal of Plant Physiology 11: 519–538.

Kirschbaum, M.U.F & G.D. Farquhar, 1987. Investigation of the CO_2 dependence of quantum yield and respiration in *Eucalyptus pauciflora*. Plant Physiology 83: 1031–1036.

Klapwijk, D., 1975. Effects of aerial humidification on the growth of young tomato plants. Annual Report Glasshouse Crops Research and Experimental Station, Naaldwijk, 1973 and 1974, pp. 52–55.

Kleinendorst, A. & B.W. Veen, 1983. Responses of young cucumber plants to root and shoot temperatures. Netherlands Journal of Agricultural Science 31: 47–61.

Klepper, B., 1991. Root-shoot relationships. In: Y. Waisel & A. Eshel (Eds), Plant roots: the hidden half. M. Dekker Inc., New York, p. 265–286.

Knight, S.L., R.B. Rogers, M.A.L. Smith & L.A. Spomer, 1992. Effects of NaCl salinity on miniature dwarf tomato 'Micro-Tom': I. Growth analysis and nutrient composition. Journal of Plant Nutrition 15: 2315-2327.

Kramer, P.J., 1988. Changing concepts regarding plant water relations. Plant, Cell and Environment 11: 565–568.

Krug, H. & H.-P. Liebig, 1980. Diurnal thermoperiodism of the cucumber. Acta Horticulturae 118: 83–94.

Kuiper, P.J.C., 1964. Water uptake of higher plants as affected by root temperature. Mededelingen Landbouwhogeschool Wageningen 64-4, Wageningen, 11 pp.

Küppers, M., 1988. Water vapour and carbon dioxide exchange of leaves as affected by different environmental conditions. Acta Horticulturae 229: 85–112.

Lang, A., 1990. Xylem, phloem and transpiration flows in developing apple fruits. Journal of Experimental Botany 41: 645–651.

Lange, O.L., P.S. Nobel, C.B. Osmond & H. Ziegler (Eds), 1982. Physiological plant ecology II. Water relations and carbon assimilation. Encyclopedia Plant Physiology New Series 12B. Springer-Verlag, Berlin, 747 pp.

Larson, R.A., 1992. Introduction to floriculture. Academic Press, New York, 636 pp.

Lee, D.R., 1990. A unidirectional water flux model of fruit growth. Canadian Journal of Botany 68: 1286–1290.

Lee, D.R., M.A. Dixon & R.W. Johnson, 1989. Simultaneous measurements of tomato fruit and stem water potentials using in situ stem hygrometers. Canadian Journal of Botany 67: 2352–2355.

Leffring, L., 1981. De bloemproduktie van Gerbera. *(Flower production of Gerbera)*. Dissertation Wageningen Agricultural University. Pudoc, Wageningen, 86 pp. (in Dutch).

Lepage, I., J. De Jong & L. Smeets, 1984. Effect of day and night temperatures during short photoperiods on growth and flowering of *Chrysanthemum morifolium* Ramat. Scientia Horticulturae 22: 373–381.

Liebig, H.-P., 1978. Einflüβe endogener und exogener Faktoren auf die Ertragsbildung von Salatgurken (*Cucumis sativus* L.) unter besonderer Berücksichtigung von Ertragsrhythmik, Bestandesdichte und Schnittmaβnahmen. Dissertation Technischen Universität Hannover, 155 pp. (in German).

Liebig, H.-P., 1988. Temperature integration by kohlrabi growth. Acta Horticulturae 230: 371–380.

Longuenesse, J.J., 1990. Influence of CO_2 enrichment regime on photosynthesis and yield of a tomato crop. Acta Horticulturae 268: 63–70.

Longuenesse, J.J., C. Gary & M. Tchamitchian, 1993. Modelling CO_2 exchanges of greenhouse crops: a matter of scales and boundaries. Acta Horticulturae 328: 33–47.

Louwerse, W., 1981. Effects of CO_2 concentration and irradiance on the stomatal behaviour of maize, barley and sunflower plants in the field. Plant, Cell and Environment 3: 391–398.

Ludwig, L.J. & A.C. Withers, 1978. Effect of temperature on single leaf photosynthesis and respiration in tomato. Annual Report of the Glasshouse Crops Research Institute for the Year 1977, pp. 51–52.

Maas, F.M., 1992. Stuurlichteffecten in de tuinbouw. *(Photomorphogenetic effects of light in horticulture).* Agrobiologische Thema's 5: 63-80 (in Dutch).

Madsen, E., 1968. Effect of CO_2-concentration on the accumulation of starch and sugar in tomato leaves. Physiologia Plantarum 21: 168–175.

Maksymowich, R., 1973. Analysis of leaf development. Cambridge University Press, Cambridge, 109 pp.

Mansfield, 1985. Porosity at a price: the control of stomatal conductance in relation to photosynthesis. In: J. Barber & N.R. Baker (Eds), Photosynthetic mechanisms and the environment. Topics in photosynthesis, Vol. 6. Elsevier, Amsterdam, pp.

Marcelis, L.F.M., 1989. Simulation of plant-water relations and photosynthesis of greenhouse crops. Scientia Horticulturae 41: 9–18.

Marcelis, L.F.M., 1991. Effects of sink demand on photosynthesis in cucumber. Journal of Experimental Botany 42: 1387–1392.

Marcelis, L.F.M., 1992[a]. The dynamics of growth and dry matter distribution in cucumber. Annals of Botany 69: 487–492.

Marcelis, L.F.M., 1992[b]. Non-destructive measurements and growth analysis of the cucumber fruit. Journal of Horticultural Science 67: 457–464.

Marcelis, L.F.M., 1993[a]. Fruit growth and biomass allocation to the fruits in cucumber. 1. Effect of fruit load and temperature. Scientia Horticulturae 54: 107–121.

Marcelis, L.F.M., 1993[b]. Fruit growth and biomass allocation to the fruits in cucumber. 2. Effect of irradiance. Scientia Horticulturae 54: 123–130.

Marcelis, L.F.M., 1993[c]. Effect of temperature on the growth of individual cucumber fruits. Physiologia Plantarum 87: 321–328.

Marcelis, L.F.M., 1993[d]. Simulation of biomass allocation in greenhouse crops: a review. Acta Horticulturae 328: 49–67.

Marcelis, L.F.M., 1994[a]. Effect of fruit growth, temperature and irradiance on biomass allocation to the vegetative parts of cucumber. Netherlands Journal of Agricultural Science 42: 115–123.

Marcelis, L.F.M., 1994[b]. A simulation model for dry matter partitioning in cucumber. Annals of Botany 74: 43–52.

Marcelis, L.F.M., E. Heuvelink & A.N.M. De Koning, 1989. Dynamic simulation of dry matter distribution in greenhouse crops. Acta Horticulturae 248: 269–276.

McAvoy, R.J. & H.W. Janes, 1988. Alternative production strategies for greenhouse tomatoes using supplemental lighting. Scientia Horticulturae 35: 161–166.

McCree, K.J., 1972. The action spectrum, absorptance and quantum yield of photosynthesis in crop plants. Agricultural Meteorology 13: 349–357.

McIntyre, G.I., 1977. Environmental control of lateral bud growth in sunflower *(Helianthus annuus).* Canadian Journal of Botany 55: 2673–2678.

McIntyre, G.I., 1987. The role of water in the regulation of plant development. Canadian Journal of Botany 65: 1287–1298.

Meeuwissen, G., 1985. Potplanten behoeven meer licht. *(Pot plants require more light).* Vakblad voor de Bloemisterij 19: 30–31. (in Dutch).

Meinzer, F.C. & D.A. Grantz, 1991. Coordination of stomatal, hydraulic, and canopy boundary layer proporties: do stomata balance conductances by measuring transpiration? Physiologia Plantarum 83: 324–329.

Michael, G. & H. Marschner, 1962. Einfluß unterschiedlicher Luftfeuchtigkeit und Transpiration auf Mineralstiffaufnahme und -Verteilung. Zeitschrift für Pflanzenernährung, Düngung und Bodenkunde 96: 200-212.

Moe, R. & R.D. Heins, 1990. Control of plant morphogenesis and flowering by light quality and temperature. Acta Horticulturae 272: 81-90.

Moe, R. & T. Kristoffersen, 1969. The effect of temperature and light on growth and flowering of Rosa 'Baccara' in greenhouses. Acta Horticulturae 14: 157-165.

Moe, R., R.D. Heins & J. Erwin, 1991. Stem elongation and flowering of the long day-plant *Campanula isophylla* Moretti in response to day and night temperature alternations and light quality. Scientia Horticulturae 48: 141–151.

Moe, R., T. Fjeld & L.M. Mortensen, 1992. Stem elongation and keeping quality in poinsettia (Euphorbia pulcherrima Willd.) as affected by temperature and supplementary lighting. Scientia Horticulturae 50: 127-136.

Molz, F.J. 1979. A circuit analog model for studying quantitative water relations of plant tissues. Plant Physiology 64: 712–716.

Monteith, J.L. & M.H. Unsworth, 1990. Principles of environmental physics. Second edition. Edward Arnold, London, 291 pp.

Mor, Y. & A.H. Halevy, 1979. Translocation of ^{14}C-assimilates in roses. Physiologia Plantarum 45: 177–182.

Morison, J.I.L., 1985. Sensitivity of stomata and water use efficiency to high CO_2. Plant, Cell and Environment 8: 467–474.

Morison, J.I.L., 1987. Intercellular CO_2 concentration and stomatal response to CO_2. In: E. Zeiger, G.D. Farquhar & I.R. Cowan (Eds), Stomatal function. Stanford University Press, Stanford, pp. 229–251.

Morison, J.I.L. & R.M. Gifford, 1983. Stomatal sensitivity to carbon dioxide and humidity: a comparison of two C_3 and two C_4 grass species. Plant Physiology 71: 789–796.

Mortensen, L.M., 1987. Review: CO_2 enrichment in greenhouses. Crop responses. Scientia Horticulturae 33: 1–25.

Mortensen, L.M. & H.R. Gislerød, 1989. Effect of CO_2, air humidity, and nutrient solution concentration on growth and transpiration of begonia x hiemalis Fotsch. Gartenbauwissenschaft 54: 184–189.

Mortensen, L.M. & S.O. Grimstad, 1990. The effect of lighting period and photon flux density on growth of six foliage plants. Scientia Horticulturae 41: 337–342.

Mortensen, L.M. & E. Strømme, 1987. Effects of light quality on some greenhouse crops. Scientia Horticulturae 33: 27–36.

Mott, K.A., 1988. Do stomata respond to CO_2 concentrations other than intercellular? Plant Physiology 86: 200–203.

Mott, K.A. & D.F. Parkhurst, 1991. Stomatal responses to humidity in air and helox. Plant, Cell and Environment 14: 509–515.

Murakami, T., M. Inayama & H. Kobayashi, 1982. Translocation and distribution of ^{14}C-photosynthates in cucumber plants. Sink strength of growing fruit. Bulletin of the National Institute of Agricultural Sciences D (Physiology and Genetics) 33: 235–275.

Nederhoff, E.M., 1992. Effects of CO_2 on greenhouse grown eggplant (*Solanum melongena* L.). I. Leaf conductance. Journal of Horticultural Science 67: 795–803.

Nederhoff, E.M. & R. De Graaf, 1993. Effects of CO_2 on leaf conductance and canopy transpiration rate of greenhouse grown cucumber and tomato. Journal of Horticultural Science 68: 925–937.

Nederhoff, E.M. & J.G. Vegter, 1994. Photosynthesis of stands of tomato, cucumber and sweet pepper, measured in greenhouses under various CO_2 concentrations. Annals of Botany 73: 353–361.

Nederhoff, E.M., A.N.M. De Koning & A.A. Rijsdijk, 1992. Leaf deformation and fruit production of glasshouse grown tomato (*Lycopersicon esculentum* Mill.) as affected by CO_2, plant density and pruning. Journal of Horticultural Science 67: 411–420.

Nederhoff, E.M., A.A. Rijsdijk & R. De Graaf, 1992. Leaf conductance and rate of crop transpiration of greenhouse grown sweet pepper (*Capsicum annuum* L.) as affected by carbon dioxide. Scientia Horticulturae 52: 283–301.

Newton, P., 1963. Studies on the expansion of the leaf surface. II. The influence of light intensity and photoperiod. Journal of Experimental Botany 14: 458–482.

Nielsen, T.H. & B. Veierskov, 1988. Distribution of dry matter in sweet pepper plants (*Capsicum annuum* L.) during the juvenile and generative growth phases. Scientia Horticulturae 35: 179–187.

Nieuwhof, M., 1976. The effect of temperature on growth and development of cultivars of radish under winter conditions. Scientia Horticulturae 5: 111–118.

Nieuwhof, M., F. Garretsen & J.C. Van Oeveren, 1991. Growth analyses of tomato genotypes grown under low energy conditions. Netherlands Journal of Agricultural Science 39: 191–196.

Nilsson, S.B., C.H. Hertz & S. Falk, 1958. On the relation between turgor pressure and tissue rigidity. II. Theoretical calculations on model systems. Physiologia Plantarum 11: 818–837.

Nilwik, H.J.M., 1981[a]. Growth analysis of sweet pepper (*Capsicum annuum* L.) 1. The influence of irradiance and temperature under glasshouse conditions in winter. Annals of Botany 48: 129–136.

Nilwik, H.J.M., 1981[b]. Growth analysis of sweet pepper (*Capsicum annuum* L.) 2. Interacting effects of irradiance, temperature and plant age in controlled conditions. Annals of Botany 48: 137–145.

O'Leary, J.W., & G.N. Knecht, 1972. Salt uptake in plants grown at constant high relative humidity. Journal of the Arizona Academy of Science 7: 125-128.

Pallas, J., 1965. Transpiration and stomatal opening with changes in carbon dioxide content of the air. Science 147: 171–173.

Passioura, J.B., 1988[a]. Water transport in and to roots. Annual Review of Plant Physiology and Plant Molecular Biology 39: 245–265.

Passioura, J.B., 1988[b]. Response to Dr P.J. Kramer's article, 'Changing concepts regarding plant water relations', Volume 11, Number 7, pp. 565–568. Plant, Cell and Environment 11: 569–571.

Pearcy, R.W. & O. Björkman, 1983. CO_2 and plants: physiological effects. In: Lemon, E.R. (Ed.), CO_2 and plants: the response of plants to rising levels of atmospheric carbon dioxide. Westview Press Inc., Boulder CO, pp. 65–105.

Peaz, A., H. Hellmers & B.R. Strain, 1984. Carbon dioxide enrichment and water stress interaction on growth of two tomato cultivars. Journal of Agricultural Science 102: 687–693.

Peet, M.M., S.C. Huber & D.T. Patterson, 1986. Acclimation to high CO_2 in monoecious cucumbers. II. Carbon exchange rates, enzyme activities, and starch and nutrient concentrations. Plant Physiology 80: 63–67.

Pell, E.J., W.E. Winner, C. Vinten-Johansen & H.A. Mooney, 1990. Response of radish to multiple stresses. I. Physiological and growth responses to changes in ozone and nitrogen. New Phytologist 115: 439–446.

Penning de Vries, F.W.T. & H.H. Van Laar, 1982. Simulation of growth processes and the model BACROS. In: F.W.T. Penning de Vries & H.H. Van Laar (Eds), Simulation of plant growth and crop production. Pudoc, Wageningen, pp. 114–135.

Picken, A.J.F., 1984. A review of pollination and fruit set in the tomato (*Lycopersicon esculentum* Mill.). Journal of Horticultural Science 59: 1–13.

Picken, A.J.F., K. Stewart & D. Klapwijk, 1986. Germination and vegetative development. In: J.G. Atherton & J. Rudich (Eds), The tomato crop: a scientific basis for improvement. Chapman and Hall, London, pp. 111–166.

Plodowska, J.W., P.H.J. Jongebloed, P.A.C.M. Van de Sanden & P.C. Struik, 1989. Effects of a short period of drought on changes in tuber volume and specific leaf weight. II. Diurnal changes. Potato Research 32: 255–266.

Powell, D.B.B. & M.R. Thorpe, 1977. Dynamic aspects of plant water relations. In: J.J. Landsberg & C.V. Cutting (Eds), Environmental effects on crop physiology, Academic Press, London, pp. 259–279.

Preissel, H.G., G. Schmidt-Stohn & O. Krebs, 1980. Untersuchungen an Codiaeum variegatum var. pictum III. Der Einfluss von Temperatur, Licht und Blattalter auf den Pigmentgehalt und die Ausfärbung der Blätter. Gartenbauwissenschaft 45 (4): 164-169.

Pytlinski, J. & H. Krug, 1989. Modelling *Pelargonium zonale* response to various day and night tempera-

tures. Acta Horticulturae 248: 75–84.
Ramos, C. & A.E. Hall, 1983. Effects of photon fluence rate and intercellular CO_2 partial pressure on leaf conductance and CO_2 uptake rate in *Capsicum* and *Amaranthus*. Photosynthetica 17: 34–42.
Raschke, K., 1970. Temperature dependence of CO_2 assimilation and stomatal aperure in leaf sections of *Zea mays*. Planta 91: 336–363.
Raschke, K., 1979. Movements of stomata. In: W. Haupt & M.E. Feinleib (Eds), Physiology of Movements. Encyclopedia of Plant Physiology (NS). Vol. 7, Berlin. pp. 383–441.
Raschke, K., 1986. The influence of the CO_2 content of the ambient air on stomatal conductance and the CO_2 concentration in leaves. In: H.Z. Enoch & B. Kimball (Eds), Carbon dioxide enrichment of greenhouse crops. Vol. II. Physiology, yield and economics, pp. 87–102.
Reid, J.B. & M.G. Huck, 1990. Diurnal variation of crop hydraulic resistance: a new analysis. Agronomy Journal 82: 827–834.
Richards, D., 1981. Root-shoot interactions in fruiting tomato plants. In: R. Brouwer (Ed.), Structure and function of plant roots. Martinus Nijhoff, The Hague, pp. 373–380.
Roberts, E.H. & R.J. Summerfield, 1987. Measurement and prediction of flowering in annual crops. In: J.G. Atherton (Ed.), Manipulation of flowering. Butterworths, London, pp. 17–50.
Rudich, J. & U. Luchinsky, 1986. Water economy. In: J.G. Atherton & J. Rudich (Eds), The tomato crop. A scientific basis for improvement. Chapman and Hall, London, pp. 335–367.
Rudich, J., E. Rendon-Poblete, M.A. Stevens & A.I. Ambri, 1981. Use of leaf water potential to determine water stress in field grown tomato plants. Journal of the American Society for Horticultural Science 106: 732–736.
Rünger, W., 1976. Licht und Temperatur im Zierpflanzenbau, 3rd ed. Parey, Berlin-Hamburg, 353 pp. (in German).
Rünger, W. & K.E. Cockshull, 1985. Camellia japonica. In: A.H.Halevy (Ed.), Handbook of flowering. Vol. II. CRC Press, Inc. Florida, pp. 115–116.
Rylski, I. & M. Spigelman, 1982. Effects of different diurnal temperature combination on fruit set of sweet pepper. Scientia Horticulturae 17: 101-106.
Sagar, J.C., J.L. Edwards & W.H. Klein, 1982. Light energy utilization efficiency for photosynthesis. Transactions of the American Society of Agricultural Engineers 25: 1737–1746.
Sage, R.F., T.D. Sharkey & R.W. Pearcy, 1990. The effect of leaf nitrogen and temperature on the CO_2 response of photosynthesis in the C_3 dicot *Chenopodium album* L. Australian Journal of Plant Physiology 17: 135–148.
Salisbury, F.B. & N.G. Marinos, 1985. The ecological role of plant growth substances. Chapter 9. In: R.P. Pharis & D.M. Reid (Eds), Hormonal regulation of development III Role of environmental factors. (Encyclopedia of plant physiology new series Vol. 11), Springer Verlag, Berlin, Heidelberg, New York, Tokyo, pp. 707–766.
Sánchez-Blanco, M.J., M.C. Bolarín, J.J. Alarcón & A. Torrecillas, 1991. Salinity effects on water relations in *Lycopersicon esculentum* and its wild salt-tolerant relative species *L. pennellii*. Physiologia Plantarum 83: 269–274.
Saxe, H., 1989. Air pollution, primary physiological responses, and diagnostic tools. PhD dissertation, Royal Veterinary and Agricultural University, Kopenhagen, 234 pp.
Schapendonk, A.H.C.M. & P. Brouwer, 1984. Fruit growth of cucumber in relation to assimilate supply and sink activity. Scientia Horticulturae 23: 21–33.
Schulte, P.J., 1992. The units of currency for plant water status. Plant, Cell and Environment 15: 7–10.
Schulte, P.J. & T.M. Hinckley, 1985. A comparison of pressure-volume curve data analysis techniques. Journal of Experimental Botany 36: 1590–1602.
Schulze, E.D., 1986. Carbon dioxide and water vapor exchange in response to drought in the atmosphere and the soil. Annual Review of Plant Physiology 37: 247–274.
Schulze, E.D., 1991. Water and nutrient interactions with plant water stress. In: H.A. Mooney, W.E.

Winner & E.J. Pell (Eds), Responses of plants to multiple stresses. Academic Press, London, pp. 89–101.

Schulze, E.D., & A.E. Hall, 1982. Stomatal responses, water loss and CO_2 assimilation rates of plants in contrasting environments. In: O.L. Lange, P.S. Nobel, C.B. Osmond & H. Ziegler (Eds), Physiological plant ecology II. Water relations and carbon assimilation. Encyclopedia Plant Physiology New Series 12B. Springer-Verlag, Berlin, pp. 181–230.

Schulze, E.D., E. Steudle, T. Gollan & U. Schurr, 1988. Response to Dr P.J. Kramer's article, 'Changing concepts regarding plant water relations', Volume 11, Number 7, pp. 565–568. Plant, Cell and Environment 11: 573–576.

Schulze, E.-D., N.C. Turner, T. Gollan & K.A. Shackel, 1987. Stomatal responses to air humidity and soil drought. In: E. Zeiger, G.D. Farquhar & I.R. Cowan (Eds), Stomatal function. Stanford University Press, Stanford, pp. 311–321.

Seginer, I., 1993. Crop models in greenhouse climate control. Acta Horticulturae 328: 79–98.

Seginer, I., D. Kantz, N. Levav & U.M. Peiper, 1990. Night-time transpiration in greenhouses. Scientia Horticulturae 41: 265–276.

Shackel, K.A., C. Greve, J.M. Labavitch & H. Ahmadi, 1991. Cell turgor changes associated with ripening in tomato pericarp tissue. Plant Physiology 97: 814–816.

Shaer, Y.A. & C.H.M. Van Bavel, 1987. Relative role of stomatal and aerodynamic resistances in transpiration of a tomato crop in a CO_2-enriched greenhouse. Agricultural and Forest Meteorology 41: 77–85.

Sharkey, T.D., 1985. Photosynthesis in intact leaves of C_3 plants: physics, physiology and rate limitation. Botanical Review 51: 53–105.

Sharp, R.E. & W.J. Davies, 1989. Regulation of growth and development of plants growing with a restricted supply of water. In: H.G. Jones, T.J. Flowers & M.B. Jones (Eds), Plants under stress. SEB Seminar series 39. Cambridge University Press, Cambridge pp. 71–93.

Shell, G.S.G., A.R.G. Lang & P.J.M. Sale, 1974. Quantitative measures of leaf orientation and heliotropic response in sunflower, bean, pepper and cucumber. Agricultural Meteorology 13: 25–37.

Sheriff, D.W., 1984. Epidermal transpiration and stomatal responses to humidity: some hypotheses explored. Plant, Cell and Environment 7: 669–677.

Sinclair, T.R. & M.M. Ludlow, 1985. Who taught plants thermodynamics? The unfulfilled potential of plant water potential. Australian Journal of Plant Physiology 12: 213–217.

Slack, G. & Hand, D.W., 1983. The effect of day and night temperatures on the growth, development and yield of glasshouse cucumbers. Journal of Horticultural Science 58: 567–573.

Slack, G., J.S. Fenlon & D.W. Hand, 1988. The effects of summer CO_2 enrichment and ventilation temperatures on the yield, quality and value of glasshouse tomatoes. Journal of Horticultural Science 63: 119–129.

Slatyer, R.O., 1958. The measurement of diffusion pressure deficit in plants by a method of vapour equilibration. Australian Journal of Biological Science 11: 349–365.

Slatyer, R.O., 1967. Plant-water relationships. Academic Press, London, 366 pp.

Slootweg, G. & U. Van Meeteren, 1991. Transpiration and stomatal conductance of roses cv Sonia grown with supplemental light. Acta Horticulturae 298: 119-125.

Smith, H., 1981. Adaptation to shade. In: C.B. Johnson (Ed.), Physiological processes limiting plant productivity. Butterworths, London. pp. 159–173.

Sondern, J.A., 1962. Grondbedekking met witte kunststoffolie. *(Covering soil with white plastic).* Mededelingen van de Direktie Tuinbouw 25: 175–180 (in Dutch).

Sonneveld, C., & G.W.H. Welles, 1988. Yield and quality of rockwool-grown tomatoes as affected by variations in EC-value and climatic conditions. Plant and Soil 111: 37-42.

Spitters, C.J.T., 1986. Separating the diffuse and direct component of global radiation and its implications for modeling canopy photosynthesis. II. Calculation of canopy photosynthesis. Agricultural

and Forest Meteorology 38: 231–242.
Spitters, C.J.T., H. Van Keulen & D.W.G. Van Kraalingen, 1989. A simple and universal crop growth simulator: SUCROS87. In: R. Rabbinge, S.A. Ward & H.H. Van Laar (Eds), Simulation and systems management in crop protection. Pudoc, Wageningen, pp. 147–181.
Squire, G.R., C.R. Black & C.K. Ong, 1983. Response to saturation deficit of leaf extension in a stand of pearl millet (Pennisetum typhoides S. & H.). II. Dependence on leaf water status and irradiance. Journal of Experimental Botany 34: 856–865.
Sruamsiri, P. & F. Lenz, 1985. Photosynthese und stomatäres Verhalten bei Erdbeeren (Fragaria x ananassa Duch.). IV. Einfluß von Kohlendioxid und Sauerstoff. Gartenbauwissenschaft 50: 221–227. (in German).
Stadelmann, E.J., 1984. The derivation of the cell wall elasticity function from the cell turgor potential. Journal of Experimental Botany 35: 859–868.
Stanghellini, C., 1985. Transpiration and temperature of greenhouse crops, in relation to internal and external resistances. Acta Horticulturae 174: 87–95.
Stanghellini, C. & J.A. Bunce, 1994. Response of assimilation and conductance to light, CO_2, temperature and humidity in tomato plants acclimated to ambient and elevated CO_2. Photosynthetica (in press).
Starck, Z., 1973. The effect of shading during growth on the subsequent distribution of ^{14}C-assimilates in *Raphanus sativus*. Bulletin de l'Academie Polonaise des Sciences Serie des Sciences Biologiques 21: 309–314.
Stitt, M., 1991. Rising CO_2 levels and their potential significance for carbon flow in photosynthetic cells. Plant, Cell and Environment 14: 741–762.
Sutcliffe, J., 1979. Plants and water. 2nd edition. Institute of Biology's Studies in Biology no. 14. Edward Arnold Publishers Ltd., London, 122 pp.
Swalls, A.A. & J.W. O'Leary, 1976. Growth, water consumption, and salt uptake of tomato plants in high humidity-high carbon dioxide greenhouse environments. Journal of the Arizona Academy of Science 11: 23–26.
Syvertsen, J.P. & Y. Levy, 1982. Diurnal changes in citrus leaf thickness, leaf water potential and leaf to air temperature difference. Journal of Experimental Botany 33: 783–789.
Tachibana, S., 1991. Import of calcium by tomato fruit in relation to the day-night periodicity. Scientia Horticulturae 45: 235-243.
Tchamitchian, M., 1990. Photosynthese d'une culture de tomates sous serre: mise au point et validation d'un modèle analytique. Dissertation, L'Institut National Polytechnique de Toulouse, 97 pp. (in French).
Tenhunen, J.D., R.W. Pearcy & O.L. Lange, 1987. Diurnal variations in leaf conductance and gas exchange in natural environments. In: E. Zeiger, G.D. Farquhar & I.R. Cowan (Eds), Stomatal function. Stanford University Press, Stanford, pp. 323–351.
Thornley, J.H.M., 1976. Mathematical models in plant physiology. Academic Press, London, 318 pp.
Thornley, J.H.M. & R.G. Hurd, 1974. An analysis of the growth of young tomato plants in water culture at different light integrals and CO_2 concentrations. II. A mathematical model. Annals of Botany 38: 389–400.
Tikhomirov, A.A., I.G. Zolotukhin, G.M. Lisovskii & F. Ya. Sid'ko, 1987. Specificity of responses to the spectral composition of PAR in plants of different species under artificial illumination. Soviet Plant Physiology 34: 624–633.
Ting, I.P, 1987. Stomata in plants with Crassulacean Acid Metabolism. In: E. Zeiger, G.D. Farquhar & I.R. Cowan (Eds), Stomatal function, Stanford University Press, Stanford, pp. 353–366.
Turner, M.A., D.W. Reed & D.L. Morgan, 1987. A comparison of light acclimatization methods for reduction of interior leaf drop in Ficus spp. Journal Environmental Horticulture 5: 102-104.
Tyree, M.T. & P.G. Jarvis, 1982. Water in tissues and cells. In: O.L. Lange, P.S. Nobel, C.B. Osmond & H.

Ziegler (Eds), Physiological plant ecology II. Water relations and carbon assimilation. Encyclopedia Plant Physiology New Series 12B. Springer-Verlag, Berlin, pp. 35–77.

Tyree, M.T. & J.D. Alexander, 1993. Plant water relations and the effects of elevated CO_2: a review and suggestions for future research. In: J. Rozema, H. Lambers, S.C. Van de Geijn & M.L. Cambridge (Eds), CO_2 and biosphere. Kluwer Academic Publishers, Dordrecht, pp. 47–62. (Vegetatio 104/105: 47–62)

Udink ten Cate, A.J., G.P.A. Bot & J.J. Van Dixhoorn, 1978. Computer control of greenhouse climates. Acta Horticulturae 87: 265–272.

Uitermark, K. & J. Benninga, 1991. Bedrijfsvergelijkend onderzoek Ficus (2), Kwaliteit is niet alles. Vakblad voor de Bloemisterij 46(4): 48. (in Dutch).

Van Henten, E.J., 1994. Greenhouse climate management: an optimal approach. Dissertation, Wageningen Agricultural University, Wageningen, 329 pp.

Van Henten, E.J. & J. Bontsema, 1991. Optimal control of greenhouse climate. In: Y. Hashimoto & W. Day (Eds), Mathematical and control applications in agriculture and horticulture. Pergamon Press, Oxford, pp. 27–32.

Van Holsteijn, G.P.A., 1988. Bij zonnig weer koelere vruchten en minder verdamping. *(At sunny weather less transpiration and cooler fruits)*. Groenten en Fruit 43(40): 28–29 (in Dutch).

Van Holsteijn, G.P.A., 1990. Sun screen in a tomato crop. Annual Report 1989 Glasshouse Crops Research Station, Naaldwijk, pp. 28–29.

Van de Geijn, S.C. & F. Smeulders, 1981. Diurnal changes in the flux of calcium toward meristems and transpiring leaves in tomato and maize plants. Planta 151: 265–271.

Van de Sanden, P.A.C.M. & H. Gijzen, 1993. Gewasverdamping, waterhuishouding en groei van kasgewassen *(Transpiration, water relations and growth of greenhouse crops)*. In: H. Van Keulen & F.W.T. Penning de Vries (Eds), Watervoorziening en gewasproduktie *(Water supply and crop production)*. Agrobiological Themes 8. DLO-Centre for Agrobiological Research, Wageningen, pp. 67–82. (in Dutch).

Van de Sanden, P.A.C.M. & B.W. Veen, 1992. Effects of air humidity and nutrient solution concentration on growth, water potential and stomatal conductance of cucumber seedlings. Scientia Horticulturae 50: 173–186.

Van den Berg, G.A., 1987. Influence of temperature on bud break, shoot growth, flower bud atrophy and winter production of glasshouse roses. Dissertation Wageningen Agricultural University, Wageningen, 170 pp.

Van der Vlugt, J.L.F., 1987. The case: roots vs. fruits in the cucumber. I. The effect of the nitrogen concentration in the recirculating nutrient solution on root death in cucumber. Plant and Soil 98: 299–301.

Van Goor, B.J., 1974. Influence of restricted water supply on blossom-end rot and ionic composition of tomatoes grown in nutrient solution. Communications in Soil Science and Plant Analysis 5: 13-24.

Van Spronsen, J.C., 1981. Invloed van teeltwijze op de houdbaarheid van *Codiaeum variegatum* 'Philip Geduldig'. *(Effects on cultivation on keeping quality of Codiaeum)*. Bloemisterij onderzoek Nederland 1981: 101-103. (in Dutch).

Vegter, J.G., 1989. Fotosynthesemetingen 1987–1989. Teeltverslag produktie en gewaswaarnemingen. *(Photosynthesis measurements 1987–1989)*. Intern Verslag 26, Proefstation voor Tuinbouw onder Glas, Naaldwijk, 48 pp. (in Dutch).

Vince-Prue, D., 1975. Photoperiodism in plants. Mc Graw-Hill Book Company Ltd., Maidenhead. 444 pp.

Vlahos, J.C., E. Heuvelink & G.F.P. Martakis, 1991. A growth analysis study on three cultivars of *Achimenes* grown under three light regimes. Scientia Horticulturae 46: 275–282.

Vogelezang, J.V.M., 1988. Effect of root-zone heating on growth, flowering and keeping quality of

Saintpaulia. Scientia Horticulturae 34: 101–113.

Vogelezang, J.V.M., 1990. Effect of root-zone and air temperature on flowering, growth and keeping quality of *Begonia x hiemalis* 'Toran'. Scientia Horticulturae 44: 135–147.

Vogelezang, J.V.M., 1993. Bench heating for potplant cultivation. Analysis of effects of root- and air temperature on growth, development and production. Dissertation, Wageningen Agricultural University, Wageningen, 115 pp.

Von Caemmerer, S. & G.D. Farquhar, 1981. Some relationships between the biochemistry of photosynthesis and the gas exchange of leaves. Planta 15: 376–387.

Vonk Noordegraaf, C., 1972. Anjerteelt en kasklimaat. *(Carnation and greenhouse cliomate)*. Horti-progress 72: 7-13 (in Dutch).

Vonk Noordegraaf, C., 1973. Influence of temperature on flowering in *Anthurium scherzerianum*. Acta Horticulturae 31: 71-76.

Vonk Noordegraaf, C., 1981. Bloemproduktie bij *Alstroemeria* 'Walter Fleming'. *(Flower production of Alstroemeria 'Walter Fleming')*. Mededeling Proefstation voor de Bloemisterij 69. Pudoc, Wageningen, 152 pp.

Vos, J., E.M. Drees & F.T.W. Penning de Vries, 1982. Grain formation and assimilate partitioning in wheat. In: F.W.T. Penning de Vries & H.H. Van Laar (Eds), Simulation of plant growth and crop production. Simulation Monographs. Pudoc, Wageningen, pp. 144–151.

Ward, G.M., 1964. Greenhouse tomato nutrition: a growth analysis study. Plant and Soil 11: 125–133.

Wardlaw, I.F., 1990. The control of carbon partitioning in plants. New Phytologist 116: 341–381.

Wareing, P.F. & J. Patrick, 1975. Source-sink relations and the partition of assimilates in the plant. In: J.P. Cooper (Ed.), Photosynthesis and productivity in different environments. Cambridge University Press, Cambridge, pp. 481–499.

Webb, J.A. & P.R. Gorham, 1964. Translocation of photosynthetically assimilated C^{14} in straight-necked squash. Plant Physiology 39: 663–672.

Welles, G.W.H., J. Janse & J.Y. Pears, 1992. L'influence des techniques modernes de production et des variétés sur la qualité analytique des légumes de serre. Revue Horticole 324: 43-54. (in French).

Widders, I.E. & H.C. Price, 1989. Effects of plant density on growth and biomass partitioning in pickling cucumbers. Journal of the American Society for Horticultural Science 114: 751–755.

Wiebe, H.-J., 1989. Vernalisation von wichtigen Gemüsearten: Ein Überblick. Gartenbauwissenschaft 54: 97–104. (in German).

Wiersum, L.K., 1966. Calcium content of fruits and storage tissues in relation to the mode of water supply. Acta Botanica Neerlandica 15: 406-418.

Wilson, J.B., 1988. A review of evidence on the control of shoot:root ratio, in relation to models. Annals of Botany 61: 433–449.

Wolff, X.Y. & R.R. Coltman, 1990. Productivity under shade in Hawaii of five crops grown as vegetables in the tropics. Journal of the American Society for Horticultural Science 115: 175–181.

Wolswinkel, P., 1985. Phloem unloading and turgor-sensitive transport: factors involved in sink control of assimilate partitioning. Physiologia Plantarum 65: 331–339.

Wolterbeek, H.Th., P.C.M. Willemse & J. Van Die, 1987. Phloem and xylem import of water and solutes in tomato fruits. Acta Botanica Neerlandica 36: 295–306.

Wong, S., I.R. Cowan & G.D. Farquhar, 1985. Leaf conductance in relation to rate of CO_2 assimilation. II. Effects of short-term exposures to different photon flux densities. Plant Physiology 78: 826–829.

Wurr, D.C.E. & J.R. Fellows, 1991. The influence of solar radiation and temperature on the head weight of crisp lettuce. Journal of Horticultural Science 66: 183–190.

Wyn Jones, R.G., & J. Pritchard, 1989. Stresses, membranes and cell walls. In: H.G. Jones, T.J. Flowers & M.B. Jones (Eds), Plants under stress. SEB Seminar series 39. Cambridge University Press, Cambridge, pp. 95–114.

Yelle, S., R.C. Beeson Jr., M.J. Trudel & A. Gosselin, 1989. Acclimation of two tomato species to high atmospheric CO_2. II. Ribulose-1,5-biphosphate carboxylase/oxygenase and phosphoenolpyruvate carboxylase. Plant Physiology 40: 1473–1477.

Yelle, S., R.C. Beeson Jr., M.J. Trudel & A. Gosselin, 1990. Duration of CO_2 enrichment influences growth, yield, and gas exchange of two tomato species. Journal of the American Society for Horticultural Science 115: 52–57.

Yoshioka, H. & K. Takahashi, 1979. Studies on the translocation and accumulation of photosynthates in fruit vegetables. II. The translocation and distribution of ^{14}C-photosynthates in tomato plants during reproductive development and effects of topping and shading. Bulletin of the Vegetable and Ornamental Crops Research Station A6: 71–84.

Zeroni, M. & J. Gale, 1989. Response of 'Sonia' roses to salinity at three levels of ambient CO_2. Journal of Horticultural Science 64: 503–511.

Zhang, L. & R. Lemeur, 1992. Effect of aerodynamic resistance on energy balance and Penman-Monteith estimates of evapotranspiration in greenhouse conditions. Agricultural and Forest Meteorology 58: 209–228.

Zieslin, N. & Y. Mor, 1990. Light on roses: a review. Scientia Horticulturae 43: 1–14.

Zimmermann, U. & E. Steudle, 1978. Physical aspects of water relations of plant cells. Advances in Botanical Research 6: 45–117.

List of symbols

A	leaf area per plant (m^2)
C	capacitance (water holding capacity) ($m^3\,MPa^{-1}$)
C_a	ambient CO_2 concentration ($\mu mol\,mol^{-1}$)
C_f	conversion efficiency dry weight per gram assimilates ($g\,g^{-1}$)
C_i	intercellular CO_2 concentration ($\mu mol\,mol^{-1}$)
C_{pf}	CO_2 production factor ($g\,CO_2\,g^{-1}$ dry matter formed)
C_s	CO_2 concentration at the leaf surface ($\mu mol\,mol^{-1}$)
C_{tissue}	apacitance of plant tissue (water holding capacity) ($m^3\,MPa^{-1}$)
c^s	concentration of solutes ($mol\,cm^{-3}$)
$c^s{}_{xyl}$	concentration of solutes in the xylem ($mol\,cm^{-3}$)
D_a	vapour pressure deficit of air (kPa)
D_s	leaf-air vapour pressure deficit (kPa)
E	transpiration rate ($g\,m^{-2}\,s^{-1}$)
E_l	leaf transpiration ($g\,m^{-2}\,s^{-1}$)
e_a	ambient vapour pressure (Pa)
e_l	vapour presure inside the leaf (Pa)
e_s	vapour pressure at leaf surface (Pa)
F_{hp}	fraction of dry weight diverted to the harvestable product (-)
f_a	quantum efficiency or quantum yield (mol CO_2 per mol photons absorbed)
G	water flux for growth ($cm^3\,s^{-1}$)
g_b	boundary layer conductance ($m\,s^{-1}$)
g_l	leaf conductance ($m\,s^{-1}$)
g_s	stomatal conductance ($m\,s^{-1}$)
g_{smax}	maximum stomatal conductance ($m\,s^{-1}$)
g_{tot}	total (leaf + boundary layer) conductance ($m\,s^{-1}$)
H	water flux to storage ($cm^3\,s^{-1}$)
I	light intensity ($\mu mol\,m^{-2}\,s^{-1}$ or $W\,m^{-2}$)
I_Σ	daily integral of PAR ($MJ\,m^{-2}\,d^{-1}$)
$I_{1/2s}$	irradiance level at half the light-saturated rate
I_{dif}	diffuse light intensity beneath partial LAI L_c ($\mu mol\,m^{-2}\,s^{-1}$)
$I_{o,dif}$	diffuse light intensity above the canopy ($\mu mol\,m^{-2}\,s^{-1}$)
J	electron transport rate (meq. $m^{-2}\,s^{-1}$)
J_{max}	maximum electron transport rate (meq. $m^{-2}\,s^{-1}$)
J_s	net solute flux density ($mol\,m^{-2}\,s^{-1}$)
J_w	volume flux density of water ($cm^3\,m^{-2}\,s^{-1}$)
K	extinction coefficient (-)
K_c	Michaelis-Menten constant for CO_2 ($\mu mol\,mol^{-1}$)
K_{dif}	extinction coefficient for diffuse light (-)
K_o	Michaelis-Menten constant for O_2 ($mmol\,mol^{-1}$)
K_m	effective Michaelis-Menten constant for CO_2 ($\mu mol\,mol^{-1}$)
L	path hydraulic conductance ($s^{-1}\,MPa^{-1}$)
L_c	partial LAI ($m^2\,m^{-2}$)
m	wall extensibility ($s^{-1}\,MPa^{-1}$)
O	intercellular O_2 concentration ($mmol\,mol^{-1}$)
P	photosynthesis (-)
P_c	limited by Rubisco carboxylation rate ($\mu mol\,m^{-2}\,s^{-1}$)
P_g	leaf gross photosynthesis per unit leaf area ($mg\,CO_2\,m^{-2}\,s^{-1}$)
$P_{gc,d}$	crop photosynthesis per unit glass-house area ($g\,CO_2\,m^{-2}\,d^{-1}$)
P_{gmax}	leaf maximal rate of gross photosynthesis per unit leaf area ($\mu mol\,m^{-2}\,s^{-1}$)
$P_{gp,d}$	daily rate of plant gross photosynthesis ($g\,CO_2\,plant^{-1}\,d^{-1}$)
P_j	carboxylation rate limited by RuP_2 generation ($\mu mol\,m^{-2}\,s^{-1}$)
P_n	net rate leaf photosynthesis per unit leaf area ($mg\,CO_2\,m^{-2}s^{-1}$)
P_{nc}	net rate crop photosynthesis per unit glasshouse area ($g\,CO_2\,m^{-2}\,h^{-1}$)
P_{nmax}	net rate leaf photosynthesis at saturating light per unit leaf area ($mg\,CO_2\,m^{-2}s^{-1}$)
Q_{10}	temperature coefficient (-)
R	liquid flow resistance ($MPa\,s\,cm^{-1}$)
R	respiration ($g\,CO_2\,m^{-2}\,h^{-1}$)
R_d	leaf dark respiration rate ($mg\,CO_2\,m^{-2}\,s^{-1}$)
R_g	growth respiration ($g\,CO_2\,m^{-2}\,h^{-1}$)
R_{gas}	gasconstant (8.31 $JK^{-1}\,mol^{-1}$)
R_m	maintenance respiration per unit leaf area ($mg\,CO_2\,m^{-2}\,s^{-1}$)

$R_{m,d}$ daily crop maintenance respiration per unit glasshouse area (g CH_2O m^{-2} d^{-1})
$R_{mp,d}$ daily rate of plant maintenance respiration (g CO_2 $plant^{-1}$ d^{-1})
R_{plant} liquid flow resistance total plant (MPa s cm^{-1})
R_{root} liquid flow resistance roots (MPa s cm^{-1})
$R_{soil-plant}$ liquid flow resistance between soil and plant (MPa s cm^{-1})
r_b boundary layer resistance (m s^{-1})
r_c carboxylation resistance (m s^{-1})
r_s stomatal resistance (m s^{-1})
r_t instantaneous relative growth rate (g g^{-1} d^{-1})
T temperature (K)
T_r flux for transpiration (cm^3 s^{-1})
t time (s)
t_h time of harvest (day)
t_p time of transition from young to production phase (-)
U water uptake (cm^3 s^{-1})
V tissue volume (m^3)
V_c rate of carboxylation (µmol m^{-2} s^{-1})
V_{cmax} maximum rate of carboxylation (µmol m^{-2} s^{-1})
V_{leaf} rate of leaf carboxylation (µmol m^{-2} s^{-1})
V_o rate of oxygenation (µmol m^{-2} s^{-1})
V_{tissue} volume of water in tissue (m^3)
W (crop) dry weight (g m^{-2})
W_{dry} dry weight (g)
W_{fresh} fresh weight (g)
W_{hp} dry weight of harvestable products (g)
W_i initial (crop) dry weight (g)
W_l dry weight of leaves per plant (g)
W_p plant dry weight (g $plant^{-1}$)
W_{ref} target crop weight (g)
W_t weight after *t* days (g)
W_{turgid} turgid weight (g)
Y minimum turgor for extension (MPa)

Greek symbols

α light use efficiency (mg CO_2 μmol^{-1} (photons))
α_c average crop light use efficiency (g CO_2 MJ^{-1})
α_i initial light use efficiency (mg CO_2 μmol^{-1} (photons)
Γ^* CO_2 compensation point (µmol mol^{-1})
ε bulk elastic modulus (MPa)
ρ reflection coefficient of the canopy (-)
τ time constant (s)
τ conductance for CO_2 transfer (g m^{-2} s^{-1} μmol^{-1} mol)
Ψ water potential (MPa)
Ψ_{leaf} leaf water potential (MPa)
Ψ^P turgor potential (MPa)
Ψ_{plant} plant water potential (MPa)
Ψ^S osmotic potential (MPa)
Ψ_{soil} soil water potential (MPa)
Ψ_{xylem} xylem water potential (MPa)
Ψ_o root environment water potential (MPa)

List of abbreviations

ABA abscisic acid
CAM crassulacean acid metabolism
DIF difference between day- and night temperature (°C)
DM dry matter (g)
DMC dry matter content (g g^{-1})
DVR rate of development (d^{-1})
DVS developmental stage (0–1)
EC electrical conductivity (mS cm^{-1})
GR crop growth rate (g d^{-1})
HU heat units (°C days)
LAI leaf area index (m^2 m^{-2})
LAR leaf area ratio: leaf area per unit plant dry weight (m^2 g^{-1})
LDP long-day plant
LWR leaf weight ratio: leaf dry weight/plant dry weight
NAR net assimilation rate (g m^{-2} d^{-1})
NLG Dutch guilders
PA plastochron age
PAR photosynthetic active radiation (400–700 nm) (µmol m^{-2} s^{-1} or W m^{-2})
PEP PhosphoEnolPyruvate
PCO photosynthetic carbon oxidation
PCR photosynthetic carbon reduction
PGA phosphoglyceric
PGlA phosphoglycollate
Pi orthophosphate
RH relative humidity (%)
RGR relative growth rate (g g^{-1} d^{-1})
RuP_2 ribulosebiphosphate

RWC	relative water content ($g\ g^{-1}$)
SDP	short-day plant
SLA	specific leaf area: leaf area per unit leaf dry weight
VPD	vapour pressure deficit (kPa)

3 Physics of greenhouse climate

3.1 Introduction

G.P.A. Bot and N.J. van de Braak

The relations between growth conditions and the processes contributing to plant production have been analyzed in Chapter 2. In this chapter follows a discussion of how growth conditions are modified by a greenhouse.

In general terms it can be understood qualitatively how the greenhouse works. The growth conditions are determined by factors including local irradiation, CO_2 concentration, temperature and water vapour pressure. All these factors are affected by the greenhouse cover. The primary reason is that the cover envelopes the air. This induces both a reduction in air exchange from the crop environment to the atmospheric air and a strong reduction in local air velocities. Neither absorbed energy nor transpired water will be released easily to the atmosphere but are trapped. Moreover the CO_2 exchange to the atmosphere is modified. Secondly, solar radiation is (partly) transmitted through the enclosure but the enclosure is opaque to the thermal radiation emitted by the crop, so the radiation is trapped and not easily released to the environment. A decreased solar radiation intensity would imply a lower plant production, so this has to be compensated for by considerably improved growth conditions to achieve a higher crop production.

The general terms may be simple but the growth conditions inside the greenhouse not only have to be understood qualitatively but have to be quantified in order to determine their impact on plant production. The various processes responsible for the transfer of energy and mass (water vapour, carbon dioxide) then have to be examined. In this chapter the various mechanisms for heat and mass transfer in the greenhouse and from the greenhouse to the environment will be discussed in general terms and then applied to the more specific situation of the greenhouse climate. In this way the greenhouse climate can be quantified in relation to outdoor conditions and the physical properties of the greenhouse and its equipment. Growth conditions inside the greenhouse can then be calculated and applied to the quantification of greenhouse crop production. Moreover the information can be used as a design tool in greenhouse engineering.

3.2 Transport phenomena

G.P.A. Bot and N.J. van de Braak

3.2.1 Basic principles

Though the qualitative description of the main physical processes is relatively simple, a careful consideration of the various heat and mass transfer processes and principles is needed for a quantitative description. In this section the basic principles of heat and mass transfer will be discussed (see for example Bird *et al.*, 1960), in the next section this will be applied to the greenhouse situation.

The first principle to examine is how flows of energy and mass to and from a body or volume work, and how they affect temperatures and concentrations. Any part of the greenhouse or the total greenhouse may act as an individual body or volume. The next principle concerns how such flows can be quantified.

When the transport of physical quantities like energy and mass is considered, for these quantities the law of conservation is valid for any volume. To establish a general rule, energy is considered first. Then an easy conversion will be made to mass.

If an amount of energy per unit of time (q_h, energy flux, J s^{-1}) enters a volume, then the amount of energy in the volume Q_h (J) will increase. Of course if an energy flux leaves a volume, Q_h will decrease. The relation in which we compare the entering and leaving fluxes is called the balance. We can set up an energy balance or a mass balance and so forth for any volume. Not only the in- and outflows of energy contribute to the increase or decrease of energy in the volume, but also the energy production in the volume p_h (J s^{-1}), if any. From these considerations the energy balance over any volume can be set up in general terms as:

$$\mathrm{d}Q_h / \mathrm{d}t = q_{in,h} - q_{out,h} + p_h \quad \text{(Eq. 3.2.1)}$$

The term on the left hand side is the change of energy per unit of time in the considered volume. This is determined by the terms on the right hand side, which are respectively the influx, the outflows and the production of energy per unit of time in the considered volume. In balance (3.2.1) it is assumed that no work is done by or on the medium in the volume, which in general is true for the greenhouse situation.

The amount of energy in the volume Q_h (internal energy) is directly related to the temperature T (K) of the volume through its thermal capacity Cap_h (JK^{-1}):

$$Q_h = Cap_h\, T \quad \text{(Eq. 3.2.2)}$$

In this way the time rate of change of the internal energy is translated to a time rate of change of temperature that can be implemented in the balance:

$$Cap_h\, \mathrm{d}T / \mathrm{d}t = q_{in,h} - q_{out,h} + p_h \quad \text{(Eq. 3.2.3)}$$

So the time rate of change of the temperature of a volume with given heat capacity Cap_h depends on the in- and outflows and the production rate of energy.

The same considerations lead to the mass balance over a volume:

$$\mathrm{d}Q_m / \mathrm{d}t = q_{in,m} - q_{out,m} + p_m \quad \text{(Eq. 3.2.4)}$$

The change in time of the mass Q_m (kg) in the volume (left side) equals the difference between in- and outflows of mass, $q_{in,m}$ and $q_{out,m}$ respectively, (kg s^{-1}), plus the mass produced per unit of time in the volume p_m (kg s^{-1}). In the greenhouse situation the water vapour and CO_2 balances are of special interest, but it is of course possible to set up the balance for any substance of interest.

The mass Q_m in the volume is directly related to the concentration c_m (kg m^{-3}) and the volume V (m^3):

$$Q_m = V c_m \quad \text{(Eq. 3.2.5)}$$

This leads to

$$V\, \mathrm{d}c_m / \mathrm{d}t = q_{in,m} - q_{out,m} + p_m \quad \text{(Eq. 3.2.6)}$$

So from the mass balance it is concluded that the time rate of change of concentration in a volume V depends on the in- and outflows and the rate of production mass.

3.2.2 Transport mechanisms

In the balance equation for energy or mass over a volume V (equation (3.2.3) and (3.2.6)), the time rate of change of temperature or concentration is given as a function of the in- and outflows and the production.

If the in- and outflows and production can be related to temperature or concentration, a differential equation in time can be set up, which can be solved for a known initial condition. To formulate the relations between the fluxes and the relevant variables a distinction has to be made between the various mechanisms responsible for the transport from one place to another (to or from the considered volume). The following transport mechanisms can be distinguished:

- *Conduction or diffusion*
  Transport through a resting medium, in this case through a solid medium or a fluid (liquid or gas) that is not flowing in the direction of considered transport.

- *Convection*
  Transport by a flowing medium in the flow direction (sometimes also called advection) or between a resting medium (a surface) and a flowing medium.

- *Radiation*
  Direct transport of energy between the surfaces of (opaque) bodies by electromagnetic waves through a transparent medium. Only energy is transported by this mechanism.

In the following sections firstly the mechanisms of conduction (diffusion) and convection will be treated separately for various physical quantities. Secondly radiation will be discussed.

3.2.3 Conduction (diffusion)

Conduction is the mechanism of transport in a resting medium. The mechanism itself is on a molecular scale but for overall calculations the process is considered at a macroscopic scale in measurable quantities. In the greenhouse, conduction does take place through the construction and cover but it is especially important in regard to the transport in the soil.

In general, energy flows from high to low temperature for all mechanisms. For conduction not only the change in temperature is taken into account but also the distance over which the temperature changes. The temperature change over the distance is called the temperature gradient. Fourier law states that the transport of energy in one direction, e.g. the x-direction, is proportional to the gradient of the temperature:

$$\Phi_h = -\lambda \, dT / dx \qquad \text{(Eq. 3.2.7)}$$

The transport is expressed as the flux per unit area perpendicular to the direction of transport, the so-called flux density, so for heat the heat flux density Φ_h (W m^{-2} K^{-1}). The proportionality factor λ is the property of the material to conduct energy and is called the thermal conductivity (W m^{-1} K^{-1}). The negative sign in equation (3.2.7) links a positive flux density in the positive x-direction to a negative gradient (temperature from high to low for increasing x). From a mathematical point of view the gradient is a vector which is in accordance with the characteristic of the heat flux density having direction and magnitude. For a two or three dimensional temperature field the gradient is extended to these dimensions to calculate direction and magnitude of the heat flux density.

For the transport of mass by diffusion the same approach can be followed, the driving force being a concentration gradient. In the greenhouse system this could be applied to water movement in the soil. However greenhouse soils are well saturated so water is not limited and the consideration of mass transport by diffusion is not particularly relevant.

3.2.4 Advection and convection

The transport of energy and mass by a flow from one place to the other in the direction of flow and the transport from a surface to a flowing medium or vice-versa are called convection. The first case is sometimes called advection. In a greenhouse the ventilative exchange of energy and mass (water vapour, CO_2) is transport by advection. The exchange between the greenhouse air and the internal surfaces such as cover, crop, heating pipes, soil surface) is transport by convection. The same holds for the exchange between the outer surface of the greenhouse and the ambient air.

The net energy flux q_h (J s^{-1}) from inside to outside due to ventilation through a ventilation opening can easily be calculated as:

$$q_h = q_v \rho C_p (T_i - T_a) \qquad \text{(Eq. 3.2.8)}$$

with q_V the volumetric flux through the opening (m^3 s^{-1}), ρC_p the volumetric specific heat (J m^{-3} K^{-1}) and T_i, T_a the inside and outside temperature respectively. In section 3.3.4 the mechanisms responsible for the volume flux will be outlined.

For the mass flux q_m by ventilation the same kind of relation can be set up:

$$q_m = q_v (c_i - c_a) \qquad \text{(Eq. 3.2.9)}$$

here c_i and c_a are the inside and outside concentrations of the water vapour or carbon dioxide.

The transfer of energy or mass from a surface to a flow field cannot be described in such an easy way. Therefore a more practical description needs to be developed. Knowing that the "driving force" for the transport due to convection is the temperature or concentration difference between surface and flowing medium, it is simply stated that for energy transfer the heat flux density Φ_h equals:

$$\Phi_h = \alpha_h (T_s - T_f) \qquad \text{(Eq. 3.2.10)}$$

and for mass transfer the mass flux density Φ_m equals:

$$\Phi_m = k_m (c_{m,s} - c_{m,f}) \qquad \text{(Eq. 3.2.11)}$$

These equations for energy transfer, already stated by Newton, define the heat transfer coefficient α_h (W m^{-2} K^{-1}) and the mass transfer coefficient k_m (m s^{-1}) from the surface to the medium as a ratio between the flux density and the driving force. In literature on crop transpiration the symbol E is often used instead of the general symbol Φ_m. Not taken into account in this equation is that the transfer coefficients depend on a great number of relevant factors, including the properties of the flowing medium, the flow conditions and on the geometry of the flow field. These dependencies have to be quantified for each particular geometry, mostly by experiment.

To do so, the great number of relevant factors is first reduced by combining them, through dimensional analysis, to a small quantity of dimensionless numbers thereby transferring a relation between n relevant factors to a relation between r dimensionless numbers, where of course $r < n$. The heat transfer coefficient a_h is then expressed in the so-called Nusselt-number (Nu). If the flow field is driven by external factors (such as the wind over the greenhouse) the convection is called forced convection and the flow condition is characterised by the Reynolds-number (Re). The relevant properties of the flowing medium for heat transfer are combined into the Prandtl-number (Pr). Then the following kind of relation proves to fit with experimental data for most forced convective heat transfer cases:

$$\mathrm{Nu} = C_1\, \mathrm{Re}^{n}\, \mathrm{Pr}^{m} \qquad \text{(Eq. 3.2.12)}$$

where: $\mathrm{Nu} = \alpha_h\, l\, \lambda^{-1}$
$\mathrm{Re} = u\, l\, \nu^{-1}$
$\mathrm{Pr} = \nu\, a^{-1}$

where u is the velocity in the flow field (m s^{-1}), l is a characteristic dimension (m) of the surface considered, ν is the kinematic viscosity of the flowing medium ($m^2\ s^{-1}$) and a the thermal diffusivity of the medium ($m^2\ s^{-1}$).

In the greenhouse situation the coefficient C_1 and the powers n and m depend on the geometry of the surface and the flow conditions as determined by the range of the Reynolds values. This is demonstrated by the relations for forced convection heat transfer along a flat surface:
For low Reynolds numbers:

$$\mathrm{Nu} = 0.664\, \mathrm{Re}^{1/2}\, \mathrm{Pr}^{1/3} \qquad \text{(Eq. 3.2.13)}$$

with: $2.10^2 < \mathrm{Re} < 10^5;\ \mathrm{Pr} \geq 0.7$

For high Reynolds values or turbulent flow:

$$\mathrm{Nu} = 0.036\, \mathrm{Re}^{0,8}\, \mathrm{Pr}^{1/3} \qquad \text{(Eq. 3.2.14)}$$

with: $10^5 < \mathrm{Re} < 10^7;\ \mathrm{Pr} \geq 0,7$

For air $\mathrm{Pr} \approx 0.72$. The length of the plate in the flow direction is the characteristic length, l. In heat transfer literature relations can be found for most geometries in practical cases. However these relations are validated experimentally under laboratory conditions, so it is difficult to translate these relations without modifications to the greenhouse situation where there are many disturbing factors.

Using the same approach the mass transfer coefficient k_m is combined in a dimensionless number called the Sherwood number (Sh) and its dependency on the flow conditions and the properties of medium is expressed again as a relation between dimensionless numbers:

$$\mathrm{Sh} = C_1\, \mathrm{Re}^{n}\, \mathrm{Sc}^{m} \qquad \text{(Eq. 3.2.15)}$$

with: $\mathrm{Sh} = k_m\, l\, \mathbb{D}^{-1}$
$\mathrm{Sc} = \nu\, \mathbb{D}^{-1}$

where $\mathbb{D}$ is the diffusivity of the gas component in the medium ($m^2 s^{-1}$) and Sc is the Schmidt number. For the same flow field and flow conditions and the same geometry the Nu, Re, Pr relation and the Sh, Re, Sc relation are analogous, so the coefficient C_1 and the powers n and m in relations (3.2.12) and (3.2.15) are equal.

Until now we have considered the convective transfer from a surface to a medium flowing with a known velocity v, given by some externally driven flow field. This we have called forced convection. In some situations however, the flow-field is driven by the temperature and concentration differences themselves, e.g. around the pipes of the heating system in the greenhouse. Consider a vertical hot plate resting in cool air. Along the surface of the hot plate, air is heated and therefore its density will decrease. Due to the density difference with the surrounding air, the hot air bubble will rise, and so will flow. On the downward side fresh air will come in and in the steady state a continuous flow of air is generated along the surface due to the temperature driven density differences. This type of convection is called free or natural convection. Again the transfer coefficient has to be known as a function of relevant parameters. For this type of convection heat transfer is of particular interest. For this mechanism it is found that the Nusselt number is a function of a dimensionless number characterising the density difference (due to the temperature difference) as driving force, called the Grashof number (Gr). Again the Prandtl number characterises the properties of the medium.

For air under normal conditions:

$$Gr = g \, \Delta T \, l^3 \, T^{-1} \, \nu^{-2}$$

where g is the acceleration due to gravity (m s^{-2}), ΔT the temperature difference between surface and medium (K) and T the absolute temperature of the medium (K).

As an example for the free convective heat transfer from vertical flat plates and horizontal cylinders (the characteristic length l equals diameter) the following relations are given:
For low Grashof numbers:

$$Nu = 0.55 \, (Gr \, . \, Pr)^{1/4} \qquad \text{(Eq. 3.2.16)}$$

with: $10^4 < Gr < 10^8$, $0.5 < Pr < 10$.

For high Grashof numbers:

$$Nu = 0.13 \, (Gr \, . \, Pr)^{1/3} \qquad \text{(Eq. 3.2.17)}$$

with: $Gr > 10^8$, $0.5 < Pr < 10$.

For other geometries empirical relations can also be found in the literature from laboratory experiments. Again, direct translation to the greenhouse situation is difficult. Inside a greenhouse a flow is mostly generated by temperature differences. However these differences are not between a single surface and the air. Several surfaces are involved such as the cold glass panes, the hot heating pipes and the soil surface. This means that there is interaction between the flow fields around these surfaces which will affect the various heat transfers. So *in situ* measurements are needed to validate the relations (3.2.12) and (3.2.15) for heat and mass transfer respectively.

3.2.5 Simultaneous heat and mass transfer

An important combination of heat and mass transfer appears if phase changes occur. This is of interest especially for consideration of the processes of evaporation and condensation. If water evaporates, the water vapour will be transported from the evaporating surface *i* to the undisturbed air *j* according to the transfer equation (3.2.11) for the vapour flux density:

$$\Phi_m = E = k_m (c_{m,i} - c_{m,j})$$

Energy is needed for evaporation, the heat of evaporation per unit mass is referred to as L ($J\ kg^{-1}$). Therefore heat has to flow to the evaporating surface given by the heat transfer equation (3.2.10) for the heat flux density:

$$\Phi_h = a_h (T_j - T_i)$$

Note that in the steady state the surface is cooler than the air due to the evaporation process. The heat balance of the surface in the steady state requires:

$$k_m (c_{m,i} - c_{m,j}) L = a_h (T_j - T_i) \qquad \text{(Eq. 3.2.18)}$$

The heat and mass transfer coefficients are given by the typical Nu, Re, Pr and the Sh, Re, Sc relations. The analogy between relations (3.2.12) and (3.2.15) leads to:

$$\mathrm{Nu} / \mathrm{Sh} = (\mathrm{Pr} / \mathrm{Sc})^m = \mathrm{Le}^m \qquad \text{(Eq. 3.2.19)}$$

With Le the Lewis Number as ratio of Pr and Sc.

A typical value for *m* in equation (3.2.19) is found to be 1/3. With $\lambda / rC_p = a$, this leads to:

$$\alpha_h / k_m = \rho C_p (D / a)^{-2/3} \qquad \text{(Eq. 3.2.20)}$$

so equation (3.2.18) will be:

$$(c_{m,i} - c_{m,j}) / (T_j - T_i) = (\rho C_p / L)(a / D)^{2/3} \qquad \text{(Eq. 3.2.21a)}$$

An evaporating surface will cool down to a temperature different from the ambient temperature ($T_j - T_i$) and at this temperature the water vapour concentration at the surface is saturated. On the right hand side of equation (3.2.21a) only physical properties are given so the ratio on the left hand side is known. So if the temperature difference is measured together with the surface temperature, the concentration of water vapour in the air can be determined. The dry and wet bulb psychrometer operates according to this principle.

In environmental physics the quantity of vapour concentration is not widely used; partial vapour pressure *p* is used instead. According to the ideal gas law, vapour concentration *c* ($kg\ m^{-3}$) can be translated to vapour pressure *e* (Pa) according to

$$c = e M (R T)^{-1} \qquad \text{(Eq. 3.2.22)}$$

where *R* is the universal gas constant (8314 $J\ kmol^{-1}\ K^{-1}$), *T* the absolute temperature (K) and *M* the molar mass of water vapour (18 $kg\ kmol^{-1}$).

Equation (3.2.20) can then be rewritten:

$$(e_i - e_j) / (T_j - T_i) = \{(R\,T\rho C_p) / (M\,L)\}(a / D)^{2/3} \qquad \text{(Eq. 3.2.21b)}$$

3.2.6 Radiation

3.2.6.1 Introduction

Radiation refers to the continual emission of energy from the surface of all bodies of a given temperature. This emitted radiation is of electromagnetic origin. The wavelength l_w (m) and the frequency f (s^{-1} or Hz) are related:

$$l_w\,f = \text{const.} \qquad \text{(Eq. 3.2.23)}$$

where const. is the velocity of light (2.9979 10^8 m s^{-1}).

The radiant energy emitted by a body depends on the nature of the surface of that body and its temperature. At low temperatures the rate of radiant energy is low. At higher temperatures total radiant energy increases rapidly. Another property of radiation is that, at any surface temperature, the radiant energy consists of waves of different wavelengths with a distribution according to Planck's law. At lower temperatures the distribution occurs at longer wavelengths. For instance, at a temperature of 300 K (about room temperature), practically all of the radiant energy is emitted in the infrared band at wavelengths between about 2.5 and 25 μm with a maximum emission at about 10 μm. At the surface temperature of the sun (6000 K) the wavelength region is between about 300 and 2500 nm (0.3–2.5 μm) with a maximum at about 500 nm so the visible region and the photosynthetic active region are included in this distribution. In Table 3.2.1 some interesting fields of the electromagnetic spectrum are shown.

3.2.6.2 Emittance and Stefan-Boltzmann's law

The radiant energy emitted by a body per unit time and per unit area, depends on the temperature of the surface of the body and on the nature of the body. In 1879 Stefan found that the total emitted radiant energy flux density, thus integrated over all wavelengths, is proportional to the fourth power of the absolute temperature:

Table 3.2.1. The electro-magnetic spectrum.

Type of radiation	Wavelength (nm)	Frequency (Hz)
gamma radiation	< 0.01	> 3 10^{19}
X-radiation	0.01 – 10	3 10^{19} – 3 10^{16}
ultra violet	10 – 390	3 10^{16} – 7.7 10^{14}
visible light	380 – 760	7.9 10^{14} – 3.9 10^{14}
solar radiation	300 – 2,500	1 10^{15} – 1.2 10^{14}
infrared (IR)	760 – 3 10^5	3.9 10^{14} – 1 10^{11}
thermal IR, ~300K	2,500 – 25,000	1.2 10^{14} – 1.2 10^{13}
micro waves	3 10^5 – 1 10^8	1 10^{12} – 3 10^9
radio waves	> 1 10^8	< 3 10^9

$$\Phi_{rad} = \varepsilon \sigma T^4 \qquad \text{(Eq. 3.2.24)}$$

where Φ_{rad}: radiated energy flux density ($W\ m^{-2}$)
ε: the emissivity of the surface (-)
σ: a universal physical constant, called the Stefan-Boltzmann constant ($5.6699\ 10^{-8}\ W\ m^{-2}\ K^{-4}$)
T: absolute temperature (K).

The number ε characterises the emitting properties of the surface of a body and is called the emissivity. It is a dimensionless number between 0 and 1. A body with an emissivity of 1 is called a black body.

From extensive measurements it can be deduced that there are roughly two groups of emissivity values. For non-metals, including white paint and also plant leaves, the value of ε at about room temperature (thermal radiation) is high (0.7 to 1). For metals (especially if the metals are polished) ε is low (0.3 to 0.05).

3.2.6.3 Absorption, reflection and transmission

If radiation falls on a body, generally part of the radiation is absorbed, a part is reflected and a part is transmitted. If the absorptivity is α, the reflectivity is ς and the transmissivity is τ, it can easily be seen from the law of conservation of energy, that

$$\alpha + \varsigma + \tau = 1 \qquad \text{(Eq. 3.2.25a)}$$

Where opaque bodies are concerned no radiation will be transmitted, and relation (3.2.25a) then reads:

$$\alpha + \varsigma = 1 \qquad \text{(Eq. 3.2.25b)}$$

If all of the radiation is absorbed by a body ($\alpha = 1$), the body is called a black body.

For the same region of wavelengths it can be proven easily that the emissivity and the absorptivity have the same value. In the greenhouse situation the various opaque parts emit and absorb thermal radiation in the thermal wavelength region. Some parts however are exposed to solar radiation, so absorb radiant energy in the short wave wavelength region but emit in the thermal band. In principle then the absorptivity for the solar light and the emissivity for the thermal radiation will differ.

3.2.6.4 Radiative energy exchange between surfaces

In a greenhouse the various surfaces of the different components (crop leaves, cover, soil surface and heating pipes) are at an absolute temperature of about 300 K. The surfaces then emit thermal radiation with a wavelength between 5,000 and 50,0000 nm and absorb radiation emitted in the same wavelength region from the other surfaces. If only two surfaces were involved, radiating and absorbing to and from each other with an emissivity (and thus also absorbtivity at the same wavelength) equal to 1, the net energy flux density $\Phi_{rad,12}$ from the surface with temperature T_1 to that with T_2 will be the difference between emitted and absorbed radiation:

$$\Phi_{rad,12} = \sigma (T_1^4 - T_2^4) \qquad \text{(Eq. 3.2.26a)}$$

If the emissivities are not equal to 1, multiple reflections between the surfaces will occur and equation (3.2.26a) will change to:

$$\Phi_{rad,12} = \varepsilon_{12}\,\sigma\,(T_1^4 - T_2^4) \tag{Eq. 3.2.26b}$$

The effective emissivity ε_{12} between the surfaces depends on the individual emissivities ε_1 and ε_2 and the geometry of the surfaces. For large parallel surfaces a relatively simple relation is found:

$$\varepsilon_{12} = (1/\varepsilon_1 + 1/\varepsilon_2 - 1)^{-1} \tag{Eq. 3.2.27}$$

Determining radiative exchange becomes more complicated however if more than two surfaces are involved, where each only intercepts part of the radiation emitted from others. The view factor, F_{ij}, determines which part of the total radiation from surface i falls directly on surface j. The fraction of the radiation coming directly from surface i absorbed by surface j equals $F_{ij} \times \varepsilon_j$. Surface j also receives radiation indirectly, coming from i, via reflections at the surfaces k. If B_{ij} is the absorption factor representing the fraction of all radiation coming from surface i which is absorbed by surface j, this absorption factor can be written as

$$B_{ij} = F_{ij}\varepsilon_j + \Sigma_k F_{ik}\varsigma_k B_{kj}, \tag{Eq. 3.2.28}$$

with $\varsigma_k = 1 - \varepsilon_k$. This equation can be rearranged and written in matrix form as

$$\underline{B} = (\underline{I} - \underline{F}\,\underline{\varsigma})^{-1}\,\underline{F}\,\underline{\varepsilon}. \tag{Eq. 3.2.29}$$

For an enclosure it can be proven that $\varepsilon_i A_i B_{ij} = \varepsilon_j A_j B_{ji}$ (Gebhart, 1961) thus the net radiative exchange between the surfaces i and j can be written as

$$q_{ij} = \varepsilon_i A_i B_{ij}\,\sigma\,(T_i^4 - T_j^4). \tag{Eq. 3.2.30}$$

The description of radiative exchange in greenhouses often takes place between surfaces with a high emissivity, which are approximated as being parallel. For these the view factor can easily be found in the literature (Gebhart, 1961).

If multiple reflections are ignored (high emissivities) the net heatflux from surface i to surface j can be written as:

$$q_{rad,ij} = A_i\,\varepsilon_i\,\varepsilon_j\,F_{ij}\,\sigma\,(T_i^4 - T_j^4) \tag{Eq. 3.2.32}$$

So, though the greenhouse situation seems to be complex, relatively simple relations allow the calculation of the thermal radiative exchange from the surface temperatures.

3.3 Energy balance

G.P.A. BOT AND N.J. VAN DE BRAAK

3.3.1 Introduction

As explained in section 3.2 the state of important greenhouse variables is determined by energy and mass balances (equation (3.2.3) and (3.2.6)). In this section the various terms of the energy balance will be discussed. Before going into detail it is necessary to determine which part(s) of the greenhouse (or in the terms of section 3.2.1 which volume) we are considering in one balance. Although for every single part of the greenhouse and its content a balance could be formulated, it is sufficient for the sake of understanding to restrict ourselves to a few main parts. We will consider the energy balances of (1) the greenhouse hull, (2) the greenhouse air, (3) the crop and (4) the greenhouse soil. For each of them equation (3.2.3) is valid. Figure 3.3.1 illustrates which components determine the energy inflow $q_{in,h}$ and the energy outflow $q_{out,h}$ of each part.

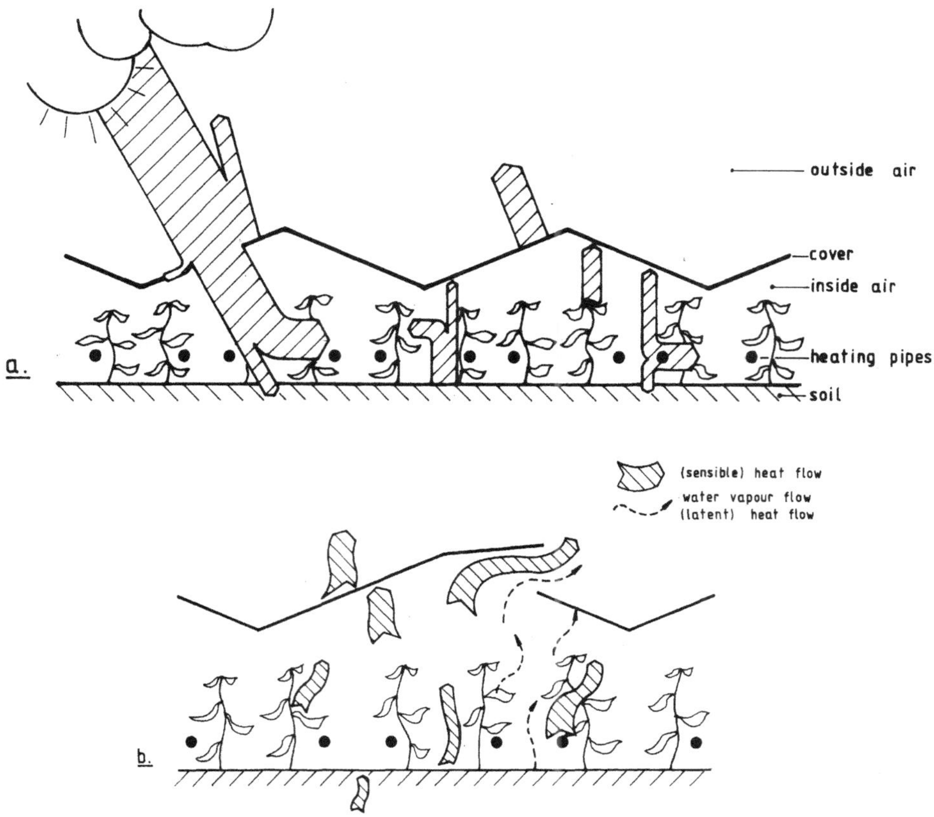

Figure 3.3.1. Terms in the energy balance over the various compartments of a simplified greenhouse. (a) radiative terms, left: shortwave or solar radiation, right: thermal or longwave radiation, (b) latent and sensible heat flows.

For the greenhouse hull the total of $q_{in,h}$ and $q_{out,h}$ is composed of absorbed solar radiation, radiative exchange (IR) with the sky, radiative exchange with the interior of the greenhouse, convective exchange between the hull and both the greenhouse air and outside air, and latent heat released by condensation of water vapour at the inside.

For the greenhouse air the composing terms are convective exchange with the hull, the crop,the soil and the heating system respectively and exchange with the outside air (advection or ventilation).

For the crop these are absorbtion of solar radiation, radiative exchange with hull, soil and heating system, convective exchange with the greenhouse air and latent heat linked to evaporation.

For the soil these are absorbtion of solar radiation, radiative exchange with the hull, the crop and the heating system, convective exchange with the greenhouse air, and conductive exchange with the underlying soil layers.

In the following sections a number of the components mentioned will be considered in more detail.

3.3.2 Solar radiation

The total solar radiation (global radiation) can be divided into direct radiation (originating from the sun at solar position) and diffuse radiation (scattered in the atmosphere and by the clouds). Most of this incoming solar energy flux (99%) at earth level is within the wavelength region between 300 and 2,500 nm. For plant growth a special part of the spectrum in the visible region between 400 and 700 nm is of interest, this part is called photosynthetic active radiation (PAR). About half of the total solar energy is irradiated within this wavelength region. Only a small part of the PAR energy is absorbed by the crop and is directly converted into the photosynthesis process. The remainder is converted into heat. Kondratyev (1972) described the spectral distribution of solar irradiance at the earth surface for various meteorological conditions. This description is accurate enough to estimate the ratio between PAR and global radiation for outside conditions.

The global radiation at crop level contributes to the energy balance of the crop and so affects crop temperature and transpiration (the latter will be treated in section 3.4). In the energy balance of the crop the energy converted in the photosynthesis process can be neglected, as stated above. It is of great importance to translate both direct and diffuse solar radiation at earth level to that inside the greenhouse at crop level. The interaction of the greenhouse cover with the solar radiation determines how much radiation is transmitted and available at crop level. This can be calculated from the basic optical laws of reflection, absorbtion and transmission of transparent layers and opaque construction parts. The optical properties of the cover and construction, the angle of incoming radiation and the geometry of the construction have to be known. For the direct component of the global radiation, the angle follows from the solar position determined by the time and date and the latitude of the observed greenhouse and by the orientation and geometry of the surfaces. For the diffuse radiation it follows from the distribution of the radiation intensity over the hemisphere. This is different for various meteorological conditions; the most striking difference is that between a clear and cloudy sky. Stoffers (1967) was the first to develop a transmission model. Bot (1983) has described the considerations outlined above in full detail. A review of the literature on greenhouse light transmission has been provided by Critten (1992). Recently De Zwart (1993) formulated a vectorial scheme to calculate light transmission in an easy way.

A comparison of this scheme with Bot's model calculations and measurements of transmission factors shows an agreement for both diffuse and direct radiation under varying conditions to within a few percent points (Heuvelink *et al.*, 1994). Therefore this model is applied in plant production and energy consumption modelling (Gijzen & Ten Cate, 1988; Houter *et al.*, 1989; Gijzen *et al.*, 1990; Houter, 1990).

Light transmitting properties of greenhouses can be compared by the experimental determination of the light transmission for diffuse radiation (high cloudiness, low solar altitude). In this way the strong effects of spatial and time dependent variations in direct radiation transmission are avoided.

3.3.3 Thermal radiation exchange

All surfaces inside the greenhouse exchange radiation with their environment. This can be calculated according to the methods described in section 3.2.6.4.

A special difficulty is the characterization of the radiative exchange between the outside of the cover and the celestial hemisphere, because this is a complex exchange between the various layers of the atmosphere and the greenhouse hull. The atmospheric temperatures at various heights determine the temperature differences with the cover and the composition of the atmosphere determines the absorbtivity and emissivity at various heights. So for a full description both the atmospheric composition and the temperature as a function of the height have to be known. To overcome this, a sky temperature T_{sky} is defined as a temperature of a black hemisphere ($\varepsilon = 1$) exchanging thermal radiation with the greenhouse cover according to Stefan-Boltzmann by the same amount as the real atmospheric exchange. Wartena *et al.* (1973) reported correlations of experimental data relating sky temperatures to standard meteorological observations such as air temperature, humidity and cloudiness, but these apply only for an average sky temperature over a long period and are valid only for regions with the same meteorological characteristics as the region in which the measurements have been performed.

For real time greenhouse performance, sky temperature is an important boundary condition for the greenhouse climate and the energy budget. Fortunately sky temperature can be measured nowadays directly using a pyrgeometer with sufficient accuracy (De Zwart, 1995).

3.3.4 Ventilation

The greenhouse hull prevents the internal air mixing with the external air. Air exchange through openings in the hull (leaks and ventilation windows) is called ventilation (advection in section 3.2.4) and can be expressed in terms of volumetric flow ($m^3\ s^{-1}$). This flow carries energy and mass according to equations (3.2.8) and (3.2.9) respectively. Because the ventilation process affects the most important greenhouse effect, i.e. the enveloping of air, a proper description of the ventilation flux as a function of external and internal factors is essential for the description of greenhouse climate.

Ventilation flux as the flow of air from inside to outside and vice-versa through openings has to be driven by pressure differences over the openings. These pressure differences can be due to the effect of an outside airflow (wind effect) or due to the density differences between internal and external air, generated by temperature differences (temperature effect) and to a much lesser extent by concentration differences. The pressure differences generated by the wind field proved to be of a fluctuating nature for large greenhouses, with vents in the roof (Bot, 1983). As a consequence the ventilation fluctuates as well, but can be described as an effective flow from inside to outside and the reverse, due to an effective pressure difference.

A flow through an opening, with a known pressure difference over it, depends on the flow resistance of the opening itself. Relations are deducted for the ventilation rate through specific openings as a function of relevant parameters in literature on the ventilation of buildings. Bot (1983) adopted the approach in which the wind and temperature effects and the flow characteristics of the openings are discussed separately. This approach was given a more sound base and validated in more detail by

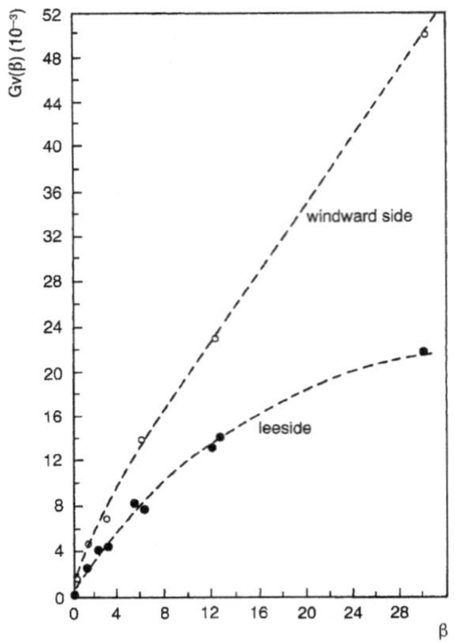

Figure 3.3.2. Ventilation number Gv as function of the window opening angle ß of a typical ventilation window with aspect ratio of 2.5 and mounted at the ridge. Relations are given for windows opened at the lee-side or the windward-side (after De Jong, 1990).

De Jong (1990). Nederhoff *et al.*, (1984) and Fernandez & Bailey (1992) presented data on specific greenhouses. The effective ventilation volume flux q_V ($m^3\ s^{-1}$) appeared to be linear proportional to the outside wind speed u ($m\ s^{-1}$) (defined at a reference height of 10 m) and the area of ventilation windows A_o for any window opening angle ß. Due to the linear proportionalities a dimensionless ventilation number Gv can be defined as:

$$Gv = (q_V u) / A_o \qquad \text{(Eq. 3.3.1)}$$

Once Gv has been derived by experiment from specific greenhouses this data can then be used for other greenhouses. The only parameter to be aware of is the window type. In Figure 3.3.2 typical ventilation characteristics are given for some commonly used window types in terms of the relation Gv(ß).

From Gv the ventilation rate or other interesting greenhouse specific ventilation figures can be deduced. Also β seems not to be so practical. However it has also been chosen as a non-specific variable. From β more specific variables often used in practice, such as the percentage of maximal opening can be deduced.

Some questions concerning the relation between ventilation rate and outside wind field still require an answer. The first one is how leeward and windward side ventilation combine. The above data are for leeward or windward side ventilation separately. De Jong (1990) proved experimentally that both separate effects can be added for small opening angles on the windward side, and for a wide range of opening angles on the leeward. Whether some shortcut occurs between windward and leeward sides when there are larger opening angles on the windward side, still has to be checked. The second question is about the effect of the dimensions of the greenhouse on the ventilation. De Jong (1990) found an effect which is not yet understood. The third question is on the variation in ventilation that can be expected due to the fluctuating nature of the wind field. For small window openings variation was found to be small, for large window openings the effect has to be quantified more accurately.

The amount of ventilation due to temperature differences between the inside of the greenhouse and the exterior can also be quantified. Bot (1983) and De Jong (1990) have presented extensive findings. In this case the vertical height of the opening, or for a number of openings, the vertical distance between these openings (chimney effect), is of importance. The ventilation due to wind effects will be dominant, however, even at low wind speeds of about 2–3 m s^{-1}. The temperature effect will therefore be relevant under special low wind, high radiation conditions.

With the known dependency of ventilation flux on wind speed, temperature difference between inside and outside, window characteristics and window opening the exchanged energy q_{vnt} (W) between the greenhouse interior and the immediate surroundings can be calculated according to equation (3.2.8). Similarly, the mass transfer (water vapour, CO_2) can then be calculated according to equation (3.2.9). As stated before, greenhouse specific figures such as ventilation rate (defined as volume changes per hour) can easily be calculated from the ventilation flux.

3.3.5 Convective exchange

3.3.5.1 Greenhouse cover

The greenhouse cover exchanges energy at the inner surface to the greenhouse air and to outside air at the other side. Moreover, water vapour is transported from the greenhouse air to the cover and condenses there. The mechanism of these exchanges is that of convection (section 3.2.2). Inside natural convection is expected due to low local air velocities generated by the prevailing temperature differences. Outside forced convection is expected due to local air velocities generated by the wind field. Following equation (3.2.10), the convective heat exchange is defined as

$$q_{cnv} = \alpha_h A_s (T_a - T_s) \qquad \text{(Eq. 3.3.2)}$$

where T_a and T_s are the ambient air and cover surface temperature (K) respectively, A_s the surface area and α_h the heat transfer coefficient (W m^{-2} K^{-1}). The transport of water vapour to the cover can be described analogously according to equation (3.2.11).

As shown in section 3.2.4, both for the forced and natural convection, α_h is dependent on fluid properties and system parameters for a particular geometry. These dependencies are expressed in relations between dimensionless numbers (equations (3.2.12) through (3.2.17)). In the greenhouse situation the fluid is air and fluid properties only vary due to temperature dependencies of these properties. These variations can be neglected in a preliminary approach.

The ratio between Gr (Grashof) and Re^2 (Reynolds) indicates whether the exchange is due to purely natural or purely forced convection (Morgan, 1975). The cover of a multispan greenhouse complex has a saw tooth surface geometry. A common Dutch greenhouse type, manufactured industrially, is the Venlo-type greenhouse (see also Chapter 4). The geometry of the saw tooth surface of this type of greenhouse has a characteristic length of about 1.75 m (ridge-gutter distance). The exchange with the cover at both the inside and outside are of interest. On the inside, local air velocities are in the order of 0.1 ms^{-1} ($Re \approx 10^4$) and the temperature differences about 10 K ($Gr \approx 10^{10}$), so it can be expected that natural convection is the prevailing form of heat transfer. Forced convection can be expected at the outside. Following section 3.2.5 the mass transfer coefficient k (equation (3.2.11)) for the transport of water vapour to the cover can be calculated according to equation (3.2.20).

Because the flow field over a saw-tooth surface will be different compared to that over a flat plate, it is not known a priori if data from the literature on the transfer to and from flat plates can be applied. Furthermore, no general data on convective heat transfer to and from saw-tooth surfaces are available. Therefore the convective heat transfer to and from the greenhouse cover has been measured and the flow field over the cover has been sampled by Bot (1983). This yielded natural convection relations for the heat transfer at the inside and the outside surfaces for low wind speeds up to 3 m s^{-1}. At higher wind speeds forced convection has been found at the outside.

3.3.5.2 Heating pipes

Dutch greenhouses are commonly heated by a heating pipe system, distributing hot water from a central boiler (Chapter 4). The mechanism of heat transfer between the pipes and the air is by convection. The same approach as indicated in section 3.2.4 can be applied here. The characteristic length is the diameter of the pipes.

With local air velocities of 0.1 m s^{-1}, Re is low (≈250), Gr will be in the order of 10^6, thus the value of $GrRe^{-2}$ is about 10 which is near to the criterion for pure natural convection around horizontal cylinders $GrRe^{-2} \geq 14$ (Morgan, 1975). However the arrangement of heating pipes in greenhouses often differs from that found in the experimental situation, on which data on the pure natural convective exchange from horizontal cylinders in the literature is based. This makes application of the data difficult.

Experiments under greenhouse conditions (Stanghellini, 1983, Nawrocki, 1985) yielded the heat transfer coefficient prevailing under conditions of natural convection. This did indeed differ from data in the literature on pure natural convection under ideal conditions. Where the heat transfer between pipes and air is described as in equation (3.2.17) the coefficient of heat transfer, being dependent on the temperature difference as determined by the Grashof number, introduces non-linearities.

3.3.5.3 The crop

The crop is a major absorber of solar energy and thereby a convertor of radiative heat into latent and sensible heat. This latent and sensible heat is transported to the greenhouse air by convection. So the energy balance over the crop can be set up from the absorbed shortwave radiation, the exchanged sensible and latent heat and the thermal radiative exchange with the various greenhouse parts. These terms have already been discussed or will be discussed in the section on the vapour balance (section 3.4).

3.3.5.4 The soil

In the energy balances of the various greenhouse parts, the exchange with the soil is of minor importance. However, the soil surface exchanges thermal radiation with the other greenhouse components and the energy storage in the soil determines the dynamics of the greenhouse system on a daily basis (Bot, 1989). So the exchange to, and the transport in the soil (for the latter see section 3.3.6) have to be represented in a proper way for the description of the greenhouse climate, especially as far as the daily rhythm is concerned. The calculation via the natural convection exchange at the surface of the soil has proved to be accurate enough. The convective exchange can be approached in the same way as discussed in section 3.3.5.1.

3.3.6 Conductive exchange

As stated in section 3.2.3 the temperature gradient in the medium is the driving force for heat transfer in conduction. For a simple representation of the greenhouse soil as infinite layers with a thickness d, equation (3.2.7) can be transformed to:

$$q_h = \alpha_h \Delta T \qquad \text{(Eq. 3.3.3)}$$

with

$$\alpha_h = \lambda A/d \qquad \text{(Eq. 3.3.4)}$$

where q_h is the heat flow through an area A due to conduction (W m^{-2}), ΔT the temperature difference between two adjacent soil layers and λ the thermal conductivity (W m^{-1} K^{-1}). To be able to calculate the conductive fluxes the thermal conductivity of the soil has to be known. This will be dependent on the character of the soil and the water content. Data on the thermal conductivity of saturated soils have been reported by Baver & Gardner (1972).

For crops cultivated in closed loop watering systems, which have been introduced recently, the soil will be dry. In these cases λ has to be determined experimentally, using, for example, a non-steady heat probe method (Van Loon *et al.*, 1989).

3.4 Vapour balance

C. STANGHELLINI

3.4.1 Introduction

One might think that the purpose of the whole business of getting so much water moving around a greenhouse is to have a fraction of it trapped within the crop, more exactly within that part of the crop one is going to sell. This, however, overlooks the main peculiarity of water, which makes it so important for energy considerations, and that is the large amount of energy that goes into the phase change of water into vapour: it takes about 2.5 kJ to make 1 g of water evaporate. Conversely, condensation of 1 g of vapour would warm up a cubic metre of air by about 2 K. It is, namely, the huge disparity between the specific heat of air ($C_p \approx 10^3$ J kg^{-1} K^{-1}) and the latent heat of vaporisation of water ($L \approx 2.5\ 10^6$ J kg^{-1}) that ensures that phase changes of water (transpiration and condensation) are conspicuous terms in the greenhouse energy balance. Hence, water use and energy consumption are more akin than one would at first suspect, as supply of energy causes evaporation of water to take place, which process removes energy from the ambient. In fact, in modulating the energy fluxes, water has a great bearing on environmental conditions, affecting thereby all growth processes.

The vapour balance of the greenhouse will be discussed in this section. Crop transpiration ($\Phi_{w,tr}$, E in the present section) is the main source of vapour. Evaporative cooling will be dealt with in Chapter 4. Vapour removal takes place through both condensation ($\Phi_{w,cnd}$, here C) and ventilation ($\Phi_{w,vnt}$, here V), so that the following balance equation holds:

$$h\frac{d\chi_a}{dt} = E - C - V \qquad \text{(Eq. 3.4.1)}$$

where h (m) is the ratio of the greenhouse volume to its ground area, that is, the mean height of the greenhouse, and χ_a (identified elsewhere also by the symbol c_w) is the mean vapour content of the air within the greenhouse (kg m^{-3}). After a short summary of the definitions useful when dealing with these matters, all the fluxes appearing in equation (3.4.1) will be quantified separately. Finally, the vapour balance of greenhouse air will be discussed, and the main factors affecting it will be highlighted.

3.4.2 Definitions

The amount of vapour contained in a parcel of air can be quantified in many ways: the mass of vapour per unit volume of air, called absolute humidity or vapour concentration, has been already introduced with the either the symbol χ or χ_w (kg m^{-3}). Specific humidity X (kg kg^{-1}), is the name used for the mass of vapour per unit mass of air. Finally, the symbol e is commonly used for partial pressure (Pa). As the amount of water vapour that can be contained in a parcel of air is always a small fraction of the total mass of the parcel (some grams per kilogram or, where the density of air $\rho_a \approx 1.2$ kg m^{-3}, some grams per cubic metre), water vapour in the air can be regarded as an ideal gas, which ensures that conversion of units among the three definitions above takes place through multiplying factors that are (almost) constant, and that the three definitions are equivalent. In particular:

$$\chi = \frac{M_w}{RT} e = \frac{\rho_a C_p}{\gamma L} e \approx 7.4\, 10^{-6} e \qquad \text{(Eq. 3.4.2)}$$

Where M_w is the molar mass of water (kg kmol^{-1}), R the gas constant (J kmol^{-1} K^{-1}), T temperature (K), γ the thermodynamic psychrometric constant (Pa K^{-1}), ρ_a air density (kg m^{-3}) and C_p specific heat content of air at constant pressure (J kg^{-1} K^{-1}). On the other hand:

$$X = \frac{\chi}{\rho_a} \quad 6.2\, 10^{-6} e \qquad \text{(Eq. 3.4.3)}$$

with the coefficients calculated, in both cases, for air at 20°C and 101.3 kPa pressure.

The maximum (or saturating, e^*; X^*; χ^*) amount of water vapour that can be contained in a parcel of air depends heavily (roughly exponentially) on the temperature of the air. Other quite common definitions of air "humidity" take this fact into account. Relative humidity ($RH \equiv e/e^* \equiv X/X^* \equiv c/c^*$) and either pressure saturation deficit ($D \equiv e^* - e$) or "*delta X*" ($\equiv X^* - X$) quantify the "drying power" of air, that is, the amount of vapour that air at a given temperature is able to absorb. Finally, dew point is also a measure of humidity. This is the temperature at which a given vapour content would be saturating.

3.4.3 Transpiration

The transpiration of a greenhouse crop, as results from prevailing microclimate conditions, can be determined as follows: first the laws governing water loss from a wet surface will be examined. After discussing the modifications necessary in the case of a non-wet surface, such as the surface of a leaf, it will be shown that similar equations can be applied to a leaf canopy, through the "big leaf" analogy. Finally, some consequences of the presence of an enclosure (the greenhouse shelter) will be listed.

3.4.3.1 Evaporation from a wet surface

For evaporation from a wet surface to take place, two conditions have to be met: first the energy necessary for the phase change has to be brought to the surface and, secondly, the layer of saturated air in contact with the surface has to be constantly removed and replaced by non saturated air. The first event is governed by a conservation law: the conservation, or balance, of energy, equation (3.2.1), that is written for the steady state:

$$R_n - H - LE = 0 \qquad \text{(Eq. 3.4.4)}$$

where R_n is the mean flux density of available radiation (W m^{-2}); H is the flux density of sensible heat transferred between the canopy and the air (W m^{-2}) and LE is the flux density of latent heat due to transpiration (W m^{-2}), L being the heat of vaporization of water (latent heat, J kg^{-1}) and E the vapour flux density (kg m^{-2} s^{-1}).

The diffusion analogy (or Dalton's) law, on the other hand, implies that the vapour flux density E leaving the evaporating surface is proportional to the vapour concentration gradient between the surface and the "free" air, equation (3.2.4), the constant of proportionality being the transfer coefficient k_m, or the "conductance", g_b (m s^{-1}), of the air layer (or boundary layer), which is a measure of the replacement rate of air alongside the surface. It is handy, for the present purpose, to define this conductance as the inverse of a resistance (r_b, s m^{-1}), accordingly:

$$E = g_b\left(\chi_s - \chi_a\right) = \frac{1}{r_b}\frac{\rho_a C_p}{\gamma L}\left(e_s - e_a\right) \qquad \text{(Eq. 3.4.5)}$$

where the subscripts s and a refer to the surface and air, respectively. As the surface is wet, obviously:

$$e_s = e^*(T_s) \qquad \text{(Eq. 3.4.6)}$$

where T_s the temperature of the surface is generally unknown. An equation for it, however, can be found by observing that transfer of heat is governed by a law formally identical to equation (3.4.5). Once one takes into account that convective transfer of heat and vapour in air has to take place at (nearly) the same rate, given the similarity of their diffusion coefficient in air:

$$H = \rho_a C_p \frac{T_s - T_a}{r_b} \qquad \text{(Eq. 3.4.7)}$$

The system formed by equations (3.4.4); (3.4.5); (3.4.6) and (3.4.7) can be solved for the transpiration rate only by numerical methods. An elegant approximate solution was calculated by Penman (1948), who observed that whenever surface temperature does not differ much from air temperature, $e^*(T_s)$ may be approximated by the first two terms of Taylor's expansion:

$$e^*(T_s) \approx e^*(T_a) + s(T_a)\,(T_s - T_a) \qquad \text{(Eq. 3.4.8)}$$

where $s(T_a)$ (Pa K^{-1}) is the slope of the saturated vapour pressure curve, calculated at air temperature, in the following simply indicated by s. If canopy temperature is within a few degrees of air temperature, the error entailed by this linearity is small indeed: Lagrange's theorem ensures that there is a temperature, in the interval contained between T_s and T_a, that, were the slope to be calculated there, would cause the linearity to be exact. Taking into account also equation (3.4.8), an analytical expression for the evaporation rate can be found:

$$LE = \frac{s}{s+\gamma} R_n + \frac{\rho_a C_p}{r_b}\frac{D}{s+\gamma} \qquad \text{(Eq. 3.4.9)}$$

The first term of equation (3.4.9), usually called the "radiative" term, represents the evaporation that would take place into saturated air ($D \rightarrow 0$) due to the radiative energy gain of the surface. The "aerodynamic" term (the second one), on the other hand, represents adiabatic evaporation, that is the evaporation rate solely caused by the air being less than saturated, D being the saturation deficit (Pa).

3.4.3.2 Transpiration from a leaf surface

For a surface which is not thoroughly wet (e.g. the surface of a leaf not wetted by dew or rain), the condition of saturation, equation (3.4.6), does not need to be met. This problem was brilliantly overcome by Monteith (1965), who observed that inside the leaf there has to be a wet surface, usually identified with the internal surface of the substomatal cavities. All equations from section 3.4.3.1 will hold for that surface, with the important proviso that vapour leaving it has to cope with an additional "leaf" resistance, r_l (also called the "stomatal resistance"), before reaching the external surface of the leaf, and the boundary layer. Accordingly:

$$LE = \frac{\rho_a C_p}{\gamma} \frac{e_s - e_a}{r_l + r_b} \qquad \text{(Eq. 3.4.10)}$$

The equation of heat transfer, equation (3.4.7), however, does not need to be re-written, since, due to the high thermal conductivity of leaf tissue, the "wet surface" inside the leaf can safely be assumed to have the same temperature as the external surface. Once the same calculations as in section 3.4.3.1 are carried out:

$$LE = \frac{sR_n + \frac{\rho_a C_p}{r_b} D}{s + \gamma\left(1 + \frac{r_l}{r_b}\right)} \qquad \text{(Eq. 3.4.11)}$$

Transpiration of a leaf, accordingly, is described by an equation formally identical to equation (3.4.9) (the evaporation of a free water surface) once $\gamma^* \equiv \gamma(1 + r_l/r_b)$ is defined.

3.4.3.3 Transpiration from a canopy

Water lost from a crop canopy is the sum of the water transpired by all individual leaves. However, there can be important differences in the microclimate around individual leaves as there are vertical profiles of radiation, temperature, humidity and wind speed within a canopy, as well as horizontal variations. How this all affects photosynthesis has been discussed in section 2.2. As equation (3.4.11) indicates, however, transpiration has a linear response to the most important climate parameters ($\boldsymbol{R}_n$ and D) and a weak one to temperature and wind speed (s and r_b). Hence it could be calculated as if it were taking place from a "big leaf" (equal in area to the sum of all individual leaves), immersed in a homogeneous, average microclimate, once the properties of the big leaf have been defined.

This is commonly done by stating that a canopy exchanges heat and mass as if all its unitary leaf surfaces were wired in parallel. As there is, by definition, twice LAI leaf surface for each square metre of ground surface, this amounts to defining, respectively, an "aerodynamic" resistance to heat transfer as $r_b/2LAI$ and a "canopy" resistance to mass transfer, similarly, as $r_l/2LAI$. Thus equation (3.4.11) can be re-written for a canopy as follows:

$$LE = \frac{sR_n + \frac{2LAI\,\rho_a C_p}{r_b} D}{s + g\left(1 + \frac{r_l}{r_b}\right)} \qquad \text{(Eq. 3.4.12)}$$

or also:

$$E = \frac{\frac{s}{\gamma}\frac{R_n}{L}}{1 + \frac{s}{\gamma} + \frac{r_l}{r_b}} + \frac{2LAI\left(\chi_a^* - \chi_a\right)}{\left(1 + \frac{s}{\gamma}\right) r_b + r_l} \qquad \text{(Eq. 3.4.13)}$$

where R_n, the mean net radiation of the canopy, is defined per unit ground area.

It may be useful to preserve the analogy with the transfer equation (3.4.5), by defining a "transpiration conductance", g_{tr}:

$$g_{tr} \equiv \frac{2LAI}{\left(1 + \frac{s}{\gamma}\right) r_b + r_l} \qquad \text{(Eq. 3.4.14)}$$

and an "effective" absolute humidity at the "big leaf" surface:

$$\chi_{eff} \equiv \chi_a^* + \frac{\delta}{\gamma}\frac{r_b}{2LAI}\frac{R_n}{L} \qquad \text{(Eq. 3.4.15)}$$

whereby equation (3.4.13) may be re-written as:

$$E = g_{tr}(\chi_{eff} - \chi_a) \qquad \text{(Eq. 3.4.16)}$$

Here we consider a fully developed tomato canopy ($LAI \approx 2$; $r_b \approx 200$ s m^{-1}; $r_l \approx 200$ s m^{-1} at day and 2000 at night). For an ambient air temperature of 20°C, s/g is about 2. It is easy to calculate, then, that g_{tr} varies from as little as 1.5 10^{-3} at night up to about 5 10^{-3} during a sunny day. The contribution of the radiative gain to the driving force for transpiration (the apparent concentration gradient), can similarly be quantified, as χ^* is about 16 g m^{-3} at 20°C. In full sunshine R_n is some 400 W m^{-2} for a greenhouse crop, then $\chi_{eff} \approx 2\chi_a^*$, whereas it is not unusual to find that $\chi_{eff} \approx \chi_a^*$ at night.

3.4.3.4 Transpiration of a greenhouse crop

In order to calculate the transpiration rate of a greenhouse crop canopy for equation (3.4.13), under given conditions, R_n, r_l and r_b have to be known. Once some parameters of the crop are known the fraction of incident PAR radiation actually absorbed by a crop canopy can be calculated, as in section 2.2.1. In section 3.3 it has been shown how this can be extended to include also long wave and near infrared radiation. On the other hand, stomatal resistance has been quantified in section 2.2.2.

In section 3.2 it has been shown that convective heat (or vapour) exchange of almost every element of a greenhouse is governed by processes collectively described as mixed convection. That means that the air movement around the surface, the physical carrier of energy, cannot be credited exclusively to the presence of a well established air stream (forced convection) nor to the buoyancy of air parcels whose temperature is affected by the temperature of the surface itself (free, or natural convection). In a regime of mixed convection, both processes contribute to the energy exchange, although their relative importance may shift, due to the fluctuating nature of the resulting air stream.

Indeed, when attempting to estimate the boundary layer resistance of leaves within a canopy, a good deal of "streamlining" is imperative. First, leaves are commonly regarded as flat plates, whose typical dimension is well-defined. Secondly, a given leaf angle distribution is assumed. Finally, the nature of air movement about the surface (whether turbulent or laminar) is regarded as exclusively determined by stability theory, once the surface has been so described, disregarding the effect of indented edges, hair, non-flatness and roughness of the leaf itself or of the presence of nearby leaves. Once one chooses to live with this, handbooks (e.g. Monteith & Unsworth, 1990) provide the appropriate values of the coefficients of the relevant equations. Accordingly, boundary layer resistance for a horizontal plate of dimension l (m), immersed in an air stream of speed u (m s^{-1}), being ΔT degrees warmer or cooler than the air is:

$$r_{b,forced} \cong 300 \left(\frac{l}{u}\right)^{1/2}$$

$$r_{b,free} \cong 1000 \left(\frac{l}{\Delta T}\right)^{1/4} \qquad \text{(Eq. 3.4.17)}$$

provided l, u and ΔT are, respectively, a few centimetres, a few centimetres per second and a few degrees as, indeed, is the case for most greenhouse crops. It is easy to see that both forms of equation (3.4.17) would give a boundary layer resistance of a few hundred s m^{-1}, though the forced convection form is likely to result in the smallest estimate, except with uncommonly large leaves such as those of a cucumber crop.

One should conclude then, that in most cases heat and mass transfer around leaves within a greenhouse canopy is driven by air movement originating elsewhere in the house, and can be somewhat enhanced by some difference in temperature between leaves and air (Stanghellini, 1993). This suits us well indeed, as:

- mean air movement in a greenhouse is fairly constant
- boundary layer resistance varies only with the square root of a variation in air movement, and
- the transpiration rate is not very sensitive to boundary layer resistance either (Stanghellini, 1988).

So it seems accurate enough for most purposes to use a constant value of the resistance as given by the forced convection form of equation (3.4.17), calculated for a prevailing mean air movement in the greenhouse. Accordingly, the influence of the ambient on the transpiration rate of a greenhouse crop takes place primarily through three variables: radiation, air temperature and humidity. This has been shown by Stanghellini & Van Meurs (1992), who were able to achieve a desired transpiration rate by controlling air temperature and humidity in a greenhouse, once available radiation was taken into account.

3.4.4 Condensation

The flux density of water condensing at the cover surface ($\Phi_{w,cnd}$, here C for simplicity) can be written similarly to equation (3.4.5):

$$C = g_{cnd}(\chi_a - \chi^*_r) \tag{Eq. 3.4.18}$$

And, of course, as no vapour can leave the dry cover:

$$C = 0 \text{ for } \chi_a < \chi^*_r \tag{Eq. 3.4.19}$$

The mass transfer conductance g_{cnd} can be calculated (section 3.2):

$$g_{cnd} = \frac{\mathbb{D}_w}{l} \text{Sh} \cong 2.49\ 10^{-5} \frac{\text{Sh}}{l} \tag{Eq. 3.4.20}$$

where $\mathbb{D}_w$ is the molecular diffusion coefficient of water vapour in air and l a typical dimension of the cover surface. In order to account for the effect of vapour concentration gradients on the density (and thus buoyancy) of air, we can here quite conveniently dispose of the of assigning l a value, as it turns out, since, equations (3.2.17) and (3.2.19):

$$Sh = Le^{1/3} Nu \approx 0.96 \cdot 0.13 Gr^{1/3} \tag{Eq. 3.4.21}$$

with:

$$\text{Gr} \cong 1.47\ 10^8\ l^3 (\tilde{T}_a - \tilde{T}_r) \tag{Eq. 3.4.22}$$

where $\tilde{T}$ is the virtual temperature (e.g. Monteith & Unsworth, 1990). As C is referred to unit ground area, one has to take into account A_r/A_g, the ratio of cover area to ground area. Accordingly:

$$g_{cnd} \approx \frac{A_r}{A_g} 1.64\ 10^{-3} (\tilde{T}_a - \tilde{T}_r)^{1/3} \tag{Eq. 3.4.23}$$

A_r/A_g is about 1.1, for a Venlo house, and the factor on the far right is a weak function of the temperature difference, commonly (that is, when there is condensation at the surface) with a value between 1 and 2.5. Hence, g_{cnd} is some $3\ 10^{-3}$ m s^{-1}.

3.4.5 Ventilation

We can preserve the same analogy, whereby the vapour flux due to ventilation ($\Phi_{w,vnt}$, here V) can be written as:

$$V = g_{vnt}(\chi_a - \chi_o) \tag{Eq. 3.4.24}$$

with χ_o the absolute humidity outdoor.

Let *n* be the volume changes per hour. Then,

$$g_{vnt} = h \frac{n}{3600} \qquad \text{(Eq. 3.4.25)}$$

We will see (section 4.4.3) that, for leakage ventilation, *n* is 1 h^{-1} or less, which would make g_{vnt} some 1 10^{-3} m s^{-1}. Fully open roof ventilators give a *n* of some 20 or more h^{-1}, depending on other conditions and g_{vnt} would be 2 10^{-2} m s^{-1} or more. With ventilators open only a crack (a common way of controlling humidity) g_{vnt} is some 10^{-3}, which is comparable with both g_{tr} and g_{cnd}.

3.4.6 Vapour concentration of ambient air

After quantifying all the vapour fluxes, the vapour balance of the greenhouse air equation (3.4.1), can be re-written as follows:

$$h \frac{d\chi_a}{dt} = g_{tr}(\chi_{eff} - \chi_a) - g_{cnd}(\chi_a - \chi^*_r) - g_{vnt}(\chi_a - \chi_o) \qquad \text{(Eq. 3.4.26)}$$

An apparent cumulative transfer conductance, g_{tot} may be defined by combining the coefficients of ambient absolute humidity χ_a in equation (3.4.26),

$$g_{tot} \equiv g_{tr} + g_{cnd} + g_{vnt} = \frac{2LAI}{\left(1 + \frac{s}{\gamma} r_b + r_l\right)} + 1.64\ 10^{-3} \frac{A_r}{A_g} \left(\tilde{T}_a - \tilde{T}_r\right)^{1/3} + h \frac{n}{3600} \qquad \text{(Eq. 3.4.27)}$$

whereas the terms independent from the latter may be regrouped in a mass flux density *G*, the water vapour "gain" of the ambient air, thus defined:

$$G \equiv g_{tr}\chi_{eff} + g_{cnd}\chi^*_r + g_{vnt}\chi_o$$

$$= g_{tr}\left(\frac{s}{\gamma}\frac{r_b}{2LAI}\frac{R_n}{L} + \chi^*_a\right) + g_{cnd}\chi^*_r + g_{vnt}\chi_o \qquad \text{(Eq. 3.4.28)}$$

Therefore, the ambient vapour concentration balance, equation (3.4.26), reads:

$$G - g_{tot}\chi_a - h\frac{d\chi_a}{dt} = 0 \qquad \text{(Eq. 3.4.29)}$$

In fact, an analytical solution to this differential equation exists only if both *G* and g_{tot} are either unaffected by χ_a or, at least, meet some requirements. If these constraints, the consequences of which will be discussed later, are met, ambient vapour concentration at any time is given by the general solution of equation (3.4.29):

$$\chi_a(t) = \chi_a(\infty) - [\chi_a(\infty) - \chi_a(0)]\, e^{-t/\tau} \qquad \text{(Eq. 3.4.30)}$$

with the vapour concentration at equilibrium ($t \rightarrow \infty$) given by:

$$\chi_a(\infty) = G/g_{tot} \qquad \text{(Eq. 3.4.31)}$$

and τ, the time constant of the system by:

$$\tau = h/g_{tot} \qquad \text{(Eq. 3.4.32)}$$

In fact, strictly speaking, all components of g_{tot} change following a variation in ambient humidity. Stomata are known to react to it (section 2.2.2), virtual temperature of air (and thus g_{cnd}) is a function of it, and the resulting variation in buoyancy will affect g_{vnt} too. None of these, however, is the primary factor determining any of the conductances and, for the sake of the present discussion, g_{tot} could be approximated by a linear function of ambient humidity. It can be shown, though it will not be done here, that in this case true ambient humidity is approximated very nearly by equation (3.4.31). As far as G is concerned, both χ^*_a and χ^*_r (that is, T_a and T_r) are independent of condensation or ventilation only in so far as the climate control system is able to deliver the desired temperature setpoint, which modern systems manage rather well. Accordingly, for the sake of the following discussion, ambient humidity will be calculated by means of equation (3.4.31).

First of all, we need to determine the size of τ, that is the time ambient humidity would take to get through 2/3 of a variation due to changing conditions. It may be observed that $g_{tr} + g_{cnd}$ is fairly constant (in full sunshine there is no condensation, after all) and roughly equal to $5\ 10^{-3}$ m s^{-1}. Hence, g_{tot} is confined, to say, between some $6\ 10^{-3}$ and $2\ 10^{-2}$ m s^{-1}. With a mean greenhouse height of 3.6 m, equation (3.4.32) gives a τ contained between 2 and 10 minutes, which is comparable with the scanning and actuation time of most climate control systems. Accordingly, for the purpose of the control of humidity, ambient vapour concentration can safely be calculated by means of equation (3.4.31).

That will be done here, for a few simple instances. We will confine ourselves to a medium-size leaf crop such as tomato or roses, kept in a typical Venlo house ($h \approx 3.6$ m), the parameters having been listed before. Transpiration rates will be determined through a model (Stanghellini, 1987) based on equation (3.4.12). In order to compute ventilation rates as a function of window opening, De Jong's (1990) model will be used, assuming a wind speed of 2 m s^{-1}. Finally, for the condensation rates, roof temperature will be calculated as a "weighted mean" (2/3 outdoor and 1/3 indoor temperature), an approximation roughly valid for single-glass covers. Ambient humidity resulting from setpoint temperature and window opening, is shown in the upper half of Figure 3.4.1, for three typical conditions, the corresponding transpiration rate being in the panels underneath.

The example on the far left refers to a fine, cold winter day (sun radiation, I=350 W m^{-2}; air temperature, T_o, and relative humidity outdoor, RH_o, 5 °C and 45%, respectively, that is χ_o=3 g m^{-3}; the crop having LAI=2). The upper panel makes clear that humidity inside the greenhouse is determined by supply (temperature), as well as removal. Whence it may be deduced that any ventilation rate does "dry" the air, at any temperature. However, crop transpiration, the true aim of humidity control, follows mainly from air temperature and ventilation appears to have an effect only for quite small rates. The story is quite different for a sunny, summer day (I=900 W m^{-2}; T_o=25°C; RH_o=45%, that is χ_o= 10 g m^{-3}; with LAI=3, which is more common in summer), pictured in the central part of the Figure 3.4.1. Both humidity and transpiration are largely determined by ventilation rate. Finally we consider that headache of most growers: control of ambient humidity during relatively warm autumn nights (the panels on the right of Figure 3.4.1). We will take T_o=12°C; RH_o=80% (χ_o=8 g m^{-3}), also with a LAI=3.

Then, χ_a is some 10–12 g m^{-3}, little affected by ventilation rate or, for that matter, by anything else, including air temperature. A small opening of the ventilators (instead of none), however, could "stimulate" crop transpiration, though one might wonder if the same effect could not be achieved by the corollary pipe heating alone, that is, by raising the temperature setpoint without ventilation.

3.4.7 Conclusion

Vapour content of air within a greenhouse is determined by many factors, of which crop transpiration, condensation at the cover surface and ventilation are the most important. In turn, transpiration and photosynthesis (the latter through stomatal reaction) are affected by ambient humidity. Humidity is also the single most important factor causing outbreak of some diseases, it is not surprising, therefore, that modern climate control systems include some means of humidity control.

As it has been illustrated here above, any attempt to modify ambient humidity should take into account the many feed-back effects that a change in conditions would set in motion. In fact, in many conditions the three processes (transpiration, condensation, ventilation) contribute almost equally to the resulting balance, although sometimes one of the processes may predominate. The most efficient means of climate manipulation, in order to achieve a desired goal, is, therefore, dictated by the prevailing conditions at any time.

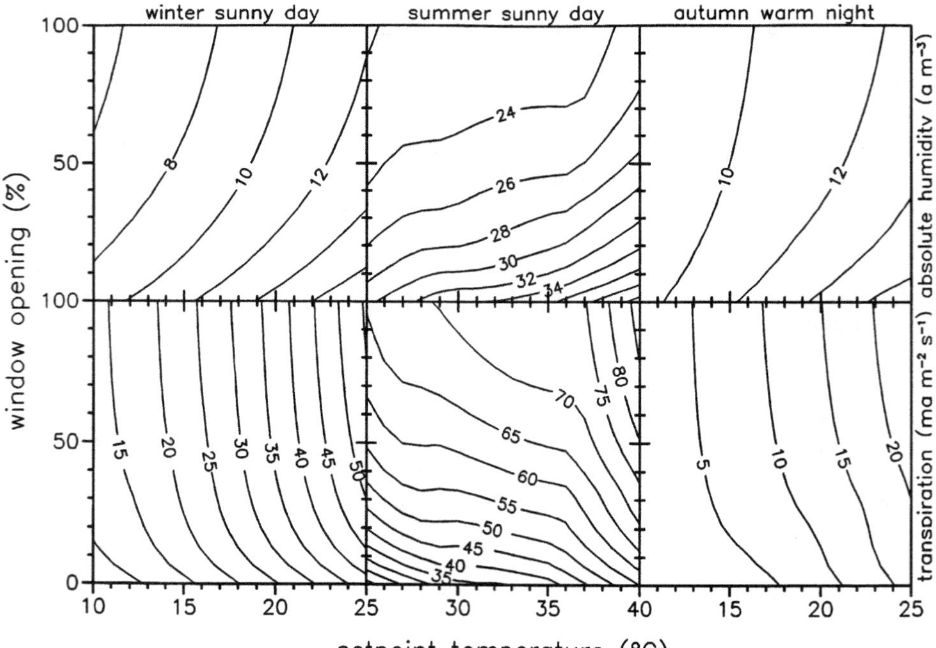

Figure 3.4.1. Ambient humidity (g m^{-3}, upper part) and crop transpiration (mg m^{-2} s^{-1}, lower part), under three typical sets of conditions (left, mid and right sections), further detailed in the text. Contours illustrate the effect of temperature within the greenhouse (horizontal axis) and ventilator opening (vertical axis). Values were calculated by means of equation (3.4.31), the variables being given by a transpiration (Stanghellini, 1987) and a ventilation model (De Jong, 1990), as explained in the text.

3.5 Carbon dioxide balance

E.M. NEDERHOFF

3.5.1 Introduction

Carbon is a principal element for life, as it comprises about 40–50% of the dry matter of living organisms (Levanon *et al.*, 1986). Carbon is acquired from the environment by green plants, in the form of carbon dioxide taken up from the air. The average CO_2 concentration of the ambient air is around 350 μmol mol^{-1} at present on an annual basis, but it shows a diurnal and an annual cycle, and moreover a rising trend, due to combustion of fossil fuels. A reconstruction of the CO_2 concentration in previous times and prognoses for future atmospheric CO_2 are presented in Figure 3.5.1.

CO_2 enters the plant through pores (stomata) in the leaf surface, where it is assimilated into carbohydrates and other plant substances. Relevant aspects of these processes are presented in Chapter 2. Normally the rate of CO_2 assimilation is limited by the amount of CO_2 present in the vicinity of the plant. Hence the assimilation, and thus the crop growth and production, can be accelerated by supply of additional CO_2 in the surrounding air (section 2.2.1).

Increasing the CO_2 concentration can be achieved relatively easily in a closed environment such as a greenhouse. When a greenhouse is ventilated to prevent excessive temperatures, a marked increase in the CO_2 level can be achieved only at the expense of much CO_2.

The present section (3.5) deals with some basic aspects of the CO_2 balance and the relevant properties of CO_2. The technical facilities for CO_2 enrichment are discussed in section 4.6 and the CO_2 control in section 5.4.5.

3.5.2 Some basic features of carbon dioxide

Under normal conditions CO_2 is a colourless, acid, non-toxic gas, with a boiling point of -78.5 °C. The molar mass of CO_2 is 44.01 kg kmol^{-1}. The gas density at normal atmospheric pressure (101.3 kPa) is 1.98 kg m^{-3} at 0 °C and 1.83 kg m^{-3} at 20 °C.

The amount of CO_2 in the air can be defined in different dimensions and expressed in different units (Table 3.5.1). Unit conversion can be made by using the ideal gas equation 3.2.22.

In this section the amount of CO_2 is expressed in vpm (which equals mmol mol^{-1}, ml l^{-1} etc.) because this unit is commonly used in horticultural practice. From this amount of CO_2 the CO_2 concentration can be calculated.

3.5.3 CO_2 in greenhouses

The CO_2 conditions in the (semi-)closed environment of a greenhouse are usually different from those outside. It is commonly observed in greenhouses that the CO_2 concentration drops below the ambient level. This so-called CO_2 depletion (Drakes, 1984; Heij & De Lint, 1984) is caused by CO_2 uptake by the crop and insufficient CO_2 influx (no supply and only little air refreshment). The CO_2 concentration then declines until an equilibrium is established, which is sometimes as low as 150 vpm. In the case of CO_2 depletion, air exchange implies influx of CO_2.

CO_2 depletion is very unfavourable for plant growth and production (Chapter 2), so the principal objective of CO_2 enrichment during ventilation is preventing CO_2 depletion (Slack & Hand, 1986). Further elevation of the CO_2 concentration is often attempted only when the ventilation rate is not

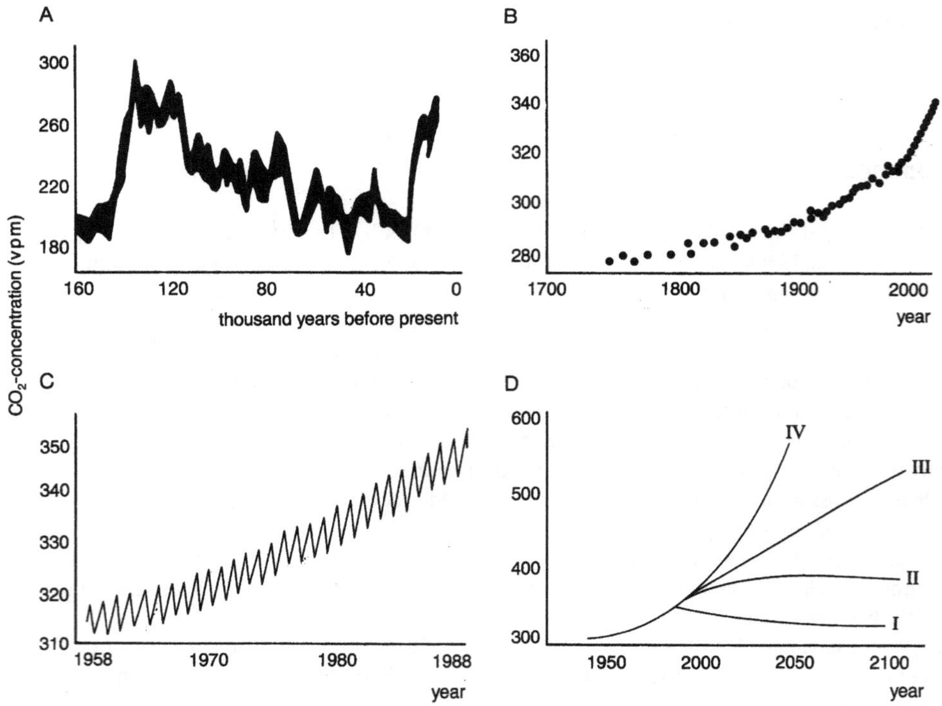

Figure 3.5.1. Atmospheric CO_2 concentration as reconstructed for the past and predicated for the future: (a) from 150,000 years ago to the present; (b) from 1650 to 1990; (c) from 1958 to 1990 and (d) from 1990 to 2100 with various levels of CO_2 emission assumed for 2100: between half (I) and eight times (IV) the current emission levels. Sources: (a), (b) and (c) after Scurlock & Hall (1991) and (d) after Langeweg (1990).

Table 3.5.1. Commonly used units for CO_2 concentration and unit conversion.

	Dimension	Units
a.	ratio of CO_2 to air, in volume:	1 vpm (ppm) = 1 ml l_{-1} = 1 µl m^{-3} =1 cm m-3 = 0.0001 % (volume)
b.	ratio of CO_2 to air, in moles:	1 ppm = 1 µmol mol^{-1} = 1 mmol $kmol^{-1}$
c.	ratio of CO_2 to air, in mass:	1 mg kg^{-1} = 0.0001 % (mass)
d.	mass of CO_2 per volume of air:	1 mg m^{-3} = 1 mg l^{-1}
e.	moles CO_2 per volume of air:	1 µmol m^{-3}
f.	partial pressure:	1 Pa = 10 mbar

Conversion (at 20 °C and 101.3 kPa):
1 vpm = 1 ppm = 1.53 mg kg-1 = 1.53 10^{-6} kg kg^{-1} = 1.83 mg m^{-3} = 1.83 10^{-6} kg m^{-3} = 41.6 mmol m^{-3} = 41.6 10^{-9} kmol m^{-3} = 0.101 Pa

too high. The optimal level is about two to three times the ambient level of 350 vpm (Mortensen, 1987).

The most simple method of reducing depletion is to increase the air exchange. It has also been suggested that fresh air be blown into the greenhouse through plastic ducts (Langer *et al.*, 1990). The CO_2 concentration used to be increased in the past by mulching the soil (Levanon *et al.*, 1986). At present the CO_2 concentration is controlled by supply of CO_2. The various methods of CO_2 enrichment are discussed in section 4.6.

3.5.4 CO_2 balance

The CO_2 level that can be maintained in a greenhouse by enrichment depends on the various terms in the CO_2 balance: the supply rate, the air exchange rate with outside, the CO_2 assimilation by the crop and the CO_2 production by degradation of organic material. This latter is assumed to be negligible. To calculate the required CO_2 supply capacity, the (dynamic) CO_2 balance of the greenhouse air must be determined according to equation (3.2.4). Here a simple method is presented for static conditions (with the assumption of a constant equilibrium CO_2 concentration). The supply rate ($F_{c,in}$) must compensate the rate of photosynthesis ($\Phi_{c,p}$) plus the rate of CO_2 exchange by air exchange ($\Phi_{c,vnt}$), so:

$$\Phi_{c,in} = \Phi_{c,p} + \Phi_{c,vnt} \qquad \text{(Eq. 3.5.1)}$$

with all terms per m^2 greenhouse area, so in kg m^{-2} s^{-1}. According to equation (3.2.9) the exchange of CO_2 depends on the volumetric air exchange rate (q_v in m^3 s^{-1}) and the CO_2 concentration difference between greenhouse air and ambient air ($c_{c,i} - c_{c,a}$ in kg m^{-3}) per m^2 greenhouse area.

The air exchange can be calculated with an appropriate model for ventilation (e.g. Bot, 1983; Nederhoff *et al.*, 1984; De Jong, 1990; Fernandez & Bailey, 1992) as given in section 3.3.4.

The rate of crop photosynthesis varies from 1 g m^{-2} h^{-1} CO_2 (units often used in horticultural practice, easy to translate into SI units) or less in dark weather, to about 4–5 g m^{-2} h^{-1} under favourable light conditions and ambient CO_2 and to about 7 g m^{-2} h^{-1} under high light and high CO_2 (Nederhoff, 1994). For a detailed estimation, a simulation model for crop photosynthesis must be applied (Acock *et al.*, 1976; Boote & Loomis, 1991). When the CO_2 concentration inside the greenhouse and outside are equal, the CO_2 loss by air exchange is zero and the supply need only compensate the net photosynthesis at the actual CO_2 concentration (equation (3.5.1)). This situation can be used to estimate the photosynthesis under the prevailing conditions.

Under specific conditions, the required CO_2 supply rate $\Phi_{c,in}$ to achieve a desired CO_2 level, or the achievable level with a given supply rate, can be approximated with equation (3.5.1) where the ventilation exchange is calculated using the ventilation model and the photosynthesis rate is estimated. For example, at 350 vpm, $\Phi_{c,in}$ must be about 3 g m^{-2} h^{-1} (practical units) while for maintaining 700 vpm at 20% window opening and 4 m s^{-1} wind speed, $\Phi_{c,in}$ must be around 20 g m^{-2} h^{-1}. With higher ventilation rates or higher CO_2 target levels, the required supply rate is even higher. The CO_2 supply will be stopped under these conditions.

A generally recommended standard (minimum) supply rate is 4.5 g m^{-2} h^{-1}, or the equivalent, the flue gas of 25 m^3 ha^{-1} h^{-1} natural gas (Hand, 1982; Van Berkel & Verveer, 1984). This rate is usually sufficient to maintain a high CO_2 level (about 1000 vpm) in a closed greenhouse and to prevent severe CO_2 depletion in a ventilated greenhouse. In many cases the supply rate is restricted to this 4.5 g m^{-2} h^{-1} for practical or economic reasons.

3.6 Synthesis

G.P.A. BOT AND N.J. VAN DE BRAAK

With the quantitative description of the main exchange processes the energy and mass balances can be calculated for the greenhouse system. Using the same approach the balances can be set up over representative, homogeneous parts of the greenhouse, such as single cover, greenhouse air, crop and layered soil (section 3.3.1). The mass balances for water vapour and carbon dioxide have to be set up over the air compartment(s) only. Where a thermal screen is used two air compartments have to be considered, one between cover and screen and one below the screen.

The energy balance for the compartment i with temperature T_i, exchanging energy to j neighbouring compartments with temperature T_j, can be represented in general as:

$$(h\,\rho\,C_p)_i\, dT_i/dt = \Sigma(-\Phi_{h,ij}) - LE_i + S_i \qquad \text{(Eq. 3.6.1)}$$

where h is the volume to area ratio of the compartment (m), $+S_i$ the absorbed solar radiation (W m^{-2}) in compartment i and $-LE_i$ the energy needed for transpiration (W m^{-2}) in compartment i (or $+LE_i$ the energy released due to condensation). The energy flux densities $\Phi_{h,ij}$ between the compartments i and j due to the various transport mechanisms can be related to the temperatures mentioned as indicated earlier. This leads to the general expression:

$$(h\,\rho\,C_p)_i\, dT_i/dt = \Sigma\, \alpha_{h,ij}(T_j - T_i) - LE_i + S_i \qquad \text{(Eq. 3.6.2)}$$

Here $\alpha_{h,ij}$ is determined by the type of mechanism and will generally contain a non-linearity. Figure 3.3.1 illustrates the various fluxes to and from each compartment to the neighbouring compartments. In this notation the boundary conditions are considered to be the temperatures of surrounding compartments.

The same approach leads to the mass balance (water and carbon dioxide) over the air compartments:

$$h_i\, dc_i/dt = \Sigma(k_{m,ij}(c_j - c_i)) \qquad \text{(Eq. 3.6.3)}$$

The water balance is linked to the energy balance through the term LE_i in which the evaporative mass flux density E_i is combined with the heat of transpiration L (equation (3.2.18)).

The energy and mass balances for all compartments result in a set of first order differential equations, describing the state variables temperature and mass concentration in the compartments as functions of time with given initial and boundary conditions. This set of differential equations for the various state variables can also be written in vector notation (Chapter 6).

There are two aspects that deserve attention. Firstly the right hand side of the equations contain the transfer coefficients a_h and k_m (k_m linked to a_h via equation (3.2.20)). For the mechanisms of natural convection and radiation these coefficients depend on both the temperature difference and level, and so contain non-linearities. For ventilation and forced convection they depend on greenhouse or outside parameters such as window opening and wind speed which vary over time. So the equations are also linked to independent varying parameters.

Another aspect is in the different time scales of the various processes involved. To argue that, for equation (3.6.2) or (3.6.3) the analogue electrical network equation can be set up:

$$Cap_i\, dv_i/dt = (1/R_{ij})\,(v_j - v_i) \qquad \text{(Eq. 3.6.4)}$$

with Cap_i the capacity, v the voltage and R_{ij} the resistance between the compartments i and j.

The combination Cap_iR_{ij} is known as the time constant (or relaxation time) of this exchange process between the compartments i and j. In equation (3.6.2) various combinations of thermal capacities and thermal resistances appear. The most striking difference is that between combinations of small capacities and small resistances such as that of the leaves and the air with time constants in the order of magnitude of one minute and combinations of large capacities and large resistances such as that of the deeper soil layers with time constants in the order of hours (Bot, 1989). When controlling a greenhouse, in order to be able to react quickly enough to the fast variations, the small time constants must be taken into account. However, in most control applications one is not interested in all details included in the complete balances, but in the dynamic behaviour of only certain state variables. Consequently for control purposes simplified balances with linear terms have proven to be sufficient (Van Henten, 1994).

This will be dealt with in Chapter 6. If only slow variations have to be followed, such as, for instance, in heat consumption studies or in general design studies, only the large time constants have to be taken into account for accurate calculations (Breuer, 1983; Houter, 1990). If one needs the full details, a complete set of balances has to be solved (Bot, 1983; Breuer & Van de Braak, 1994; De Zwart, 1994).

References

Acock, B., D.W. Hand, J.H.M. Thornley & J. Warren Wilson, 1976. Photosynthesis in stands of green peppers, an application of empirical and mechanistic models to controlled-environment data. Annals of Botany 40: 1293–1307.

Baver, L.D. & W.R. Gardner, 1972. Soil Physics. Wiley, New York, 489 pp.

Bell, G., J. Jung & H. Dehnz, 1990. CO_2-Gewinnung durch Nutzung von Rauchgasen aus mit Rohbraunkohle gefeuerten Heizkesselanlagen. Gartenbau 37(3): 77–78. (in German).

Bird, R.B., W.E. Stewart & E.N. Lightfoot, 1960. Transport Phenomena. John Wiley and Sons, New York, 780 pp.

Boote, K.J. & R.S. Loomis (Eds), 1991. Modeling crop photosynthesis from biochemistry to canopy. Crop Science Society of America. Madison, CSSA special publication 19.

Bot, G.P.A., 1983. Greenhouse climate: from physical processes to a dynamic model. PhD thesis, Wageningen Agricultural University, Wageningen, 240 pp.

Bot, G.P.A., 1989. Greenhouse simulation models. Acta Horticulturae 245: 315–325.

Breuer, J.J.G., 1983. Rekenmodel energiebehoefte in kassen *(Computer model energy requirement of greenhouses)* (2de uitgave, 2 delen). IMAG Rapport 49, IMAG-DLO, Wageningen. (in Dutch).

Breuer, J.J.G. & N.J. Van de Braak, 1994. Beschrijving van een statisch en een dynamisch simulatiemodel voor kassen *(Description of a static and dynamic simulation model for greenhouses)* IMAG-DLO Rapport 94-9, IMAG-DLO, Wageningen (in press). (in Dutch).

Chalabi, Z. & J.E. Fernandez, 1992. Spatio-temporal responses of a glasshouse to gaseous enrichment. Journal of Agricultural Engineering Research 51: 139–151.

Critten, D.L., 1992. Greenhouse transmission models. Proceedings International Workshop on greenhouse crop models. Avignon, August 1991. Acta Horticulturae. (in press).

De Jong, T., 1990. Natural ventilation of large multispan greenhouses. PhD thesis, Wageningen Agricultural University, Wageningen, 116 pp.

De Zwart, F., 1993. Determination of direct transmission of a multispan greenhouse using vector algebra. Journal of Agricultural Engineering Research 56: 39–49.

De Zwart, F., 1995. A model to evaluate the performance of heating devices in horticulture (prel. title). PhD thesis, Wageningen Agricultural University, Wageningen, (in press).

Drakes, G.D., 1984. Prevention of CO_2 depletion in tomatoes. Acta Horticulturae 162: 245–247.

Feodorow, S. & U. Jacobi, 1990. Einsatz von CO_2-Flüssiggas bei der Produktion von Tomaten in der LPG Gemüseproduktion "Edwin Hoernle" Berlin-Marzahn. Gartenbau 37: 70–71. (in German).

Fernandez, J.E. & B.J. Bailey, 1992. Measurement and prediction of greenhouse ventilation rates. Agricultural and Forest Meteorology 58: 229–245.

Gebhart, B., 1961. Heat transfer. Mc Graw-hill, New York, 764 pp.

Gijzen, H. & J.A. ten Cate, 1988. Prediction of the response of greenhouse crop photosynthesis to environmental factors by integration of physical and biochemical models. Acta Horticulturae 229: 251–258.

Gijzen, H., J.G. Vegter & E.M. Nederhoff, 1990. Simulation of greenhouse crop photosynthesis: validation with cucumber, sweet pepper and tomato. Acta Horticulturae 268: 71–80.

Goeijenbier, P., 1986. Gebruik recirculatie-ventilatoren. Tuinderij 66(24): 46–47. (in Dutch).

Hand, D.W., 1982. CO_2 enrichment, the benefits and problems. Scientific Horticulture 33: 14–43.

Hand, D.W., 1990. CO_2 enrichment in greenhouses: problems of CO_2 acclimation and gaseous air pollutants. Acta Horticulturae 268: 81–101.

Heuvelink, E., L.G.G. Batta & T.H.J. Damen, 1995. Transmission of solar radiation by a multispan Venlo-type glasshouse: validation of a model. Agricultural and Forest Meteorology, 19pp. (in press).

Heij, G. & P.J.A.L. De Lint, 1984. Prevailing CO_2 concentrations in glasshouses. Acta Horticulturae 162: 93–99.

Houter, G., H. Gijzen, E.M. Nederhoff & P.C.M. Vermeulen, 1989. Simulation of CO_2 consumption in greenhouses. Acta Horticulturae 248: 315–320.

Houter, G., 1990. Simulation of CO_2 consumption, heat demand and crop production of greenhouse tomato at different CO_2 strategies. Acta Horticulturae 268: 157–164.

Jacobi, U., M. Zschoche, W. Recker & A. Vierig, 1990. Einsatz von CO_2-Generatoren zur CO_2-Düngung auf der Basis von Erd-, Bio- und anderen Gasen. Gartenbau 37(3): 74–77. (in German).

Kiel, A.J., 1990. CO_2 enrichment with natural gas fired hot-air heaters. Acta Horticulturae 268: 111–120.

Kondratyev, K.Ya., 1972. Radiation processes in the atmosphere. WMO report 309, World Meteorological Organization, Geneva, 214 pp.

Langer, K.-H., S. Schmidt & W. Dietrich, 1990. Einrichtung und Nutzung eines bodennahen Verteilungssystems für CO_2 in Gewächshäusern. Gartenbau 37(3): 78–79. (in German).

Langeweg, F. (Ed.), 1990. Zorgen voor morgen, RIVM. Samson H.D. Tjeenk Willink, Alphen aan de Rijn. (in Dutch).

Levanon, D., B. Motro & U. Marchaim, 1986. Organic materials degradation for CO_2 enrichment of greenhouse crops. In: H.Z. Enoch & B.A. Kimball (Eds), Carbon dioxide enrichment of greenhouse crops. Volume I, Status and CO_2 sources. CRC Press Inc., Florida, p. 123–145.

Monteith, J.L., 1965. Evaporation and environment. In: The State and Movement of Water in Living Organisms. 19th Symposium of the Society of Experimental Biology: 205–234.

Monteith, J.L. & M.H. Unsworth, 1990. Principles of environmental physics. Edward Arnold, London, 291 pp.

Morgan, V.T., 1975. The overall convective heat transfer from smooth circular cylinders. Advances in heat transfer 11: 199–264.

Mortensen, L.M., 1987. Review: CO_2 enrichment in greenhouse crops: Crop responses. Scientia Horticulturae 33: 1–25.

Nawrocki, K.R., 1985. Meting warmteoverdrachts-coëfficient voor convectie van verwarmingspijpen in kassen. IMAG rapport 73, IMAG-DLO, Wageningen. (in Dutch).

Nederhoff, E.M., J. Van Vooren & A.J. Udink ten Cate, 1984. A method to determine ventilation in greenhouses. Acta Horticuturae 148: 345–350.

Nederhoff, E.M. & J.G. Vegter, 1994. Photosynthesis of stands of tomato, cucumber and sweet pepper measured in greenhouses under various CO_2 concentrations. Annals of Botany (in press).

Penman, H.L., 1948. Natural evaporation from open water, bare soil and grass. Proceedings of the Royal Meteorological Society, London, A 193: 120–145.

Scurlock, J. & D. Hall, 1991. The carbon cycle. New Scientist 132.1793 (2 Nov.): 1–4.

Slack, G. & D.W. Hand, 1986. Weighing up the cost-effectiveness of summer enrichment. Grower 105(15): 15–18.

Stanghellini, C., 1983. Calculation of the amount of energy released by heating pipes in a greenhouse and its allocation between convection and radiation. IMAG research report 83-3, IMAG-DLO, Wageningen.

Stanghellini, C., 1987. Transpiration of greenhouse crops: an aid to climate management. PhD thesis, Wageningen Agricultural University, Wageningen, 150 pp.

Stanghellini, C., 1988. Microclimate and transpiration of greenhouse crops. Acta Horticulturae 229: 405–410.

Stanghellini, C., 1993. Mixed convection above greenhouse crop canopies. Agricicultural and Forestry Meteorology 65: 111–117.

Stanghellini, C. & W.T.M. Van Meurs, 1992. Environmental control of greenhouse crop transpiration. Journal of Agricultural Engineering Research 51: 297–311.

Stoffers, J.A.,1967. Lichtdoorlatendheid van met vlakke materialen bedekte warenhuizen. ITT publikatie 14. IMAG, Wageningen, 35 pp. (in Dutch).

Van Berkel, N., 1975. CO_2 from gas-fired heating boilers: its distribution and exchange rate. Netherlands Journal for Agricultural Science 23: 202–210.

Van Berkel, N. & J.B. Verveer, 1984. Methods of CO_2 enrichment in the Netherlands. Acta Horticulturae 162: 227–231.

Van Loon, W.K.P., I.A. Van Haneghem & J. Schenk, 1989. A new model for the non-steady probe method to measure thermal properties of porous media. International Journal of Heat Mass Transfer 32: 1473–1481.

Van Henten, E.J., 1994. Greenhouse climate management: an optimal control approach. PhD thesis, Wageningen Agricultural University, Wageningen, 329 pp.

Wartena, L., C.L. Palland & G.H.L. Vossen, 1973. Checking of some formulae for the calculation of long wave radiation from clear skies. Archive für Meteorologie, Geophysik und Bioklimatologie B 21: 335–348.

List of symbols

a	thermal diffusivity (m s^{-2})
A	area (m^2)
B	absorbtion factor (-)
c	concentration (kg m^{-3})
C	condensation (kg m^{-2} s^{-1})
Cap_h	thermal capacity (J K^{-1})
C_p	specific heat (J kg^{-1} K^{-1})
C_1	coefficient (-)
d	thickness (m)
D	vapour pressure saturation deficit (Pa)
$\mathbb{D}$	diffusivity (m^2 s^{-1})
e	vapour pressure (Pa)
E	evaporative water vapour flux density (kg m^{-2} s^{-1})
f	frequency (s^{-1})
F	view factor (-)
g	acceleration due to gravity (m s^{-2})
$g_{...}$	... conductance (m s^{-1})
G	water vapour gain (kg m^{-2} s^{-1})
h	height (m)
H	flux density of sensible heat (W m^{-2})
I	solar irradiation (W m^{-2})
k	transfer coefficient (m s^{-1})
l	characteristic dimension (m)
l_w	wavelength (m)
L	heat of evaporation (J kg^{-1})
M	molar mass (kg $kmol^{-1}$)
n	ventilation volume changes per hour (h^{-1})
p	produced physical quantity (..s^{-1})
q	flux of physical quantity (..s^{-1})
Q	amount of physical quantity (..)
r	resistance (s m^{-1})
R	universal gas constant (J $kmol^{-1}$ K^{-1})
R_n	net radiation (W m^{-2})
s	slope of the saturated vapour pressure curve (Pa K^{-1})
S	absorbed solar radiation (W m^{-2})
t	time (s)
T	temperature (K)
$\tilde{T}$	virtual temperature (K)
u	velocity (m s^{-1})
V	volume (m^3)
V	vapour flux due to ventilation (kg m^{-2} s^{-1})
x	x-direction (m)
X	specific humidity (kg kg^{-1})

Dimensionless numbers

Gr	Grashof = $g\beta \Delta T l^3 \nu^{-2}$
Gv	ventilation number = $q_v (u A_o)^{-1}$
Nu	Nusselt = $\alpha_h l \lambda^{-1}$
Re	Reynolds = $u l \nu^{-1}$
Pr	Prandtl = $\nu \mathbb{D}^{-1}$
Sc	Schmidt = $a \mathbb{D}^{-1}$
Sh	Sherwood = $k_m l \mathbb{D}^{-1}$
Le	Lewis number = Pr Sc^{-1}

Greek symbols

α_h	heat transfer coefficient (W m^{-2} K^{-1})
α	absorptivity (-)
β	window aperture (-)
λ	thermal conductivity (W m^{-1} K^{-1})
Δ	difference (-)
γ	thermodynamic psychrometric constant (Pa K^{-1})
ρ	density (kg m^{-3})
χ	concentration (kg m^{-3})
ε	emissivity (-)
ς	reflectivity (-)
τ	transmissivity (-), time constant (s)
σ	Stefan-Boltzmann constant (W m^{-2} K^{-4})
Φ	flux density of physical quantity (..m^{-2} s^{-1})
ν	kinematic viscosity (m^2 s^{-1})

Subscripts

a	ambient
b	boundary layer
c	CO_2
cnd	condensation
cnv	convection
eff	effective
f	flow
g	ground
h	heat
i	compartment or surface in consideration
ij	from surface or compartment i to j
in	entering
j	neighbouring compartment or surface

k	other compartment or surface
l	leaf
m	mass
o	opening
out	leaving
p	photosynthesis
r	cover
rad	radiation
s	surface
st	stomata
sky	sky vault
tot	total
tr	transpiration
vnt	ventilation
v	volume
w	water

Superscript

*	saturation value

Underlining

symbol matrix with considered symbol

List of abbreviations

LAI	leaf area index (m^2 m^{-2})
PAR	photosynthetic active radiation (400–700 nm) (µmol m^{-2} s^{-1} or W m^{-2})
RH	relative humidity (%)

4 Greenhouse construction and equipment

4.1 Introduction

N.J. van de Braak

This chapter deals with various types of greenhouse construction and the equipment used inside them in order to meet the requirements posed by the processes described in Chapters 2 and 3. Although the description of these processes have general validity, the Dutch conditions are used as an example in this chapter for reasons of conciseness. Construction and equipment for other than Dutch conditions are described amongst others by Aldrich & Bartok (1989), Baille & Von Elsner (1988), Müller (1987), Tantau (1983) and Von Zabeltitz (1988).

As shown in Chapter 2, the physiological processes in plants are influenced by microclimatic conditions such as temperature, air humidity, air velocity, CO_2-concentration and light intensity.

According to data from the Royal Dutch Meteorological Institute, KNMI (1982), the average minimum outside air temperature in January in The Netherlands is minus 0.7 °C and the average maximum in August is 21 °C. On average each year there are three days with a temperature below -10 °C and three days with a temperature above +30 °C. The average daily solar radiation varies between 1.8 MJ m^{-2} in December and 18.6 MJ m^{-2} in June, the total yearly incoming solar energy being 3.5 GJ m^{-2}. The windspeed depends on the location and the yearly averages vary between 3 m s^{-1} and 6 m s^{-1}, with about four days per year with a higher windspeed than 19 m s^{-1}. Hail occurs on about 19 days per year. As a consequence various cultivars need to be protected in The Netherlands by means of a greenhouse construction from "hostile" weather conditions such as low temperatures, high wind speeds and rain or hail. On the other hand the light intensities are fairly low most of the time. This means that the Dutch greenhouse not only needs to protect the crop, but also has to fulfil the requirement of having a high light transmittance in order to maintain sufficient crop growth. Greenhouse construction is discussed in section 4.2.

Depending on the outside temperatures and the crop to be cultivated, a heating system may be required, in order to maintain the greenhouse air temperature above a minimum level and thus prevent crop loss. Most of the time even higher temperature levels will enhance growth. Heating systems are considered in section 4.3.

Variation in the outside conditions may necessitate heating during certain periods and cooling during others. Cooling can be achieved by means of various techniques, ranging from simple ventilation through vents to mechanical coolers (section 4.4).

Screens in greenhouses (section 4.5) are used to prevent unnecessary heat loss during colder periods especially at night, or during the periods of high insolation to reduce the cooling load and decrease the air temperature inside the greenhouse.

The protecting hull of the greenhouse enables the provision of higher levels of carbon dioxide concentration near the plants (section 4.6), thus according to Chapter 2 enhancing the growth of the crop (an increase in the CO_2-concentration of 350 vpm with respect to the outside level may increase crop production by 20 to 30%).

Another measure that is beneficial to crop growth is the addition of artificial light (section 4.7);

stimulating photosynthesis if the natural light levels are low, or influencing the crops reaction to photoperiodism.

4.2 Construction

D. WAAIJENBERG

4.2.1 Introduction

The most important function of the greenhouse construction and cover is the protection of the crop against hostile weather conditions (low temperatures, precipitation and wind), disease and pests. The construction also provides means for suspension of crop and equipment.

In countries with low natural light levels (such as The Netherlands) it is important to develop greenhouses with a maximum intensity of natural light inside. Therefore greenhouses are designed with a minimum of structural parts (which can cause shadow in the greenhouse) and the covering materials should have the highest possible light transmittance.

At the same time greenhouses should be strong enough to resist loads that occur caused by snow, wind, crops and installations (e.g. enough safety to prevent storm damage).

The costs of greenhouse construction consist of material and labour costs. Cheap materials will have in general a short lifetime. In countries with high labour costs such as the Netherlands, durability of the construction is important and will often lead to the utilisation of more expensive materials.

The requirements regarding protection and structural stability work against the requirement of maximum light transmittance. Consequently the ultimate design of a greenhouse has to be a compromise incorporating the specific properties of structural and cladding materials and the specific sensitivity to light and temperature of the crop to be grown in the greenhouse.

Greenhouses can be built using covering materials, such as single or double glazing, single or double plastic sheets and plastic films. Combinations of these materials are also used. Each covering material creates specific demands of the structure and covering used for of the greenhouse.

The greenhouses covered with glass and plastic sheets can be divided into two categories: the wider span houses and the Venlo-type houses. About 85% of newly built greenhouses in The Netherlands is of the Venlo-type and about 10% of the wider span-type. Both types use glass as covering material. Up to now these form the best compromise for the Dutch conditions, taking into account the structural requirements, low light intensities, and labour costs (durability). The national production for glass-covered greenhouses varies between 200 and 500 ha per year, and for film-covered greenhouses (including tunnels) the figure is between 15 and 20 ha per year. Where film-covered greenhouses are concerned, a distinction can be made between tunnels and film-greenhouses.

4.2.2 The requirement for standardisation

In the past the development of greenhouses was mainly based on experience. Strength calculations were rarely carried out. Consequently structural elements were often too light or too heavy and failed to carry out their functions effectively. Too little attention was paid to the safety of the greenhouses. These inadequacies were clearly revealed by the severe storms in The Netherlands of 1972 and 1973. As a result of this experience, the initiative was taken to establish a Standard for greenhouse construction. This was finally published in NEN 3859 "Tuinbouwkassen" (Standard Committee 351 37, 1978) and NPR 3860 "Tuinbouwkassen" (Standard Committee 351 37, 1985).

At the same time as the introduction of NEN 3859 a testing authority (TNO-Bouw, Delft) was set up to verify design calculations and to test specific construction details experimentally. The combined effect of the Standard and the testing authority forced the greenhouse builder to make a complete static calculation of every newly developed type of greenhouse. As a result a minimum market quality was introduced.

The Standard NEN 3859 contains information on:

- The definition and the economic lifetime of a greenhouse (fixed at 15 years);
- Wind and snow load (overall snow load 250 N m^{-2});
- Load caused by crops (150 N m^{-2} for tomatoes and cucumbers);
- Load caused by installations and other service loads;
- Permissible material stress, deformation values, etc.;
- The load combinations (under Dutch circumstance, for example, wind and snow load do not act on a structure at the same time).

4.2.3 Single glass greenhouses

A wider span greenhouse is conventional in construction, i.e. with steel or aluminium purlins attached to steel trusses. These purlins together with the steel or aluminium gutter support the glazing bars, on which the glass is placed (Figure 4.2.1). The span width of a wider span house is standardised on a multiple of 0.80 m, and thus may be 6.40 m, 8 m, 9.60 m or 12.80 m. A characteristic feature of the wider span house is the fact that there are more than one glass pane on top of one another from gutter to ridge (in the example of Figure 4.2.1 there are 5 panes in the roof plane). Another characteristic is a continuous ventilation-window over the entire length of the roof. The advantages of a wider span house are the bigger area without columns (better mechanisation-possibilities), and the better ventilation (see also section 4.2.5).

A disadvantage is the high investment-cost for a wider span house, which is higher than that for a Venlo-type house.

The Venlo-type greenhouse is the most popular type of greenhouse in The Netherlands. Here only one glass pane is placed on glazing bars covering the height from gutter to ridge (Figure 4.2.2). The standard span width is 3.20 m, consisting of two glass panes of 1 m × 1.65 m or 1.125 m × 1.65 m. joined in the ridge. The ridge is not supported by extra trusses, but the glass panes and the glazing bars support themselves and the ridge. The latest development is the use of glass panes of 0.80 m × 2.08 m or 1.00 m x 2.08 m, creating a span width of 4 m. Trellis girders (trusses) are used to support the roof and gutters. The lengths are respectively 6.40 m (2 times 3.20 m) and 8 m (2 times 4 m) (Figure 4.2.3). The centre to centre distance of the columns in the direction parallel to the gutter is 4 m or 4.50 m.

Recently the height of the Venlo houses was increased from 3–3.50 m up to 4 and 4.50 m, in order to create enough space for crops, thermal screens and light fittings for artificial lighting.

The Dutch greenhouse industry has undergone an enormous technical development over the past decade under the pressure of the need to save energy and maximise the availability of natural light inside the greenhouse. Three measures to increase the light entry were introduced in Venlo houses as follows:

- the use of windowpanes 1 m wide instead of 0.73 m;
- reduction in the width of the gutters from 0.22 m to about 0.16 m;
- increase in the spacing between trusses from 3 m to 4 m.

The total entry of diffuse light has increased from 65% to 72% as a result of these measures. Additional measures including further reducing the sizes of structural parts and the stowing away of energy-

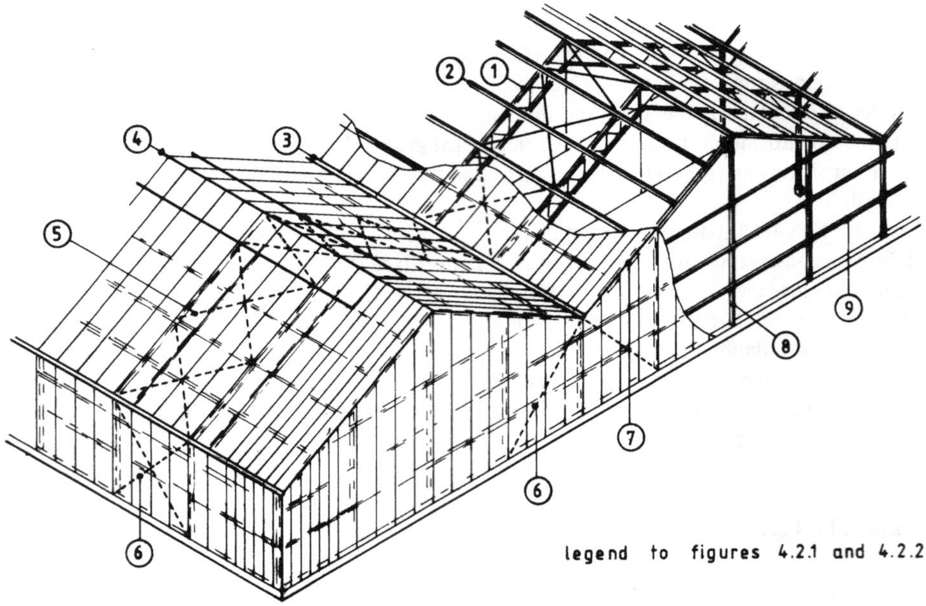

legend to figures 4.2.1 and 4.2.2

1. truss
2. purlin
3. gutter
4. ridge
5. roof braces
6. stability braces
7. top bar in front wall
8. front wall stanchion
9. front wall rail

Figure 4.2.1 – View of a wider span greenhouse.

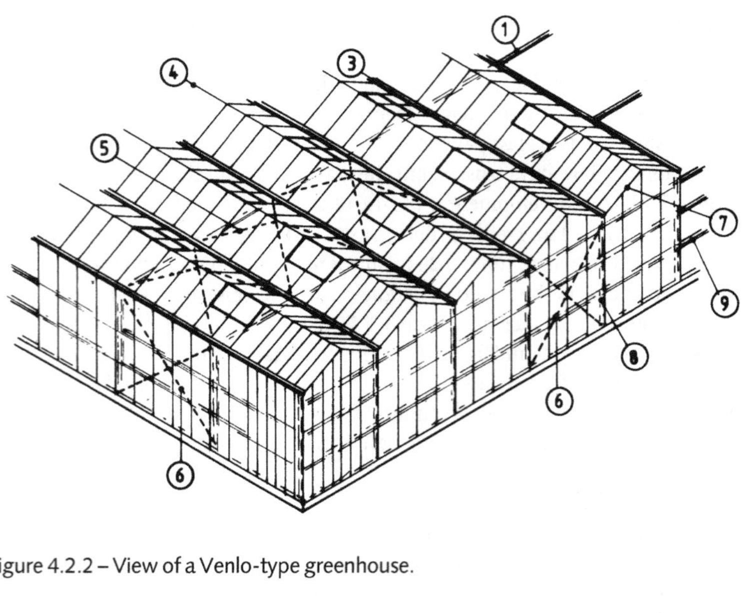

Figure 4.2.2 – View of a Venlo-type greenhouse.

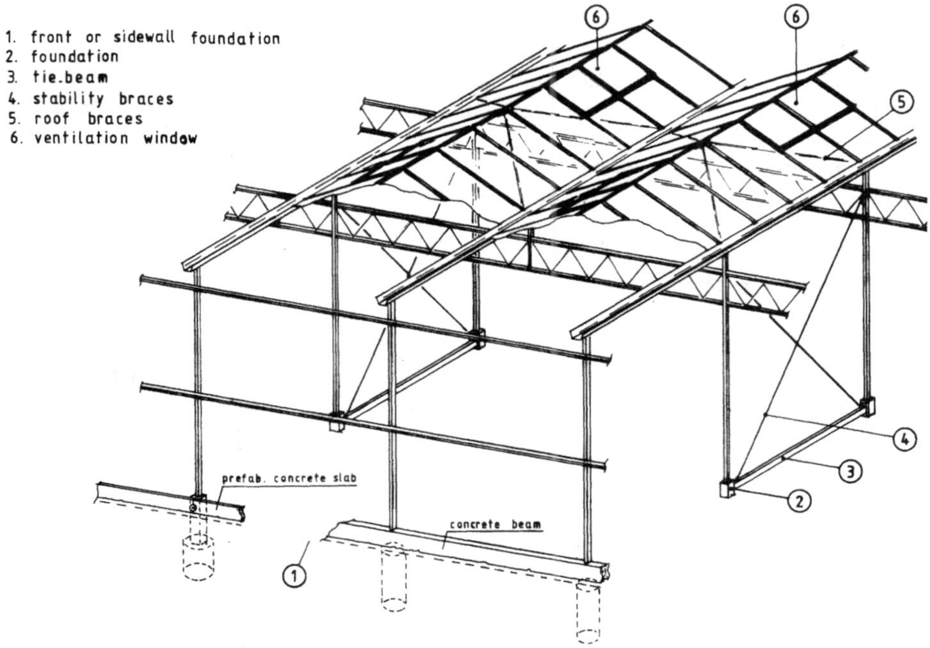

Figure 4.2.3 – Structure of a Venlo-type greenhouse with trellis-girders.

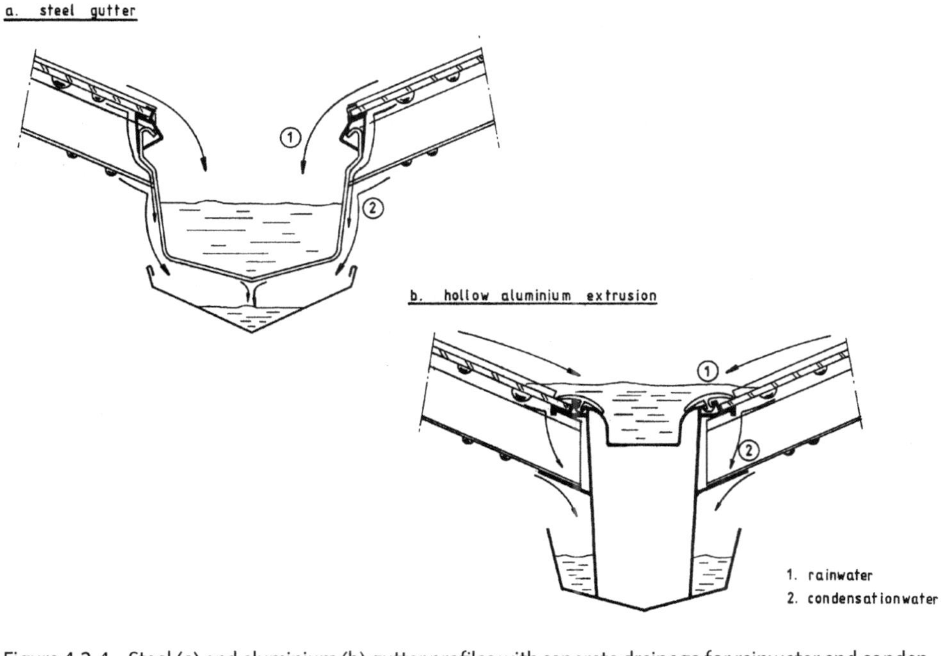

Figure 4.2.4 – Steel (a) and aluminium (b) gutter profiles with separate drainage for rainwater and condensation water

saving screens inside the house, have resulted in the total diffuse light transmittance reaching 75%.

To improve the reflection of light inside the houses structural components such as gutters and trusses are now sometimes coated with white paint.

Figure 4.2.4 shows another aspect that became much more important during the last years in order toprevent pollution of the environment: the separate drainage of rainwater and condensation water (which is polluted with pesticides and zinc). These steel (Figure 4.2.4a) and aluminium (Figure 4.2.4b) gutter profiles keep the different water flows separate.

4.2.4 Insulating cover materials

Improving the insulation values of cover materials is one of the methods to conserve energy in greenhouses. This can be achieved by applying composite materials, such as double glazing or synthetic double-web sheets or coated glazing with a low emissivity (to reduce the radiation to the sky) or by applying a second layer to an existing single cladding.

In determining which material or construction should be used the various properties should be taken into account, such as:

- Light transmittance;
- Strength and deflection under wind and snow loads;
- Resistance to hail load;
- Insulation value;
- Thermal transmittance (infrared above 3000 nm);
- UV transmittance (ultraviolet up to 400 nm);
- Resistance to ageing, soiling and chemical products such as pesticides;
- Condensation behaviour;
- Sizes in which materials can be obtained.

The light transmittance of cladding materials is measured at wavelengths between 400 and 700 nm. The interception of light by structural elements can be calculated with a computer programme. The interception of diffuse light by a double glazing system means about 10% light loss compared to single glazing. For this reason double glazing systems are no longer used for the roof in newly built greenhouses in The Netherlands. Single or double thermal screens are used instead for insulation (section 4.5).

Beside the double glazing systems, several types of plastic sheets are used for greenhouse cladding. Materials for rigid plastic sheets include: PMMA (polymethylmethacrylate), PC (polycarbonate) and PVC (polyvinylchloride). The transmittance of several of these materials both for direct and diffuse light are given in Table 4.2.3 (Waaijenberg, 1984). Use of these materials has decreased recently due to bad light transmittance.

Coatings for glass such as metal oxide with a low emissivity (e.g. Hortiplus) have been introduced recently in the quest to combine energy savings with adequate light transmittance.

4.2.5 Ventilation windows

Most greenhouses in The Netherlands are ventilated by natural ventilation through different types of windows. The number and size of ventilation windows and the mechanism of movement vary. The exact opening-angle of the ventilation windows is becoming increasingly important, because of the higher requirements of climate control in modern greenhouses.

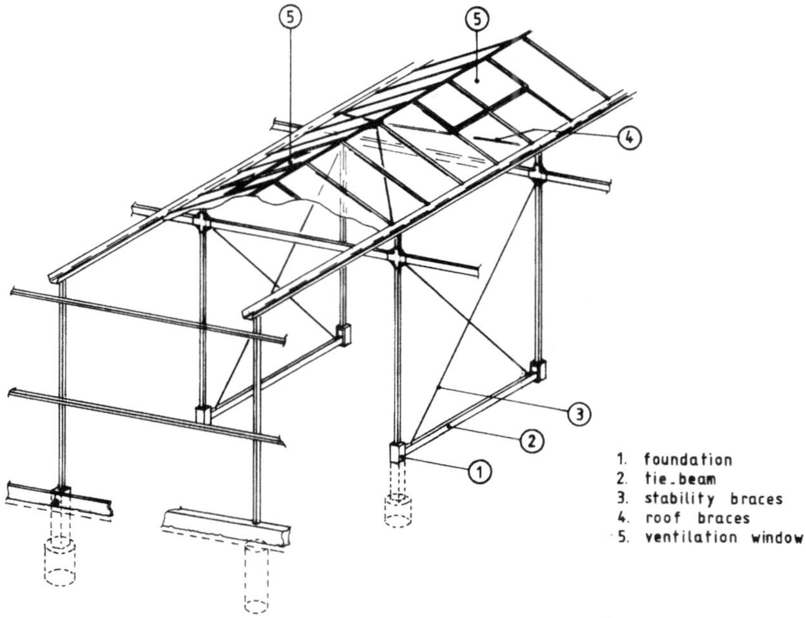

Figure 4.2.5 – The roof of a Venlo-type greenhouse with a "two half glass pane" ventilation-window.

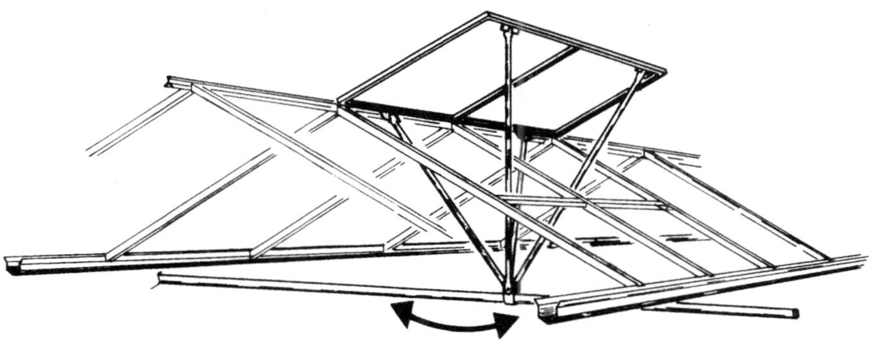

Figure 4.2.6 – "Swing mechanism" for the operation of ventilation-windows in a Venlo-type greenhouse.

For Venlo-type greenhouses the "one glass pane window" is used, (one windowpane per bay with dimensions 0.73 m × 1.65 m) along with the "two- or three half glass pane window" (windows with two or three glass panes across the width and a height which is half that of a normal glass pane i.e. 0.825 m) (Figure 4.2.5). Ventilation windows 1 m in height are also used.

The ventilation effect of the various windows is discussed in section 4.4.4. The windows can be opened at a level above the greenhouse ridge. In Table 4.2.1 the ratio of the ventilation window area to the greenhouse area is given for different types of Venlo-type windows driven by a "swing mechanism" (Figure 4.2.6). In the swing mechanism the main rod of the system is located between the trellis girders, while in the "truss rail mechanism", the main rod is situated above the trellis girder (Figure 4.2.7). This rail mechanism has been developed to reduce the shadow-casting elements in a greenhouse (shadow of pipe and trellis girders fall together).

Table 4.2.1 – Ratio of the standard ventilation openings of Venlo-type greenhouses to the greenhouse area (with swing mechanism).

Greenhouse type * (span × bay) (m)	Glass width (m)	Window dimensions (m)	Ratio window to greenhouse area (%)
$^{3.2}/_{6.4} \times 3$	0.73	1.50 × 0.825	12.75
$^{3.2}/_{6.4} \times 3$	0.73	2.25 × 0.825	19.10
$^{3.2}/_{6.4} \times 4$	0.997	2.00 × 0.825	12.90
$^{3.2}/_{6.4} \times 4$	0.997	3.00 × 0.825	19.30
$^{3.2}/_{6.4} \times 4$	0.997	2.00 × 1.00	15.60
$^{3.2}/_{6.4} \times 4$	0.997	3.00 × 1.00	23.40
$^{3.2}/_{6.4} \times 4.5$	1.12	2.25 × 0.825	12.70
$^{3.2}/_{6.4} \times 4.5$	1.12	2.25 × 1.00	15.40
$^{4.0}/_{8.0} \times 4$	0.797	1.60 × 1.04	10.30
$^{4.0}/_{8.0} \times 4$	0.797	2.40 × 1.04	15.50

** The first dimension in the table under "greenhouse type" is the span of the roof, the second figure is the span of the trellis girder and the third figure is the mutual distance of the trusses.*

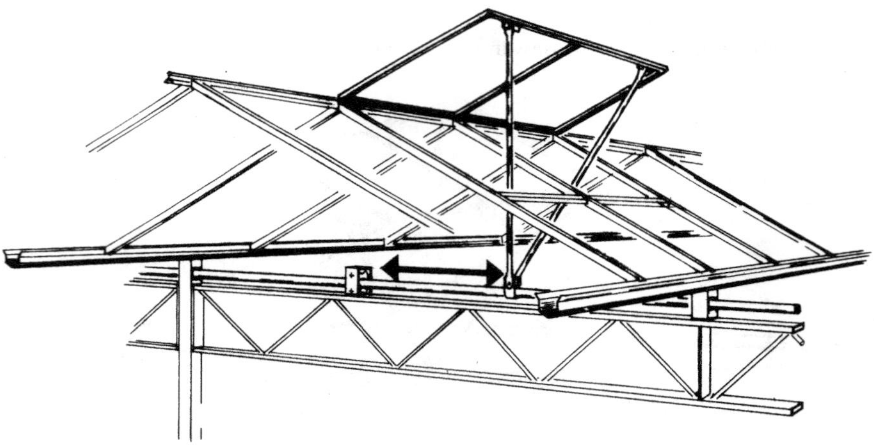

Figure 4.2.7 – "Truss rail mechanism" for the operation of ventilation-windows in a Venlo-type greenhouse.

In wider span houses there are continuous ridge-ventilation windows over the entire length of the roof (Figure 4.2.8).
For positioning of these ventilation windows on both sides of the ridge, a mechanism with a turning shaft and toothed bars is used. The height of these ventilation windows varies between 1 m and 1.6 m. In Table 4.2.2 the relation is given between the window area and the greenhouse area for different types of wider span greenhouses.

To reduce the use of insecticides and pesticides inside greenhouses there is a tendency to close the openings for insects when the ventilation windows are open by using different kinds of nets that are fixed to the structure and move together with the windows.

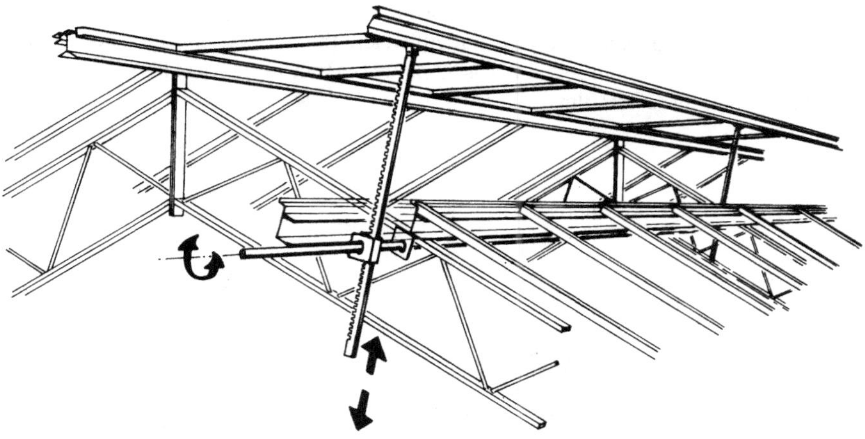

Figure 4.2.8 – The roof of a wider span greenhouse with continuous ventilation-windows

Table 4.2.2 – Ratio of window area to greenhouse area in wide span greenhouses.

Greenhouse span (m)	Window height (m) *	Ratio window to greenhouse area (%)
8	1.4	33.6
9.6	1.4	28
12.8	1.4	21

** Other window heights are 1.0, 1.2 and 1.6 m; the corresponding ratios are depending linearly on the height.*

4.2.6 Plastic film greenhouses and tunnels

Plastic film-covered greenhouses and tunnels form a minor proportion (less than 5%) of the total greenhouse area in The Netherlands. However, demand for this type is increasing. In particular, growers of crops with low heat requirements such as ornamental trees, strawberry and summer flowers, are interested in plastic greenhouses.

Unfortunately film-covered greenhouses are more susceptible to damage, above from wind. To improve the quality of film houses the initiative has been taken to draft a special Standard for this category, following the Standard for glass-covered greenhouses. Some of the preliminary rules are summarised by Waaijenberg (1990) and in the first draft for a European Greenhouse Standard (Working Group, 1991).

Tunnel greenhouses generally consist of bent trusses (hoops) which are secured to the ground by means of screw anchors or cast in concrete (Figure 4.2.9). The framework has to be able to cope with all loads and has to convey these loads to the soil.

To establish the wind coefficients for different tunnels and other types of plastic greenhouses with curvilinear roof types research has been done in wind tunnels and in real practice (Van Koten, 1974; Richardson, 1985). These coefficients are also given in the new draft of NEN 3859 (Standard Commitee, 1988).

For the covering of plastic greenhouses and tunnels many different plastic films are available. The important properties of light transmittance, durability, and thermal transmittance determine whether a film can be used for this purpose. Table 4.2.3 shows the transmittance of IR-radiation and light incident at rightangle for several films.

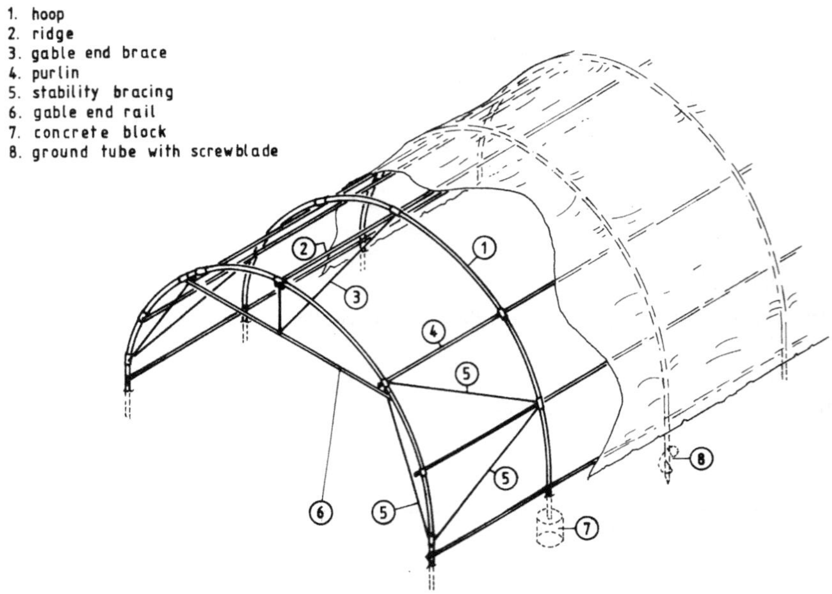

Figure 4.2.9 – Structure of an 8 m wide tunnel greenhouse with stability bracing.

Table 4.2.3 – Transmittance for long-wave heat radiation (wavelength 5000–14000 nm), for direct light (perpendicular) and diffuse light (wavelength 400–700 nm) of several greenhouse covering materials (Waaijenberg, 1988).

Material	Thickness (mm)	IR-trans-mittance	Light transmittance	
			diffuse	direct
Single glass	4	0	0.83	0.89
Double glass, Cavity 9 mm	2×3	0	0.72	0.82
PMMA acrylic Double-web Sheet (Röhm)	16	d.n.a. [1]	0.75	0.84
Polycarbonate Double-web Sheet (Qualex)	10	d.n.a.	0.65	0.74
PE, UV-stabilized	0.20	0.56	d.n.a.	0.89–0.92
EVA (ethylenevinyl-acetate)	0.18	0.22	d.n.a.	0.91
Teflon FEP	0.05	0.57	d.n.a.	0.96

[1] *d.n.a. = data not available.*

4.3 Heating equipment

N.J. VAN DE BRAAK

4.3.1 Requirements

In order to determine the capacity of a heating system for a greenhouse, the heat balance of the greenhouse system is considered for design conditions. The terms in the heat balance are the radiative, conductive and convective heat losses of the greenhouse on the one side and the supply of heat by the heating system and the captured solar radiation, if present, on the other side. The basic concepts of these terms are explained in Chapter 3.

In a simplified model the heat demand of a greenhouse can be described as

$$q_h = U A_{gg} (T_{in} - T_{ex}) - (1-f) \tau S A_{gg} \quad \text{(Eq. 4.3.1)}$$

in which A_{gg} is the ground surface of the greenhouse, T_{in} and T_{ex} the desired greenhouse air temperature and the prevailing outdoor temperature respectively. U is the effective heat transfer coefficient of the greenhouse, which is composed out of the various heat transfers through greenhouse construction, cladding material, air infiltration, etc. S is the insolation, τ the transmissivity of the greenhouse and f a factor representing the fraction of incoming solar radiation converted into latent heat by evaporation from the crop.

For a given greenhouse, the heating capacity is determined by the difference between the desired greenhouse temperature and the design outside air temperature, which generally occurs at night. In the western part of The Netherlands with an outside design temperature of -8 °C a crop with a required air temperature of 20 °C will need a heating capacity of 246 W m^{-2} in a greenhouse with a single glass cover. In Table 4.3.1 the U-values of greenhouses with various covering materials are given at a windspeed of 4 m s^{-1}. The U-value of greenhouses can be reduced by the using thermal screens, which are discussed in section 4.5.

The distribution of the heat demand according to the time of year has been investigated by several researchers. Breuer (1990) determined the sensitivity of the cumulative frequency distribution of the heat demand for various parameters such as U-value and the application of ventilation for dehumidification shortly after sunrise. Figure 4.3.1 shows the typical heat duration curve for the cultivation of tomatoes in a greenhouse located in the western part of The Netherlands.

The location of the heat input into a greenhouse influences the temperature distribution in the greenhouse. Temperature differences will cause, according to the processes discussed in Chapter 2,

Table 4.3.1 – U-values of a square greenhouse of 0,5 ha at a wind velocity of 4 m s^{-1}.

Covering material	U-value (W m^{-2} K^{-1})
Single glass	8.8
Double glass in sidewalls	7.9
All double glass	5.2
Double acrylic	5.0
Double polycarbonate	4.8
Single PE film	8.0
Double PE film	6.0

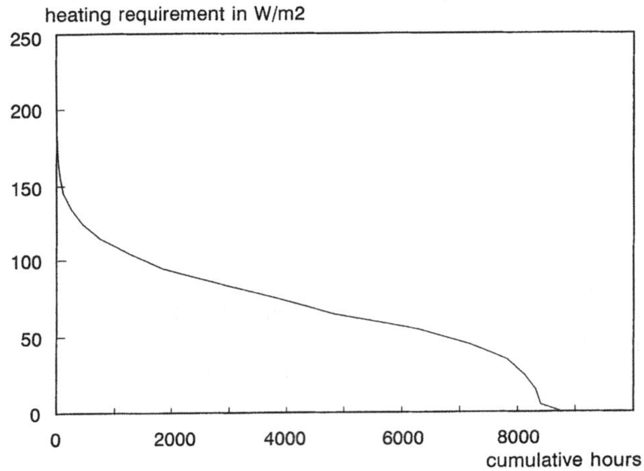

Figure 4.3.1 – Cumulative frequency curve of the heating requirement of a Dutch greenhouse with tomatoes.

undesirable differences in growth rate. Special attention has to be paid in order to ensure that the heat is supplied in such a way that losses through the sidewalls of the greenhouse are compensated for and temperature differences are avoided.

4.3.2 Traditional heating systems

If it is required, heat can be introduced into the greenhouse, through central or local heating systems. In central heating systems two main parts can be distinguished, the heat source and the heat distribution system, while in local systems the distribution component is absent.

In the orangeries of the 17th century a number of stoves without any distribution system served as heating system. Peat and wood were used as fuel. In the 18th century the flue gases of stoves were led through ducts in the walls or the floors of lean-to greenhouses, and coal made its entry as a fuel. In that period the utilisation of steam through ducts was introduced, a technique which was further developed in the following century. In the 19th century a hot water boiler with a closed circuit of hot water pipes was used for the first time. Coal and coke provided the energy at the time.

In this century initially oil and later natural gas (methane) have become important energy sources. The most commonly used heating system in the period before the oil crisis of the 1970's consisted, in The Netherlands, of a central hot water boiler, fired by natural gas or in some cases oil, and a closed circuit of steel pipes with an inner diameter of 51 mm through which the hot water circulates to distribute the heat in the greenhouse. The design temperatures of the supply and return water are 90 °C and 70 °C respectively. The pipes are attached under the growing benches or to the posts of the greenhouse, overhead and near the sidewalls. The majority of the Dutch growers still uses this heating system although the location of the pipes has often been adapted to new insights concerning energy conservation and local crop heating. Various pipe arrangements are in use, for crops such as tomatoes, cucumbers and green peppers, the most common one being four 51 mm pipes per bay of 3.2 m at a small distance (5–10 cm) from the soil surface. The pipes are combined in two pairs, with a mutual distance in the pair of about 0.4 m and the pairs at a mutual distance of 1.6 m. In this way the heating pipes can be used as a rail system for internal transport as well (Figure 4.3.2). For crops like lettuce as well as for roses, the heating pipes are often mounted overhead (Figure 4.3.3). Crops that are cultivated in beds, such as chrysanthemums, often have two small heating pipes per bed, which are adapted frequently to the crop height.

A small group of growers in The Netherlands uses free discharge air heaters with an integrated burner; these can be either of the type where the flue gases are blown directly into the greenhouse (Figure 4.3.4) or of the type with a heat exchanger between flue gases and the greenhouse air.

4.3.3 Alternative heating systems

Three levels can be distinguished on which alternatives have been investigated and/or introduced to achieve a reduction of energy consumption: the type of fuel or energy source, the energy conversion equipment, and the heat distribution system. In this section these will be considered in some detail.

4.3.3.1 Energy sources

Various fuels can be used for heating greenhouses. Table 4.3.2 shows the heat content of a number of these. Natural gas is the major fuel, it provides 99% of the total heating energy in The Netherlands.

Figure 4.3.2 – Pipe heating system functioning as transport rail.

Figure 4.3.3 – Heating pipes mounted overhead

Figure 4.3.4 – Free discharge air heater.

Oil is much less used; it provides about 0.8% of the total heating energy.

The typical problems associated with the use of alternative energy sources investigated (geothermal, solar and wind energy, industrial thermal effluents, refuse derived fuel (RDF) and coal) are presented in Table 4.3.3.

Due to these drawbacks as well as high investment costs and decreasing energy prices in the 1980's, the alternatives are seldom applied. Only a few coal and refuse derived fuel (RDF) installations and one greenhouse district of about 20 hectares connected to a electricity power plant are operational today in The Netherlands.

4.3.3.2 Energy conversion equipment

The central hot water boiler is the standard for greenhouse heating in The Netherlands. Much effort has been put into the improvement of the efficiency of the boilers during the last 15 years. By means of an additional heat exchanger the flue gases of the boiler can be cooled down further, leading to an improved efficiency. If no corrosive elements are present in the exhaust gases, which is the case when firing natural gas, the temperature can even be brought below the condensation point of the water vapour in the flue gases, thus considerably increasing the amount of extracted heat. Energy savings of up to 15% can be obtained by means of these flue gas condensors (Meijndert, 1982); as a result they are commonly used by Dutch growers.

Table 4.3.2 – Heat content of various fuels.

Fuel	Heat content (MJ kg^{-1})
Natural gas	35.17 (MJ m^{-3})
Propane	50.29
Butane	49.55
Light oil	39.03
Heavy oil	43.17
Coal	20–30
Dried wood pellets	19.8

Table 4.3.3 – Typical drawbacks of various alternative energy sources.

Drawback	Geothermal energy	Solar and wind energy	Industrial thermal effluents	Coal and RDF
Distance source to greenhouse	×		×	
Temperature level	×		×	
Salinity/corrosion	×			
Available if needed		×		
Space constraint		×		
Separate CO_2 supply	×	×	×	×
Investment costs		×	×	
Warranty on delivery			×	
Automatic control				×
Stockroom				×
Ashes				×
Air pollution				×
Requirement on return temperature			×	

The growing interest in alternative fuels has led to a revival of coal fired boilers. Automatic feeding and ash removal systems have been developed to tackle the problem of time consuming handling. Filter systems and cyclones are installed to clean exhaust gases and reduce air pollution.

The fluidised bed technique has been adopted by boiler manufacturers in the greenhouse industry to prevent pollution and obtain better levels of efficiency. In a fluidised bed burner, air is blown through a bed of sand; small coal particles are brought into it and float surrounded by air in the sand bed. In this way the coal is burned more efficiently. A number of techniques can be used for coal fired burners as well as for RDF installations.

A special type of heating system to be mentioned here is the infrared radiant heater. It consists of a number of small burners, generally fired by natural gas, attached to a long steel tube, through which the flue gases are led. Depending on the length of the tube behind the last burner, the exhaust gases can be cooled to a temperature as low as 60 °C. The temperature of the radiating part of the tube between two burners varies between 375 °C and 250 °C. The heaters are installed overhead in the greenhouse. The steel pipe emits infrared radiation which is directed towards the crop by means of a reflector attached above the pipe. The absorbed radiation is converted into heat. In this way the crop is heated directly and in principle the air temperature can be lower than in systems where the crop is heated via the greenhouse air. Due to this lower greenhouse air temperature and the lower exhaust gas temperature of the infrared heater than of the conventional boilers energy can be saved. Knies *et al.* (1983) reported savings of between 10% and 12%. They also found that the distribution of radiation at crop level is rather uneven which leads to significant temperature differences in the crop. Another disadvantage of the radiant heaters is their interception of light. The application of radiant heaters in The Netherlands is therefore rare.

Another way to improve the efficiency of energy conversion is through the use of heat pumps. Heat can be transported from a source with a certain temperature level to a sink with a higher temperature level by means of an energy input which is in general lower than the transferred heat. Theoretically the Carnot factor, defined as the ratio of the transferred heat q_h over the supplied power P, can be derived by this process:

$$C_f = q_h / P = T_1 / (T_1 - T_2) \qquad \text{(Eq. 4.3.2)}$$

with T_1 the high temperature level and T_2 the low temperature level. In practical settings a Coefficient of Performance (COP) which is defined as the ratio of the amount of heat brought into the greenhouse and the driving energy supplied, of 1.4 to 1.7 have been achieved (Telle *et al.*, 1987). In view of the high investment costs this often makes the economic feasibility doubtful.

Where growers are using both large amounts of electricity and heat, cogeneration (section 4.7.4) might be an option to reduce energy costs. A cogenerator consists of a combustion engine coupled to a generator. The heat from the cooling water and the exhaust gases can be used to heat the greenhouse when electricity is generated for artificial lighting or other purposes. In this way an overall efficiency of 85% to 90% can be obtained. In general a cogenerator produces heat and electricity at a ratio of 2 to 1. Unfortu-nately the concurrent need of those two forms of energy is often not in that ratio. In practice this leads to a loss of efficiency and a negative effect on the reduction of energy costs. The growing interest in artificial light supplementation, as well as the increasing possibility of coupling local cogenerators to the public electricity network might prove to be a turning point. At present about 500 Dutch growers are using the heat or electricity of a cogenerator.

Energy storage systems can also be used in order to reduce energy consumption for heating purposes in greenhouses. The principle is based on the storage of energy during a period where there is excess, and utilisation of that energy during a period where there is an energy need. Energy storage can be based on a short time period (day to night shift) or on a longer period (summer to winter shift).

In The Netherlands only small storage systems in combination with CO_2 supply units have proved to be economically efficient (see also section 4.6.3.5).

4.3.3.3 Heat distribution systems

The primary purpose of the heating system in a greenhouse is to raise the temperature of the crop. To obtain this against the lowest energy costs, the obvious thing to do would be to bring the heat as close as possible to the plants and heat their direct environment only. In fact a lot of growers have done so by lowering overhead pipes. Energy savings up to 20% have been reported by Heijna (1985) as a result of this measure.

If the use an alternative energy source or conversion equipment causes no changes in the supply or return temperature with respect to the traditional systems, there are no consequences for the distribution systems. However in many cases both the supply temperature and the required return temperature are lower than usual. This immediately affects the amount of heat that can be transferred by the distribution system according to the heat transfer formulae 3.2.10 given in Chapter 3. There are two ways to compensate for this: enlargement of the heat exchanging surface and improvement of the heat transfer coefficient.

An increase of the exchanging surface can be obtained in various ways:

- More of the same heating elements;
- Considerably more smaller heating elements;
- Heating elements with a larger surface such as finned tubes;
- Buried heating elements or heated concrete floors, using the entire ground surface as exchanging surface;
- Growing benches with heated bottoms.

To obtain an improvement of the transfer coefficient the following means are available:

- Utilisation of materials with a high conductivity such as aluminium;
- Application of forced instead of free convection;
- Utilisation of vapour transport.

Three groups can be distinguished when considering heat distribution systems for greenhouses: pipe heating systems, soil and floor heating systems and air heating systems. We shall discuss some examples of each of these groups.

Pipe heating systems

The conventional heating element in Dutch greenhouses is a steel pipe with an inner diameter of 51 mm, which can also be used as transport rail. During the last decade interest has grown in alternative sizes and alternative locations such as nearer to the crop growing point and near substrates for root zone heating. Nijeboer & Van Holsteijn (1981) reported the heat transfer of various sizes of steel pipe and plastic tube of 25 mm outer diameter (Table 4.3.4).

Recently a new droplet-shaped pipe was developed. This pipe has a smaller water content then a circular steel pipe with the same outer surface and possesses a high mechanical stiffness. Its convective heat transfer is equal to that of a steel pipe with the same perimeter and its radiative exchange with the roof of the greenhouse is, depending on its location with respect to the crop, up to 15% lower (Stoffers, 1989).

The application of fins on steel pipes provides a larger exchange surface. However the temperature drop over the fins reduce the effect to a certain extent. The use of aluminium, which has a conductivity four times that of steel, copes with this problem so well that an aluminium tube with a diameter of 22 mm and two fins of 24 mm each, transfers as much heat as a steel pipe with a diameter of 51 mm under the same circumstances. Moreover such a tube has a smaller water content causing its response to control actions to be quicker than that of the traditional steel pipes.

When using several different metals in one circuit one has to be aware of possible problems of corrosion. The aluminium finned tubes have been successfully applied in research projects concerning the utilisation of waste heat (Knies, 1992) and are applied by growers in commercial greenhouses as well.

Soil, floor and bench heating systems

If the entire ground area of a greenhouse acts as heating element it provides a considerably larger heat exchanging surface than the traditional pipe heating systems. Moreover the heat is brought homogeneously into the greenhouse, if the temperature differences at the ground surface are kept small.

The interest in soil heating is also a result of the beneficial effects of root zone heating for some crops. In general plastic tubes with outer diameters varying between 15 and 30 mm are buried at a depth of 20 to 50 cm depending on the supply water temperature. A temperature rise of a few degrees at the soil surface is obtained in this way, resulting in fairly low heat fluxes (10 to 40 W m^{-2}).

Other forms based on the same principle are the heated concrete floor (Knies, 1992), where the heating pipes are embedded in concrete, and the heated growing benches (Vogelezang *et al.*, 1988; Vogelezang, 1993) where the heating pipes are attached to the bottom surface of the bench. Because of the small distance of the pipes to the surface and the often better conductivity of the concrete and bench material compared to soil the temperature difference between pipes and surface will be smaller and the heat flux larger than in the case of soil heating. However the surface temperature and consequently the amount of heat transferred to the greenhouse is limited by the maximum temperature requirement of the roots of the crop.

A significant disadvantage of these heating systems is the slow response to control actions due to the large thermal mass.

In The Netherlands heated concrete floors are mainly used in seedling nurseries and heated benches by growers of potplants.

Table 4.3.4 – Heat transfer in Watts per metre pipe length at various temperature differences between pipe and greenhouse air.

Temperature difference	Steel pipe diameter (mm) 51	33.2	26.4	Plastic tube diameter (mm) 25
10	15	10	8	6
20	34	23	18	14
30	55	38	31	24
40	77	53	44	35
50	101	71	58	46
60	128	90	73	- [1]
70	156	108	90	-
80	185	129	107	-

[1] Not applied at this temperature.

Air heating systems

Considering the mechanisms of heat transfer in Chapter 3 we have seen that the coefficient of heat transfer increases with the velocity of the air passing the heat exchanging surface (equation (3.2.12)). In air heating systems this is achieved by forcing the air along a heating element, such as a finned coil, by means of a fan. The heated air can be discharged freely into the greenhouse, but it is preferable to distribute it evenly by a system of ducts to prevent undesirable temperature differences. Perforated polyethylene tubes are often used for this purpose as they are relatively cheap.

The air distribution takes place through one duct per air heater, a system which can be applied either in the attic of a greenhouse bay or under growing benches or between crop rows. Another option (less realistic for large greenhouses) is the utilization of a central air conditioning unit branched onto a network of ducts located between crop rows or under benches. Definite advantages of air heating systems are the quick response to control actions and the possibility of applying a wide range of water temperatures. A disadvantage is formed by the additional consumption of electrical power by the fan, which for a system with ducts often will be in the order of 10% of the required heating energy (Knies, 1992). For this reason this system is adopted mostly in countries with relatively low electricity prices.

4.4 Ventilation and cooling

J.J.G. BREUER AND P. KNIES

4.4.1 Introduction

Two types of ventilation can be distinguished: natural and forced. In the case of natural ventilation the pressure difference over the openings, the driving force for ventilation, is caused by wind effects and by a difference in air temperature between the inside and outside air. Energy for the forced ventilation process is supplied by fans.

As has been described extensively in Chapter 3, each of the following processes contribute to the energy and water vapour balance in a greenhouse:

- Heat transfer by radiation (solar heat input and long wave radiation exchange);
- Heat transfer by conduction through the greenhouse hull;
- Condensation against the hull;
- Air exchange by infiltration and ventilation;
- Crop evaporation and evaporation from other surfaces;
- Heat gain through the heating system.

Energy storage, either intended or unintended, in the soil for instance, may in some cases play an important role in the energy balance. This is the case when the heating system fails and in unheated greenhouses. Nevertheless, for reasons of conciseness, it is not taken into consideration in this section.

The purpose of ventilating with outside air is the discharge of water vapour (latent heat) or sensible heat. In some cases ventilation is applied to admit outside air either in order to increase or maintain the CO_2 level in the greenhouse. Ventilation also plays a role in the discharge of gaseous pollutants.

4.4.2 Requirements

Cooling load

The energy associated with ventilation, necessary to maintain the desired temperature or relative humidity is arbitrarily called the cooling load.

If the global radiation intensity is high, a substantial amount of this energy input is used by the crop for transpiration; sensible solar heat is transferred into latent heat. 50 to 80% of the incoming solar radiation is used for (evapo)transpiration (Van der Post *et al.*, 1974).

Screens may reduce incoming solar radiation substantially and can therefore be used as a tool to control the climate inside the greenhouse.

Pollutants

Interest in gaseous pollutants in greenhouses is restricted; knowledge about the detrimental effects is limited and detection is cumbersome and expensive. Pollutants known to cause injuries to (greenhouse) crops are: ozone, ethylene, sulphur dioxide, mercury vapour and phenolics (Aldrich, 1986). Sources of gaseous pollutants are: the crop itself (ethylene), photo-chemical reactions, burning of fuels (see also section 4.6.3.2), coatings, fungicides, pesticides and wood preservatives.

Air infiltration plays an important role in reducing the concentration of gaseous pollutants originating from sources inside the greenhouse under low ventilation conditions. Some sources of gaseous pollutants are located outside the greenhouse. In this case ventilation plays no role in reducing the concentration inside the greenhouse.

Air velocity

Air velocity and direction of the air flow are inextricably linked to ventilation. Air velocity affects a number of factors related to plant growth: transpiration, respiration and photosynthesis (through transportation of CO_2). Aldrich *et al.* (1983) observed that an air velocity of 0.5–0.7 m s^{-1} is generally accepted as optimum for plant growth. Above 1.0 m s^{-1} growth is inhibited and above 4.5 m s^{-1} physical damage is likely to occur. ASAE (1984) recommends a velocity of less than 1.0 m s^{-1}. The effect of direction of the air flow is hardly dealt with in the literature.

4.4.3 Air infiltration

In almost all greenhouses uncontrolled infiltration of outside air occurs. The air infiltration flux depends on: the area, the position and the geometry of chinks and the pressure difference between inside and outside the greenhouse due to wind- and temperature effects.

Many researchers quoted by De Jong (1990) found a linear dependency between the rate of air infiltration in h^{-1} (defined as the exchange of greenhouse and outside air divided by the greenhouse volume) and wind velocity and temperature; no dependency was found for the wind direction. Temperature difference appears to play a role at a low wind speed only.

Some values of air infiltration for common greenhouse types are presented in Table 4.4.1 (after ASAE, 1984; Wind speed not specified).

The sensible energy loss caused by air infiltration, or by ventilation in general, may be quantified following equation (3.2.8) of section 3.2 as:

$$q_{h,s} = V_g \, vv \, \rho \, C_p (T_{in} - T_{ex}) / 3600 \qquad \text{(Eq. 4.4.1)}$$

The latent energy loss caused by air infiltration, or by ventilation in general, with:

$$q_{h,l} = V_g \, w \, \rho \, r \, (x_{in} - x_{ex}) / 3600 \tag{Eq. 4.4.2}$$

where: $q_{h,s}$ = sensible heat loss by infiltration (W)
$q_{h,l}$ = latent heat loss by infiltration (W)
V_g = greenhouse internal volume (m^3)
w = rate of air infiltration (h^{-1})
ρ = density of air (kg m^{-3})
C_p = specific heat of air (J kg^{-1} K^{-1})
r = latent heat of vaporisation (J kg^{-1})
T_{in} = inside temperature (K)
T_{ex} = outside temperature (K)
x_{in} = specific humidity inside (kg kg^{-1})
x_{ex} = specific humidity outside (kg kg^{-1}).

4.4.4 Cooling by ventilation

4.4.4.1 Natural ventilation

Much research has been devoted to natural ventilation. However in most cases the outcome is not generally applicable. Results with a broader field of application have been reported by Kozai & Sase (1978), Kozai *et al.* (1980), Bot (1983) and De Jong (1990). De Jong has experimentally derived an expression for two pane ventilation windows used in Venlo-type greenhouses:

$$q_v = (c_1 \, \beta \exp(-\beta / c_2)) \, u \, A_w \tag{Eq. 4.4.3}$$

where: u = outside wind speed (m s^{-1})
q_v = ventilation flux (m^3 s^{-1})
A_w = total window area (m^2)
β = window opening angle (°)
c_1 en c_2 = constants related to type of vent; see Table 4.4.2.

Equation (4.4.3) is valid for "quasi infinite" greenhouse covers; this means that the equation can be applied for the calculation of natural ventilation in greenhouse compartments located relatively far from the outside walls. In a finite greenhouse the average ventilation rate is affected by the presence of the side walls. De Jong (1990) observed that the ventilation rate in a finite greenhouse is higher than in a "quasi infinite" greenhouse, but did not quantify the effect.

Combined leeward side and windward side ventilation has also been investigated by De Jong (1990). The results show that for window opening angles on the windward side up to about 12° the total flux can be regarded as the sum of the separate lee side and windward side ventilation. No measurements

Table 4.4.1 – Natural air exchanges for greenhouses (ASAE, 1984).

Greenhouse construction system	Air infiltration rate (h^{-1})
New construction, glass or fibreglass	0.75–1.5
New construction, double layer plastic film	0.5–1.0
Old construction, glass, good maintenance	1–2
Old construction, glass, poor maintenance	2–4

Table 4.4.2 – Values of constants of a two pane ventilation window with a length of 1.46 m and a height of 0.8 m (De Jong, 1990).

Condition	c^1	c^2
Lee side vents open, windward side vents closed	0.00103	54.6
Windward side vents open, lee side vents closed	0.0012	211.1

were taken for larger opening angles of windward side windows. The most common sizes of vents are given in Table 4.2.1and Table 4.2.2 in section 4.2.

4.4.4.2 Forced ventilation

Calculations carried out by Van de Braak & Breuer (1991) show that, in The Netherlands, a ventilation capacity of about 120 $m^3\ m^{-2}\ h^{-1}$ is necessary in order to achieve desirable conditions in a greenhouse with mechanical ventilation. The yearly number of calculated ventilation hours has been determined to be between 1900 and 6000 depending on the method of control. As the cost of electricity is high in The Netherlands this method of ventilation is applied in research projects only.

4.4.5 Other cooling systems

4.4.5.1 Direct evaporative cooling

Direct evaporative cooling systems are based on the principle of cooling greenhouse air by the evaporation of water. The evaporation takes energy (about 2260 kJ kg^{-1} water) which is withdrawn from greenhouse air, which in turn will cool down. In its basic form evaporative cooling will decrease the temperature of the treated air but will increase the humidity because the treated greenhouse air is in direct contact with the evaporating water. Three types of this cooling principle are discussed in the following.

Fan and pad cooling

A fan and pad system consists of the following components: fans, cooling pads and a system delivering water to the top of the pads. Van de Muyzenberg (1980) notes that the fan and pad system was used for the first time in California USA at the beginning of the 1950's. The pads are usually constructed of cross-fluted cellulose material 5 to 300 mm thick placed in the wall facing the prevailing winds (Figure 4.4.1).

The cooling pad should extend the entire length of the wall of the greenhouse in which it is installed to ensure that all plants receive cooled air. It is best to place the fans on the leeward side (Figure 4.4.2) of the greenhouse opposite the wall containing the pads. Outside air drawn by the fans over the wetted pads, cools down and flows across the greenhouse, toward the fans and leaves the greenhouse.

Effectiveness of direct evaporative cooling depends on the wet bulb depression. In The Netherlands usually no larger depression than 5 to 6 K can be obtained whereas 14 K is attainable in many places with an arid climate (Meerman, 1989). Disadvantages of fan and pad cooling are fouling and growth of algae on the pad and the occurrence of unacceptable temperature and (relative) humidity differences in the greenhouse. For these reasons, this system is hardly used at all by Dutch growers.

Figure 4.4.1 – Cooling pad at windward side of greenhouse.

Figure 4.4.2 – Fans at leeward side of greenhouse.

Fog cooling

A system applying the same principle as fan and pad cooling is fog cooling. Water is forced through nozzles placed above the crop in a greenhouse producing a fog. If the size of the droplets is less than 10 microns they stay suspended in air while they are evaporating and precipitation on crop and people will be avoided.

The effectiveness of fog cooling does not distinguish itself from fan and pad cooling with the exception that under ideal conditions with fog cooling the wet bulb temperature can be obtained whereas with fan and pad cooling the maximum depression is 1 to 2 K less. Only a few Dutch growers use fog cooling.

Roof cooling

In a roof cooling system the roof of a greenhouse is flooded with water by means of irrigation or sprinkling (Figure 4.4.3). The roof material will be cooled by the water flow. In addition water flowing down the roof will absorb solar radiation and thus reduce the heat load of the greenhouse. The extent of this effect however is rather small: the absorbtion coefficient of water is fairly low. If the vents are open there is an additional effect. The air layer immediately above the roof will be cooled by evaporation. This relatively cool air will enter the greenhouse through the vents and contribute to the cooling of the greenhouse. Under Dutch weather conditions a cooling effect of the greenhouse air of about 3 K can be achieved.

4.4.5.2 Indirect evaporative cooling

The controllability of the climate under summer conditions leaves much to be desired. The possibility to discharge large quantities of energy by natural ventilation is limited. With evaporative cooling it is not possible to control the humidity of the greenhouse air sufficiently, and mechanical cooling is very expensive. A system which more or less combines the advantages of evaporative cooling (relatively cheap) and mechanical cooling (removal of sensible and latent heat) is indirect evaporative cooling. In an indirect evaporative cooler outdoor air passes along one side of an air to air heat exchanger. This air (the secondary airstream) is cooled by evaporation by one of the following methods: direct wetting of the heat exchanger surface, passing through a wetted pad medium or water spraying into the air-

Figure 4.4.3 – Roof cooling by sprinkling water.

stream. The surfaces of the heat exchanger are cooled by contact with the secondary airstream. On the other side of the heat exchanger surface, the primary airstream (conditioned air to be supplied to the greenhouse) is sensibly cooled by contact with the heat exchanger surfaces. To the authors knowledge the principle has not been applied in greenhouse cooling yet. De Jong (1992) and De Jong *et al.* (1993) describe a design of an indirect evaporative cooler for closed greenhouses. Computer simulations of the heat transfer process in the design showed that large quantities of both sensible and latent heat can be transferred.

4.4.5.3 Mechanical cooling

Mechanical or absorbtion cooling is generally not feasible due to high investment and exploitation costs. The cultivation of some crops however, such as freesia, require soil cooling. With a system of buried plastic tubes transporting cool water produced by the mechanical cooling device, the soil is chilled in these cases.

4.5 Screens

J.C. BAKKER AND G.P.A. VAN HOLSTEIJN

4.5.1 Introduction

Screens have been used for many years in greenhouse horticulture. Traditionally, their use was restricted to black-out and shading but in the late 1970's energy saving became an important motive for the use of screens. Depending on the material used, screens can have a large impact on the energy balance of the greenhouse through the reduction of ventilation, infra-red radiation and convection. During the 1980's the introduction of screens in commercial practice was promoted by rising energy costs and the simultaneous introduction of computers for environmental control. In this period due to the combined effort of research and industry the screens developed from simple fixed screens into sophisticated moveable systems with minimal dimensions. The screens became an integral part of the greenhouse construction whereby light interception can be reduced to minimal levels.

In contemporary Dutch greenhouse horticulture about 70% of the total glasshouse area is equipped with a type of fixed or movable screen for energy saving, shading or black-out. However, large differences exist in use between the various crops.

During the last decade, the need for energy saving has decreased to be replaced by the improvement of the glasshouse climate as one of the most important motives for their use. However, in the framework of the further reduction of environmental pollution, energy saving is expected to become once again a major motive.

4.5.2 Reasons for screening

The type of screen and material used depend on the major purpose of the screen. Basically there are four reasons for the use of screens:

Black-out (darkening)

The major purpose of this type of screen is to prevent light entering the glasshouse in cases of daylength treatments (see section 2.3.1). Its light transmission should be as low as possible (< 0.1%) as even very low light intensities may disturb the short-day treatment. Recently, the use of black-out screens has been discussed to prevent light emission from glasshouses equipped with artificial lighting. For such purposes the light transmission does not have to be zero but should reduce the light emission below a certain accepted maximum value. In all cases moveable screens are used mostly with (black or aluminized) materials which have such properties that they also can be used for energy saving.

Shading
These screens are used to reduce direct radiation and the overall light level in the glasshouse to prevent water stress, heat stress and quality reduction (see sections 2.2.1, 2.3.1 and 2.3.3). The most simple ways of achieving these effects are the use of shading paint (fixed screening) and rolling mats. More sophisticated systems are roller blind systems outside, and horizontal and roof shading inside. The use of these screens in The Netherlands, where moderate climate conditions prevail, until now has generally been restricted to a small group of crops (several pot plants, cut flowers and for plant propagation). However, during the last few years more and more special summer screens have been introduced. (Yates, 1986; Andersson, 1991).

Energy saving
With these screens the reduction of energy loss is the major aim (see also Meyer, 1982 and Müller, 1987). Several solutions are possible, largely dependent on the use of the screen. For example, if the use is restricted to the dark period only, the materials used don't need to have a high light transmission. For screens which are to be used during daytime as well, a high light transmission has to be combined with a high insulating effect and anti-condensation properties.

Environmental control
The use of this screen is primarily aimed at the improvement of the glasshouse climate. Post & Maaswinkel (1984) and Van Holsteijn (1987) reported better temperature distribution under screens, and Goeijenbier & Van Holsteijn (1986) reported about control of humidity by means of screens. Generally a screen for environmental control can be regarded as a combination of a screen for shading and energy saving (Plaisier, 1992). As it is hard to combine the different demands in one type of material, these screens are sometimes constructed as a double screen (Okada, 1985).

Beside these major four reasons, recently a specific application of screens has been introduced in The Netherlands in order to prevent insects entering the glasshouse (Berlinger, 1983).

4.5.3 Ways of screening

Apart from the classification based on the above mentioned reasons for screening, screens can be divided in two major groups: (1) permanent screens (fixed systems) and semi-fixed systems (partly moveable) and (2) moveable installations.

Fixed systems for energy saving have several disadvantages of which the continuous light interception and increased humidity of the greenhouse air are of major importance (Starkey, 1985). Generally a (perforated) film with a high light transmission is used and the period of screening is restricted to the early part of the growth period in winter. To partly overcome the disadvantages of fixed screens for energy saving, semi-fixed screens can be used. With these screens the material is partly pushed aside to create a so called "moisture gap" (De Graaf, 1985). However, the major disadvantage of the continuous light interception still remains.

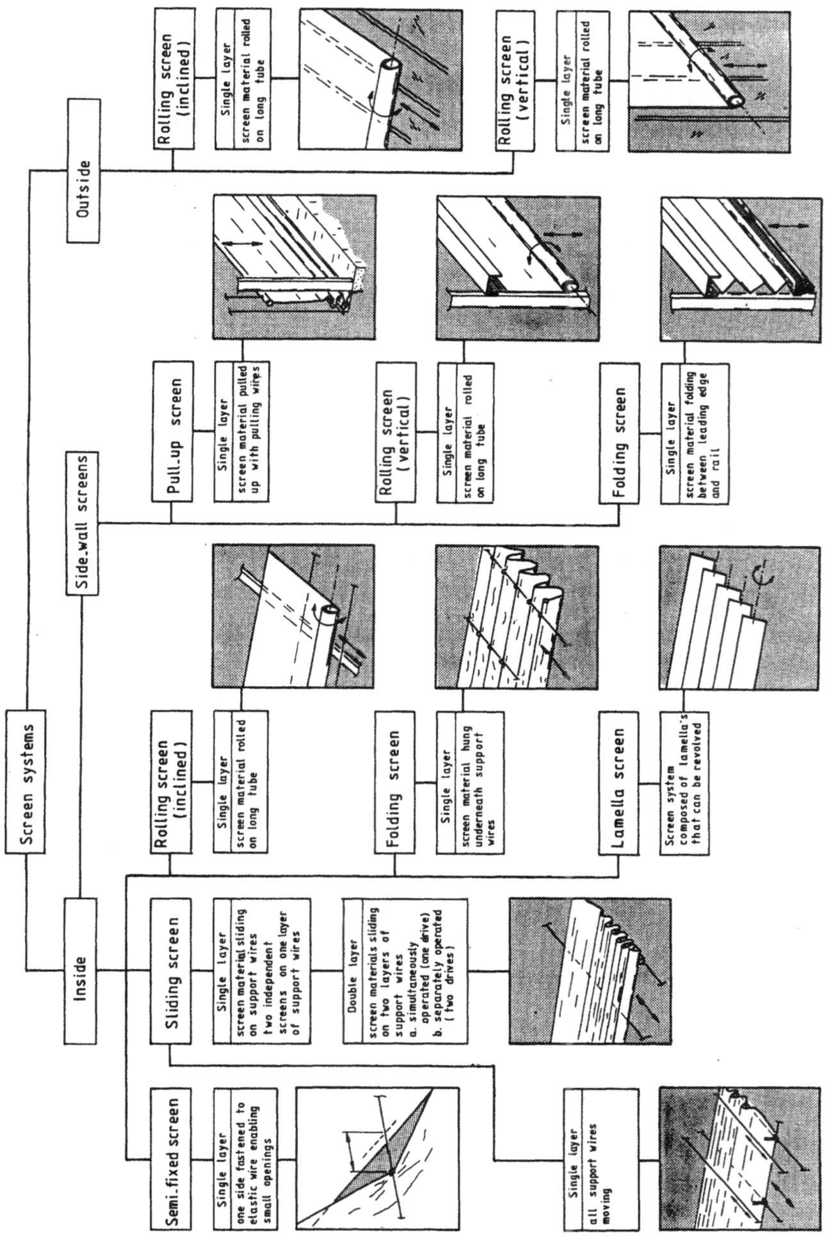

Figure 4.5.1 – Schematic overview of screens.

Shading paint can be regarded as a fixed screen, used during the summer period, of which the (continuous) light reduction depends on the amount of paint used. Some materials have a different light transmission under wet or dry conditions (e.g. T74), and in combination with a roof cooling system (see also section 4.4.5.1) this creates the possibility of on/off control of this type of fixed screen.

Because of the disadvantages of fixed and semi-fixed systems, their use (in The Netherlands) is restricted and moveable systems are preferred for most purposes.

Movable installations are the most widespread and they can be equipped with different materials depending on the major application. These systems are usually installed inside but for shading outside mounted screens are also used. In multispan glasshouses most screens are installed horizontally although in some cases (generally for shading) the screens are (partly) parallel to the cover. Systems used for moving the screening material are in order of importance: sliding, rolling, folding and lamella (Meijnders *et al.*, 1984). Vertical screens (sidewall screening), which are controlled independently from the screen overhead, are gradually being introduced. Beside the energy saving aspect these systems are generally used for black-out and constructed as rolling, folding or sliding (rumple) screens.

The different systems are schematically presented in Figure 4.5.1. In Figure 4.5.2 the different ways of opening and closing the screens (inside the glasshouse) are presented in more detail.

Generally the screen when opened should intercept as little light as possible. The folding or rolling system are the best in this respect (Wilkin & Bailey, 1985). However, in practice, sliding systems are used most widely because of easier and cheaper construction. With sliding systems, carriers or leading edge profiles are used to move the material uniformly and reduce the size of the screen package. The best solutions for reducing the light interception are folding the screen into or against a truss or under the gutter following the leading edge profiles (Figure 4.5.3). To estimate light interception of the total screen package a relatively simple rule of thumb can be used.

The light interception (for diffuse light conditions and without correction for reflection) of a construction part or screen package can be estimated by:

$$\text{light loss } (\%) = 100\% \times (0.5\, o)/s \qquad \text{(Eq. 4.5.1)}$$

with o = total perimeter of cross-section
s = mutual distance between construction parts.

Reflection of light by the construction parts reduces the total light interception, depending on the reflection coefficient and the height of the construction part. In Table 4.5.1 the light interception of the three most commonly used construction parts and screen packages are presented.

For the opening and closing of screens (sliding and folding) the drive mechanisms can roughly be divided in three groups (Figure 4.5.4):

- Pulling wires driven generally by a pipe shaft;
- Pulling and pushing bars driven generally by a rack and pinion;
- Tube motors or external motors for rolling screens.

The pulling wires are generally made of steel while for the supporting wires (to lay or hang the screen on) monofilament nylon (diameter 2.5 mm) is used. To prevent the screen from being lifted by air movement, top wires (nylon) are used.

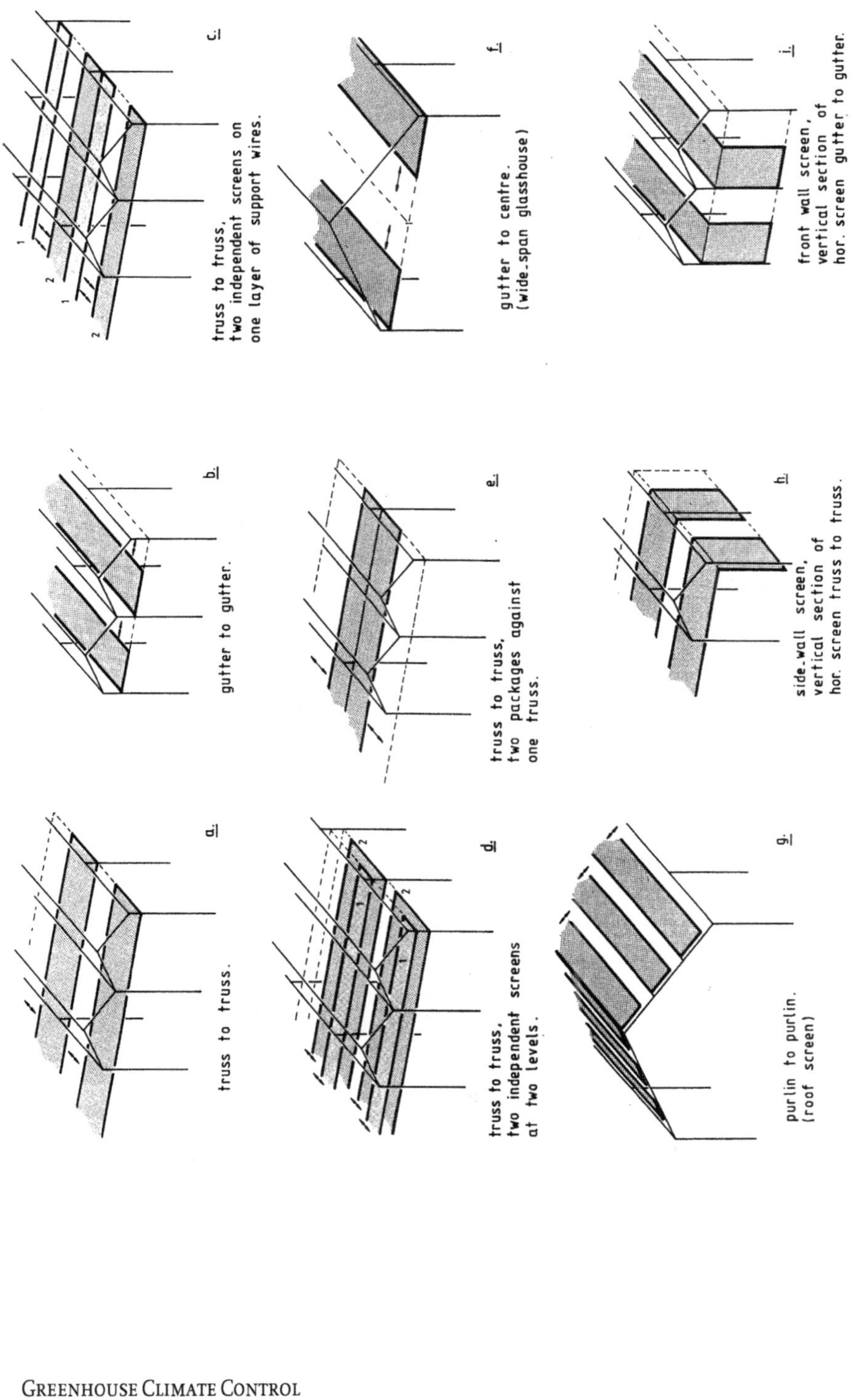

Figure 4.5.2 – Position and moving direction of screens.

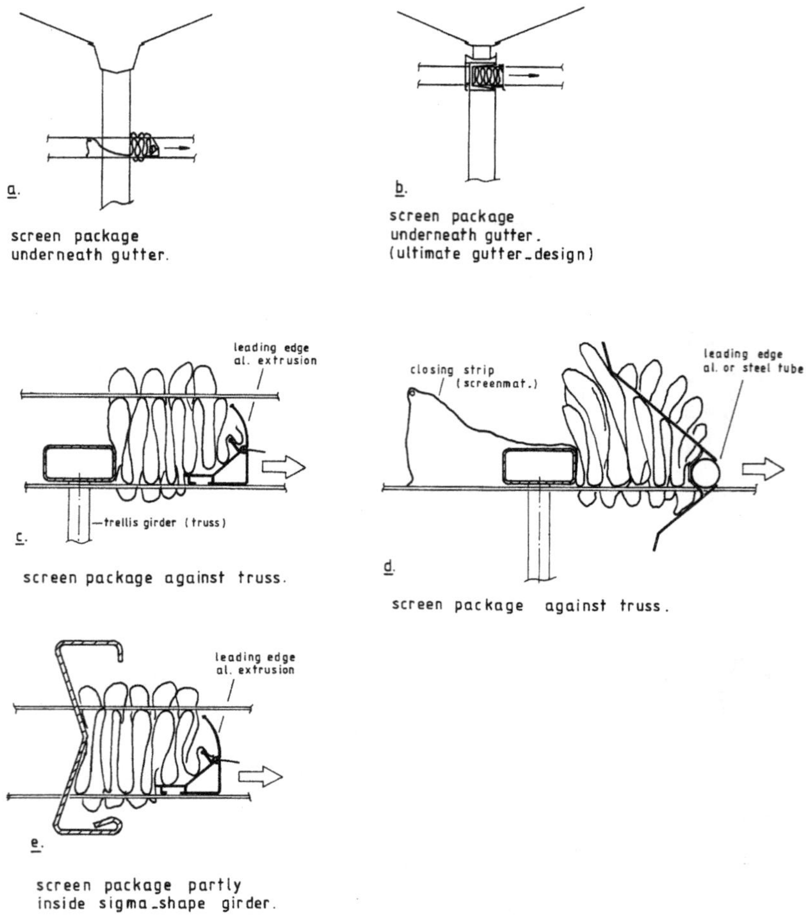

Figure 4.5.3 – Integration of screen and greenhouse construction.

Table 4.5.1 – Light interception by construction parts and screen packages.

Cross-section	Light interception	Reflection [1]	Total
Rectangle	$(w+h)/s$	$(R \times h)/s$	$(w+(1-R)h)/s$
T-profile	$1.62\,w/s$	Because of shape almost no reflection	$1.62\,w/s$
Circle	1.57 w / s	Difficult to estimate	Estimated in practice for one white pipe in each bay at: 2%

[1] *No reflection: R = 0, maximum reflection: R = 1.*

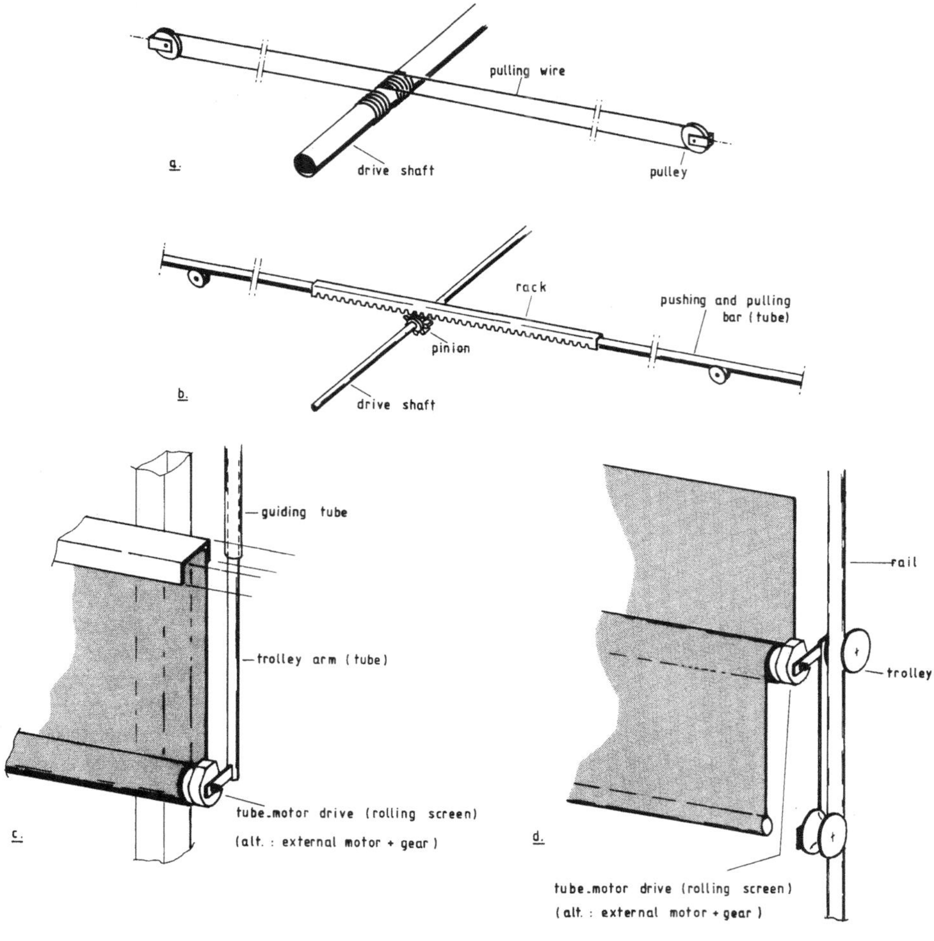

Figure 4.5.4 – Driving mechanisms for opening and closing greenhouse screens.

4.5.4 Screen materials

Beside specific criteria for the different applications, some general criteria for screening materials can be given.

The materials should be strong and have a high resistance to wear and tear. Furthermore they should be resistant to ageing from temperature, humidity, chemicals and ultra violet (UV) radiation. To increase the technical life-span special chemical stabilizers are sometimes added. With films in particular anti-fog additives are used to prevent dust adhesion.

A general criterion is also dimensional stability. Material for rolling screens especially should be virtually non-shrinkable. The maximum shrinkage for practical applications is about 2%. Suppleness and thickness are of major importance with sliding systems, materials should fold easily. Rolling systems are less demanding with respect to suppleness but more critical with respect to wrinkling and shrinkage. Finally for all screens, materials with a low inflammability are preferable.

Table 4.5.2 – Properties of different screening materials.

Type	Materials	Examples	Transmission for diffuse light (%)	Energy saving (%)[1]	Mechanical properties su/st/li [2]	Applications [3]
Film	PE	transparent	82(dry) 68(wet)	35	++/++/<1	c/e
		black	<0.1	45	++/++/3	b
		anti-fog/condens	82(dry) 82(wet)	35	++/+/<1	c/e
Fabric/yarns	polyester	Verzu-black	<0.5	55	+/+/4–5	b
		Verzu-white	55	35	+/+/4–5	s/e
Fabric/tapes	PE/aluminium	Phormilux	72	35	+/+/3–5	c/e
	PE/aluminium	PH 77–O	23	15	x/+/4–6	s
	PE/aluminium/acrylic	PH 20	70	35	x/x/3–5	s/e
Knitting/yarns	polyester	LS–100	<0.5	35	++/+/4–6	b
	polyester	LS–55	45	35	++/+/4–6	s
Knitting/ tapes + yarns	polyester	LS–10	68	40	+/+/3–5	c/e
	polyester/aluminium	ULS–17F	30	15	+/+/4–6	s
	polyester/aluminium	LS–13	65	45	+/+/5–8	e/s
	polyester/aluminium	LS–16	35	50	x/+/4–5	s/e
	polyester/aluminium	LS–11	2	60	-/+/4–5	e
	polyester/PE	LS–obscura	<0.5	35	++/+/4–6	b
	polyester/PE	LS–transforma	25	20	+/+/3–5	emission reduc.

[1] Estimated for Dutch conditions.

[2] su = suppleness; st = steadiness; li = lifetime (years); ++ = very good; + = good; x = moderate; - = little.

[c] b = black out; c = climate control; e = energy saving; s = shading.

Apart from these demands more and more attention is being paid to an environmentally conservative production process, extended lifetime and warranty period and the possibility of recycling the screen materials.

In Table 4.5.2 examples of the most widely used materials with their specific characteristics and applications are presented. The data represent estimates or indications for Dutch conditions.

4.5.4.1 Raw materials and different forms

Today's screening materials are made out of three raw materials: polythene, polyester or acrylic. Materials used previously, such as polypropylene, polyamide and cellulose have major disadvantages, especially with respect to durability. Acrylic fibre is the most resistant to ageing. Polythene without UV-stabilizers has a theoretical lifetime in greenhouses of about 1.5 years while that of polyester is around 5 years. In practice the lifetime of transparent PE-film screens is about a half to one year, for woven and knitted materials it is much longer (3 to 8 years) and the leading producers provide a warranty period of 3 to 5 years (see also Table 4.5.2).

The final form of the screen is the result of different techniques used. For the application in greenhouses four major groups of texture can be distinguished: film, fabric, knitting and nonwoven (Figure 4.5.5) or a combination of these techniques.

4.5.4.2 Specific criteria

Black-out

The light transmission of materials for black-out should be 0.1% or less which can be achieved with black PE-film, black weave or knitting. The latter materials reduce the risk of condensation forming on the screen. To combine black-out with a high energy saving, materials can be coated to reduce the emissivity (e.g. with aluminum), this is most effective with film. These films are used either directly or in woven or knitted materials.

Shading

If shading is the only application of the screen, generally woven or knitted materials with aluminium or white bands and a relatively "open" structure are used to minimize the effect on air exchange. The amount of light reduction varies from about 20 to 80% depending on the techniques and materials used (Yates, 1986; Andersson, 1991). Light distribution under these screens may be uneven, especially if the light reduction is relatively low (e.g. 20 to 30%). For purposes where relatively little light reduction is needed, open knitted materials (without aluminium bands) give a more uniform light distribution.

Open materials have much less effect on air exchange and radiation exchange than closed materials. The effects on energy consumption are therefore small to negligible. If the screen is also to be used for energy saving during the winter, a more closed material is necessary (Goebertus, 1989). At the same time this has of course disadvantages for the exchange of air and vapour during the summer. In practice this is generally compensated for by maintaining gaps.

Energy saving

While the insulating effect is of primary importance, the glasshouse climate should only be minimally affected. If the screen is used only at night, light transmission is unimportant but the screen package should be as small as possible. Film (whether or not coated to reduce the emissivity) or tightly woven or knitted aluminized materials meet these demands. The vapour exchange through PE-film is less than with the other materials but the transmittance for IR-radiation is higher which reduces the insulating effect (Nijskens, 1985; Balemans, 1989). In practice, however, the IR-radiation is absorbed by the condensation droplets on the film. To prevent very high levels of humidity and

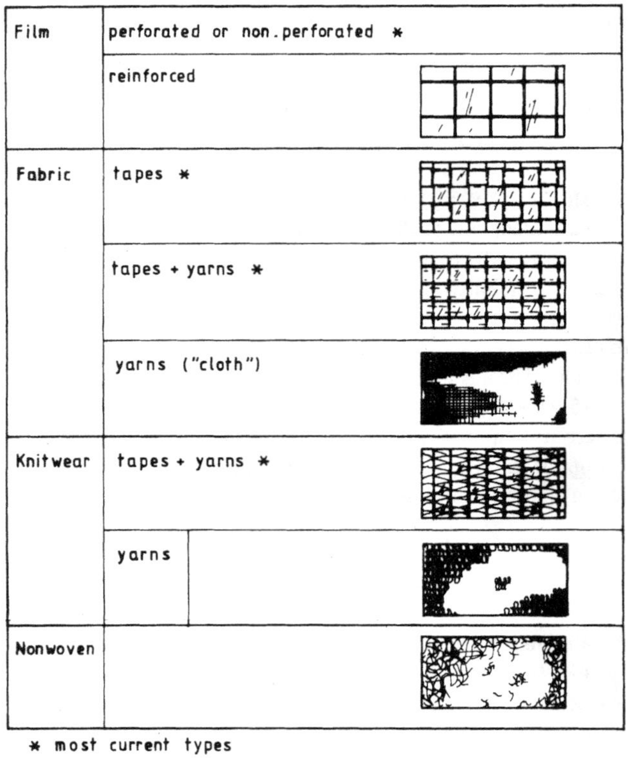

Figure 4.5.5 – Different screen textures.

increase the vapour transport, PE-films are sometimes perforated (6 mm diameter over a 10 cm or 20 cm grid which is equal to 0.25% or 0.07% opening). Energy savings of the various screen materials are given in Table 4.5.2.

In practice many energy saving screens are also used at daytime, so a high light transmission has to be combined with a high insulating effect. For these applications transparent materials (film or band weave) are used. The light transmission of PE-film is strongly reduced by condensation droplets. For example, the light transmission of clear dry PE-film is about 83% but falls to 66% with droplets on it. To overcome this problem, perforation can be used but anti-condensation additives are more effective. A disadvantage of these additives is that some films become adhesive. Furthermore, the anti-condensation effect gradually decreases with time and the maximum effective period is limited to about one year. As the price of these materials is relatively low compared to woven or knitted materials, film is usually removed after one growth season.

Environmental control

Screens for the improvement of the glasshouse climate are becoming more and more popular. The screens are mostly of woven or knitted "open" materials based on polythene or polyester each with specific properties to meet the various demands. Examples are screens with aluminium tapes used for various percentages of shading (Plaisier, 1990), screens with coloured film tapes used to modify the light spectrum within the greenhouse (Mortensen & Moe, 1992) and screens with opaline film tapes to disperse direct radiation.

4.6 Techniques of CO_2-enrichment

E.M. Nederhoff

4.6.1 Introduction

Supply of extra carbon dioxide (CO_2) is a commonly applied method to increase the yield of greenhouse crops (section 2.2.1.6). CO_2 enrichment can be achieved by different methods: (1) supply of pure (liquid) CO_2; (2) combustion of fossil fuel with small burners in the greenhouse; (3) combustion of fuels with a central burner outside the greenhouse, with the option to add a heat storage tank.

Supplying CO_2 may lead to local variations in CO_2 concentration, because the concentration declines from source to sink. Horizontal gradients in environmental conditions are generally disadvantageous, because they decrease the homogeneity of plant growth and crop production. Significant differences in tomato fruit production were found in a large greenhouse, where a systematic horizontal gradient in CO_2 concentration occurred (Van Holsteijn, 1992). Therefore special attention has to be paid to the lay out of the CO_2 distribution equipment.

CO_2 supply also causes vertical gradients. For instance with a distribution net, a high CO_2 concentration is found near the distribution tubes and a low level inside the canopy or near the opened ventilation windows. If the tubes are laid inside the plant beds or crop rows, the CO_2 enriched air first passes through the canopy before reaching the windows in the roof. This gradient is natural and not disadvantageous.

The pros and cons of the different supplying methods and the main technical characteristics of the installations are discussed in this section. The equipment for measuring and control is described in section 5.2.4.

4.6.2 Enrichment with pure CO_2

Enrichment of the greenhouse air with pure CO_2 is considered as most ideal, because supply is possible at any moment in the desired amount, restricted only by the available capacity and by costs. Pure CO_2 is obtained from the chemical industry (e.g. as a by-product of the manufacture of ammonia fertilizer), from biochemical processes (brewing) or from natural CO_2 sources. Unfortunately, in most countries, pure CO_2 cannot compete with CO_2 from fossil fuel, with respect to the price.

Pure CO_2 is supplied to the greenhouse from small refillable steel cylinders (bottles), or from a bulk storage tank refilled by road tankers, and currently also from a large central storage tank with a distribution net to several users. Bottles and small tanks contain CO_2 in liquid and gas phase, under relatively low pressure (about 1.8 MPa, depending on the surrounding temperature). In recent years small high pressure (6 MPa) tanks were quite common. Modern tanks nowadays are of larger capacity (containing 3 to 30 tons CO_2), at a constant pressure (2 MPa) and either vacuum sealed or actively cooled (below -18 °C).

There are two possible methods for the distribution of pure CO_2. Either the CO_2 passes through an (electric) evaporator (of about 1 kW per 10 kg h^{-1} CO_2) and a pressure reducer (output 100–300 kPa) and the CO_2 gas flows by its own pressure through a main duct (about 20 mm width), mostly provided with small (6–8 mm) PVC or nylon tubes at regular intervals. The alternative method is to insert the liquid CO_2 at unreduced pressure into the air flow of a fan, connected to a distribution net with lay-flat ducts. Such a system is used normally for distribution of flue gas CO_2 (section 4.6.3.4).

A special form of CO_2 supply is achieved by adding CO_2 to the irrigation water. A carbonator (e.g. trade mark "Carborain") is used to dissolve 0.6 to 0.8 gram CO_2 per litre irrigation water. The general opinion is that carbonated irrigation water has no significant effect on crop photosynthesis, but can have favourable effects on root growth and nutrient uptake. This might be due to a lower pH or to an influence on the ion exchange. There is some evidence that CO_2 can be absorbed by the roots (Baron & Gorski, 1986). It is also possible that hormones or hormone-like effects play a role (Enoch & Olesen, 1990).

4.6.3 Enrichment with CO_2 from flue gases

4.6.3.1 Combustion of fossil fuels

The oldest and still most common method of CO_2 enrichment is combustion of fossil fuel. The physiological and technical principles were researched in great detail as long ago as the 1930's (Bolas & Melville, 1935) and the method has been widely applied in practice since the 1960's.

The flue gases to be used for CO_2 enrichment must not contain dangerous amounts of injurious components (see also sections 2.2.1.3 and 4.6.3.2). Hence, appropriate fuels for CO_2 enrichment are low-sulphur oils, natural gas, premium kerosene (paraffin) and propane. The different fuels and the consequences of noxious gases are discussed by Hand (1982 and 1986). Availability and price dictate the use of a particular fuel in a region. In The Netherlands, 99% of the glasshouse area is heated by natural gas (almost sulphur free). In other countries this fuel is also important, as it is particularly appropriate for combustion for CO_2 enrichment.

Dutch natural gas consists of 81% methane (CH_4), 14% nitrogen (N_2) and 0.89% CO_2 and a number of hydrocarbons and some minor components. The gross heating value (i.e. including the latent heat in the water vapour) is 35.17 MJ m^{-3} natural gas (Nederlandse Gasunie, 1988). Natural gas produces approximately 1.8 kg CO_2 per m^3 (at 20 °C and normal atmospheric pressure).

Combustion of 1 m^3 natural gas requires 1.77 m^3 oxygen, which is equal to about 8.5 m^3 air (at equal temperature and pressure). Combustion of exactly this ratio of air and fuel is termed stoichiometric combustion, i.e. the air factor (n) equals 1. Normally the burners are adjusted to a higher air factor, e.g. $n = 1.6$, indicating a surplus of 60% air. Table 4.6.1 presents the composition of the exhaust gases for different air factors and different temperatures of the flue gas. Where $n = 1$ the CO_2 content of the flue gas mixture is 9.5% and where $n = 1.6$ it is 6.2%. A fraction of the N_2 of the air might be oxidized to NO_x, but this is omitted in Table 4.6.1.

The most direct method of CO_2 enrichment is with small burners located in the greenhouse. This method has several disadvantages, as will be explained below, and it has been superseded in The Netherlands by the method of fuel combustion in a central burner outside the greenhouse. There is a tendency to installing gas or oil fired engines for co-generation of electricity and heat, e.g. for glasshouses with artificial lighting (see also section 4.7). A drawback is that the flue gases contain large amounts of phytotoxic pollutants, making these gases unfit for CO_2 enrichment. Many attempts are being made to improve the combustion in these engines and to develop purification systems, for example by using catalysts.

4.6.3.2 Incomplete combustion and noxious gases

With all heating installations, the air supply to the burner must be properly adjusted. Insufficient air supply may lead to incomplete combustion, and hence the production of harmful gases, including carbon monoxide (CO), the highly toxic ethylene (C_2H_4) and other unsaturated hydrocarbons. The production of NO_x is also related to the air supply. In a modulated burner, the valve for the air supply is connected to the gas inlet. It is possible to install a special instrument, that measures the oxygen content in the burner, and adjusts the air supply when necessary. However, this is applied only in a few cases, due to the high costs.

It is highly advisable to monitor the flue gas of a central heater for incomplete combustion, in order to prevent the injection of ethylene into the greenhouse. There is no reliable, cheap analyzer for ethylene currently available. Based on the assumption that there is a direct relation between the production of ethylene and of carbon monoxide (CO), it is common practice to monitor CO in the flue gases. If a certain level of CO is detected (30 ppm in undiluted flue gases with 10% CO_2), the CO_2-fan is turned off, and the flue gases are led into the stack. This system can only be applied with a central boiler.

All burners produce some nitric oxide (NO), which can be oxidized to nitrogen dioxide (NO_2). The amount produced depends on the flame temperature (more NO at higher temperature) and other combustion factors. Effects of NO_x after different exposure times can be found in Hand (1990) and Hand & Hannah (1990). Gas monitoring in practical situations has demonstrated that in closed greenhouses the concentration of NO_x may easily reach a level of 0.5 to 1 vpm, at which level injury may occur (Hand, 1990; Kiel, 1990).

It is fairly unpredictable at what concentration the noxious gases will actually damage the crop. Many factors affect the incidence of damage, e.g. the duration of exposure to the gas, condition of the plant, environmental conditions (radiation, temperature, humidity, CO_2) during exposure and before, the presence of other gases, etc. The reported maximum acceptable concentrations (MAC) for humans and plants of some noxious gases are given in Table 4.6.2.

If the threshold concentration for plants of noxious gases is known (see Table 4.6.2), the level of concentration of this gas $c_{x,f}$ allowed in the flue gas can be calculated. Therefore a dimensionless dilution factor (F_d) must be known which can be derived from a simple steady state mass balance of the gas component considered without any absorbtion of the noxious gas:

$$q_{v,f} c_{x,f} = q_{v,vnt} c_{x,in} \quad \text{(Eq. 4.6.1)}$$

or:

$$F_d = q_{v,vnt} / q_{v,f} = c_{x,f} / c_{x,in} \quad \text{(Eq. 4.6.2)}$$

Table 4.6.1 – Composition of flue gas (in volume-%) from combustion of 1 m³ Dutch natural gas for four different air factors (n, with n = 1 indicating stoichiometric combustion) and two flue gas temperatures (one below and one above the dew point). The relative humidity of the combustion air is assumed to be 50% at 20 °C. The volume (m³) is total flue gas volume at 0 °C and normal atmospheric pressure.

n	1.0		1.3		1.6		1.9	
T	40	200	40	200	40	200	40	200
Volume	9.49	9.49	12.04	12.04	14.60	14.60	17.81	17.81
N_2	80.95	70.91	78.84	72.24	77.57	73.11	76.72	73.73
H_2O	7.27	18.78	7.27	15.04	7.27	12.60	7.27	10.90
CO_2	10.83	9.49	8.16	7.48	6.55	6.17	5.47	5.26
O_2	0	0	4.8	04.40	7.70	7.25	9.63	9.26
Ar	0.94	0.83	0.92	0.85	0.91	0.86	0.90	0.87

Table 4.6.2 – Maximum acceptable concentration (in vpm) for humans and plants of some noxious gases.

Gas	Humans	Plants	Plants [3]
Carbon dioxide (CO_2)	5000 [1]	4550 [1]	1600
Carbon monoxide (CO)	47 [1]	100 [1]	
Sulphur dioxide (SO_2)	3.5 [1]	0.1 [1]	0.015
Hydrogen sulphide (H_2S)	10.5 [1]	0.01 [1]	
Ethylene (C_2H_4)	5.0 [1]	0.01 [1]	0.020
Nitrous oxide (NO)	5.2 [1] / 5.0 [2]	0.5 [1] / 0.01–0.1 [2]	0.250
Nitrogen dioxide (NO_2)	5.0 [2]	0.2-2.0 [2]	0.100

[1] *Langer et al. (1990b).*
[2] *Döring (1987).*
[3] *Long-term exposure, Rijsdijk (1989).*

where $q_{v,f}$ is the volume flux of flue gas and $c_{x,in}$ the concentration of the considered component in the greenhouse.

The CO_2 balance (equation (3.5.1), Chapter 3) can be written as:

$$q_{v,f}c_{c,f} = q_{c,p} + q_{v,vnt}(c_{c,in} - c_{c,ex}) \quad \text{(Eq. 4.6.3)}$$

If $q_{c,p} << q_{c,vnt}$ $(= q_{v,vnt}(c_{c,in} - c_{c,ex}))$, (photosynthesis uptake much lower than ventilation flux of CO_2), it can be seen from equation (4.6.3) that

$$F_d = q_{v,vnt} / q_{v,f} = c_{c,f} / (c_{c,in} - c_{c,ex}) \quad \text{(Eq. 4.6.4)}$$

So then the dilution of the noxious gases is equal to that of CO_2. If $c_{c,in}$ equals $c_{c,ex}$, the CO_2 flux by ventilation will be low and comparable to the photosynthesis uptake, and equation (4.6.3) has to be applied completely. Also if the CO_2 concentration in the greenhouse is low, $c_{x,f}$ must be determined from F_d found from the complete balance (equation (4.6.3)). The results of such calculations for a number of cases are presented in Table 4.6.3.

In future the emission of exhaust gases will be bound by regulations, particularly with respect to NO_x. Hence, the (Dutch) burner manufacturers put much effort into improving the burners (e.g. "low-NOX burners"). Reduction of more than 80% in NO_x emission of hot-air heaters has been achieved (Kiel, 1990). A seal of approval is given to hot-air heaters that meet certain requirements (Kooiman, 1990). These developments greatly reduce the risk of NO_x insertion with flue gas CO_2 enrichment. To avoid "invisible injury", it is highly recommended that regular maintenance of the burner(s) takes place and that old burners are replaced by modern clean burning types.

4.6.3.3 CO_2 enrichment with small burners in the greenhouse

CO_2 enrichment can be done by the use of small burners, releasing the flue gases directly into the greenhouse. The burners range from simple natural-draught open-flame burners solely for CO_2 generation, to relatively large forced-draught direct-fired units, primarily meant for hot-air heating. Many different manufacturers and designs are known over the world. Most small burners are suspended from the greenhouse structure, but some (older) models hot-air heaters are standing. Small burners provide an easy and direct method for enrichment, but contain a number of disadvantages, as discussed below.

Table 4.6.3 – Concentration (vpm) of noxious gases allowed in flue gases used for CO_2 enrichment under winter conditions, calculated for various CO_2 concentrations (vpm) in the greenhouse air. Assumptions are: CO_2 in the flue gas 10%, photosynthesis 0.5 g m^{-2} h^{-1}, air exchange 2 m h^{-1}, threshold levels as in Table 4.6.2 after Rijsdijk (1989).

CO_2 concentration in greenhouse air	350	700	1000	2000	5000
NO	182	51	32	14	5.2
NO_2	73	21	13	5.6	2.1
C_2H_4	15	4.1	2.5	1.1	0.4

Two types of burners are distinguished, with respect to air supply for combustion: (1) atmospheric burners, with natural draw of ambient air and (2) fan assisted burners. In the latter, a fan is built in for attraction of air, either greenhouse air or outside air, inhaled through a duct. The air supply is critical and requires special attention, particularly in winter. If the burners are activated continuously and the air refreshment is low (windows closed, leaks frozen), the oxygen concentration in the greenhouse may drop by some percent, which may give rise to incomplete combustion in atmospheric burners. A fan burner might receive insufficient air in winter, because of low air den-ity at low outside air temperatures. Most burners are adjusted for a sufficiently large air factor (e.g. $n = 1.6$) to avoid incomplete combustion at all times.

Generally with small burners, the control of the CO_2 level in the greenhouse is limited, for two reasons. Usually the small burners can only operate at full capacity, which makes the control very inaccurate (on/off). There are some burners with a low and high flame, and a CO_2 generator with modulated flame has been introduced. If the burners are primarily used for heating, the CO_2 level in the greenhouse is considered of secondary importance. Hence, in practice, the CO_2 level in winter is often far too high and even injurious, while on warmer days no CO_2 is supplied at all.

The extremely high CO_2 levels (> 10,000 ppm), which are frequently observed during severe heating in winter, are above the maximum acceptable concentration (MAC) for humans and above threshold values for plants (Holländer & Krug, 1991). Therefore, ventilation windows are partly opened or side wall fans are activated, whenever the CO_2 concentration exceeds a certain level. Also special small heat-exchangers are available, to achieve loss of excessive CO_2 without loss of heat.

The flue gases of small burners are released freely or blown into the greenhouse by a built-in fan. This results mostly in a non-homogeneous CO_2 concentration. The extent of horizontal and vertical gradients depends on the number, location and emission of the burners. It is possible to connect CO_2 distribution tubes to the burners, but this is not a satisfactory solution (Jacobi *et al.*, 1990; Langer *et al.*, 1990a). The distribution of CO_2 (and heat) can occasionally be improved to a certain degree by additional, separate fans in the greenhouse (Goeijenbier, 1986).

4.6.3.4 CO_2 supply from a central burner

In a greenhouse equipped with hot water pipe heating from a central boiler, most of the flue gases from this burner can be made available for CO_2 enrichment, at least if the flue gases are pure enough (sections 2.2.1.3 and 4.6.3.2). This has been practised for many years on a large scale in The Netherlands (Van Berkel, 1975). The most important advantage of central CO_2 enrichment over enrichment with burners in the greenhouse is, that CO_2 and heat are produced outside the greenhouse itself and can be inserted separately into it. This enables enrichment to take place for many more hours in spring and summer. Further, the maintenance and also the monitoring for incomplete combustion is easier with one single burner than with a number of small burners. An additional advantage is that the water vapour produced by combustion can be removed in a flue gas condenser.

When CO_2 is produced by combustion, a considerable amount of heat is also formed, which is sometimes needed, but sometimes undesirable. Thus various possible destinations for the heat can be distinguished, as it can be (1) used directly for greenhouse heating, or (2) released into the greenhouse as a method of releasing the heat (e.g. by maintaining a minimum heating pipe temperature); or (3) stored during daytime and brought into the greenhouse at night. Theoretically, the produced heat can be rejected, i.e. transferred to the surrounding air or canal water, possibly by means of a heat exchanger. However, this is mostly not a feasible option and moreover it is not recommended from an environmental point of view.

As long as heating is needed, plenty of CO_2 will be available and the excess CO_2 can be eliminated through the chimney stack. If heating is not really needed, but not harmful to the crop either, CO_2 enrichment is still possible, using the burner at a low flame. In this case a "minimum heating pipe temperature" is usually applied, even during ventilation. On warm days, the CO_2 supply must be stopped because it is virtually impossible then to get rid of the produced heat. Thus this system does not satisfy under summer conditions either, because heat and CO_2 are not needed at the same time in equivalent amounts.

The different demands for heating and CO_2 are illustrated by the following figures, valid for a moderate climate as in the Netherlands. In winter up to 300 m^3 ha^{-1} h^{-1} of natural gas is combusted for heating. On a winter day a small fraction of the flue gases, equivalent to 25 m^3 ha^{-1} h^{-1} natural gas, is required for CO_2 enrichment. On a summer day, the heat demand is often nil, whereas the demand for CO_2 is considerable. Because the heat produced must be stored then and the storage capacity will be limited, the supply must be set to a minimum level (in practice about 4.5 g m^{-2} h^{-1} CO_2 or 25 m^3 ha^{-1} h^{-1} natural gas). In that case it must be possible to reduce the burner to about 1/8 or 1/10 of its full capacity.

An adjustable (or modulated) burner usually has a low and a high flame, with a corresponding low and high rotation speed of the burner fan and a low and high gas intake. The low flame, also called "CO_2 flame", is meant for continuous supply of CO_2 at a low level. Some burners allow for continuously variable modulation, for both low and high flames. In some cases a special small burner is built into or added to the large burner, instead of a low flame.

CO_2 enrichment from a central boiler requires a properly designed transport and distribution system for CO_2. The capacity has to be at least large enough for the transport of the flue gases from the CO_2-flame of the burner. With a modulated burner also larger amounts, up to a certain maximum, must be distributed. If less than the maximum is to be supplied, the flue gas is diluted with air. Therefore a T-shaped connection pipe is mounted in the flue gas duct near the stack. A CO_2-unit is used, consisting of a fan, to extract the flue gases from the stack, and a valve to control the inlet of air for the dilution of the flue gases. The fan is a centrifugal type of approximately 1 to 2.5 kW per ha greenhouse area.

The amount of flue gases to be transported depends on the desired CO_2 supply, and on the temperature and the CO_2 content of the flue gas. The flue gas temperature can be as high as 200 °C, or cooled down by a flue gas condenser to about 40 °C. If the flue gas temperature is cooled below the dew point, it will loose a considerable amount of water vapour. The volume and the composition of flue gas are given in Table 4.6.1 for different air factors and flue gas temperatures.

The flue gas with CO_2 is transported to the greenhouse through a duct of PVC, or of aluminum in case the flue gases are not cooled down. The main duct in the greenhouse tapers off towards the end, to maintain an equal gas pressure. The flue gas is sometimes released from holes equally spaced along the main duct, but this results in a very unequal CO_2 distribution. A good and cheap system is a net of perforated PE-film lay-flat ducts of 50 mm diameter with four 1 mm diameter perforations per 20 to 120 cm. It is recommended that one such lay-flat duct is connected to the main duct every few metres, e.g. one duct in each 3.2 m bay of a Venlo-greenhouse. The length of the ducts should not exceed 40 m. The recommended pressure at the beginning of the duct is about 750 Pa. In order to adjust the pressure distribution in the net, the opening of each duct is sometimes reduced to a certain extent by a fixed throttle at the conjunction site.

4.6.3.5 Heat storage for prolonged CO_2 enrichment

Short term heat storage to enable CO_2 enrichment, is applied on a large scale in modern glasshouse holdings in The Netherlands. The area of tomato glasshouse equipped with heat storage is estimated to be 50% of the total at present and is still increasing. Generally a well insulated water tank with a volume between 30 and 130 m^3 per ha glasshouse is used.

CO_2 enrichment particularly in the warmer seasons is greatly improved by the use of such a storage tank for heat from the central boiler. In spring and summer, if the demand for CO_2 exceeds the demand for heat, fossil fuel can be burned at day time for CO_2 enrichment, and the produced heat can be stored for utilization at night. The daily amount of CO_2 supply from a central burner is then limited by the capacity of heat storage or by possibilities to use or release the stored heat at night.

At daytime, natural gas is burned for CO_2 and warm water is pumped into the storage tank. The inlet for hot water is on the upper side, while at the same time cold water flows from the lower part of the tank to the boiler. Due to the difference in density between hot and cold water, the water temperature remains layered. Heat retrieval, mostly at night, is done by pumping hot water from the upper part of the tank to the greenhouse, in some cases via the boiler. Heat storage systems are also described by Van Berkel and Verveer (1984), Bell *et al.* (1990) and Langer *et al.* (1990b).

The amount of natural gas (Q_v in m^3) that can be combusted for enrichment if all heat produced is stored in a heat storage tank, depends on the volume of the tank (V_t in m^3), on the increase of water temperature in the tank (ΔT in K), the efficiency of the burner including a condensor (η_b) and the efficiency of the heat storage tank (η_t), as follows:

$$Q_v = (V_t \, \Delta T \, \rho C_p) \, / \, (H \, \eta_b \, \eta_t) \qquad \text{(Eq. 4.6.5)}$$

ρC_p is the volumetric heat capacity of water (4.2 MJ m^{-3} K^{-1}) and H is the gross heating value of natural gas (35.17 MJ m^{-3}). Assuming V_t is 75 m^3, ΔT is 60 K, η_b is 0.88 and η_t is 0.90, the tank can store the heat of 680 m^3 natural gas, or 68 m^3 per hour on a day of 10 hours. The average CO_2 supply is then 12.5 g m^{-2} h^{-1} (assuming one ha greenhouse area), which is almost three times the recommended minimum amount.

Short term (less than 24 h) heat storage is a feasible option for glasshouses with central heating, in a climatic region where heating is needed in summer at night. During periods when no heat is needed at all, day-to-night heat storage is not useful. Thus the capacity of the tank should be suited to the local climatic conditions. If too much heat is stored (e.g. because too large a tank is used at full capacity), not all stored energy will be utilised at night, and hence the storage water will not have cooled down sufficiently at the beginning of a new day.

A heat storage tank requires a special control programme for storage and retrieval of heat. Usually just simple algorithms are applied for this purpose. However, there is a computer algorithm commercially available, that optimizes the CO_2 enrichment and the heat utilization, on the basis of simulated photosynthesis and calculated air exchange rate, taking into account the local weather forecast for the next 24 hours.

4.7 Supplementary lighting

J.P.G. HUIJS

4.7.1 Introduction

In greenhouses artificial light is used in several ways. The most important application is the use for supplementing daylight in greenhouses to increase the irradiance level for photosynthesis. Another application is to increase the daylength (photoperiodism). The main crops in The Netherlands for the application of artificial light are roses and chrysanthemums, and the first fase of growing young plants. In this section the equipment for supplementary lighting in greenhouses will be described.

4.7.2 Lamps and fittings

The choice of lamp for an irradiation installation depends on the purpose of the lighting. A variety of interrelated factors have to be considered, such as time and level of irradiation, spectral distribution of the lamp, photoperiodic characteristics and environmental requirements of the plants, supplementation or substitution of the daylight, available space and switching cycles. The selection of the best lamp for a specific application is often difficult (Philips, 1987).

Table 4.7.1 shows the characteristics of lamps used in horticulture. A high-pressure sodium lamp (e.g. SON-T) combines a high radiant efficiency, small size, long technical life, low depreciation and a constant radiant flux with a spectral energy distribution that is suited to several crops. For these reasons the SON-T lamp is for supplementary lighting in greenhouses the most popular type.

Table 4.7.1 – Characteristics of lamps used in horticulture.

Lamp type	Rated power (W)	Luminous flux (lm)	Conversion factor (mW lm^{-1})	Radiant flux (mW)	Radiant eff. (mW W^{-1})	Economic life time (h)	Application
Incandescent GLS (150 W)	150	2,220	4.2	9,320	62	1,000	photoperiodism
Fluorescent 'TL'D 33 (58 W)	70	4,800	2.9	14,000	194	7,500	photoperiodism photosynthesis
Metal halide HPI-T (400 W)	413	31,500	2.8	88,200	214	8,000	photosynthesis
Compact gas-discharge SL (25 W)	25	1,200	2.8	3,360	134	6,000	photoperiodism
High-pressure sodium SON-T (400 W)	436	47,000	2.3	108,100	250	12,000	photosynthesis

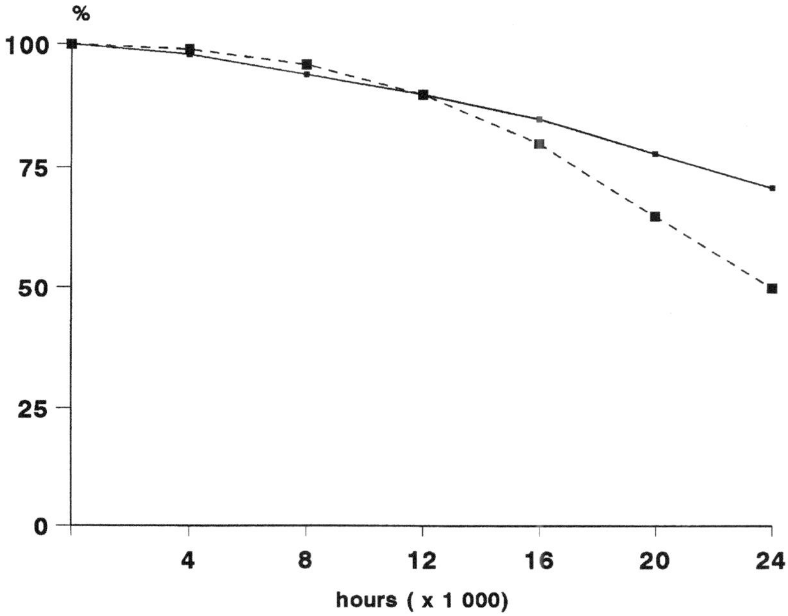

Figure 4.7.1 – Depreciation (•) and lamps alive (▫) of a high pressure sodium lamp as a function of operating time.

Figure 4.7.1 shows the depreciation and lifetime of a high pressure sodium lamp. The economic life of the lamp is 12,000 hours. After this time radiant efficiency and number of lamps still working are both about 95%.

Fittings used in greenhouses have to be dust- and waterproof. For photosynthesis lighting a special range is available. These fittings combine a wide angle of light distribution with good lighting uniformity. This is necessary to allow a wide spacing between the fittings at a restricted mounting height. All fittings are equipped with built-in control equipment, and have to be as small as possible to minimise interception of daylight. A mirror reflector ensures a radiant efficiency of approximately 90%.

4.7.3 Sources of electricity supply

The electricity required for the lighting system can be provided by the public network or by a cogenerator (Huijs, 1988). The advantage to the grower of using electricity from the public network is that no capital outlay is required for electricity generating equipment and that heat surpluses are usually low. The disadvantages are the high price of the electric energy, the large financial contributions when cables have to be laid for higher capacity and the loss of thermal energy produced in the generating process.

The advantage of generating electricity with a cogenerator is that the total energy efficiency (electrically and thermally) can be 20 to 30% higher than if electric energy is produced by the public network. This is due to the fact that a part of the heat produced in the generating process is used to

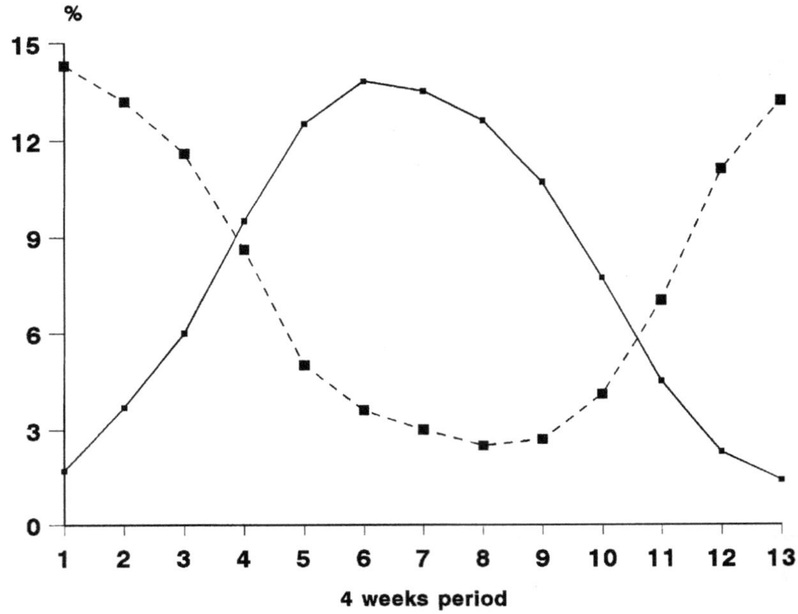

Figure 4.7.2 – Percentage of the annual radiation (•) and the heat consumption of roses (▫) during a 4 weeks period.

heat the greenhouse. Disadvantages are that a high capital outlay is required and that heat surpluses can occur through fluctuating demands of electricity and heat.

The choice between electric power supply through the public system and a cogenerator on the nursery is predominantly an economic issue. What is chosen is mainly determined by the capital outlay required cogenerator, the ratio between energy prices of both concepts and specific aspects of the nursery. A major aspect as regards the feasibility of a private generating unit is the number of operating hours.

4.7.4 Supplementary lighting and cogenaration

The need for supplementary lighting exists when the light level inside the greenhouse is too low as a result of inadequate global radiation. When assessing the economic feasibility of a total energy generating system it is of a great importance to determine whether the needs for electric power and heat coincide to a considerable degree (IKC-AT, 1990; 1991).

Figure 4.7.2 shows the relationship between the global radiation and the heating requirement for roses (cultivar Madelon). The radiation level in period 6 appears to be ten times as high as in period 13. The heat consumption in period 1 appears to be about six times as high as in period 8. This figure demonstrates that in winter there is a clear minimum for the global radiation, whereas for the rose crop the highest heat consumption is found.

Therefore, the use of cogeneration combined with supplementary lighting is a promising option from the energy point of view. Even if the needs for additional light and heat over the year run fairly well in parallel, the demand for heat will not always be met (peaks), so that the use of a boiler remains necessary. For a cogeneration the ratio between electricity and heat generated is fairly constant. The need for electric energy and light tends to be subject to short-term fluctuations. These fluctuations can be bridged by temporary storage of heat.

References

Aldrich, R.A., 1986. In: M.L. Esmay & J.E. Dixon (Eds), Environmental control for agricultural buildings (Chapter 16). The AVI Publishing Company Inc., Westport, Connecticut, 287 pp.

Aldrich, R.A. & J.W. Bartok, 1989. Greenhouse engineering. Northeast Agricultural Engineering Service, Ithaca, 203 pp.

Aldrich, R.A., R.J. Downs, D.T. Krizek & L.E. Campbell, 1983. In: M.A. Hellickson & J.N. Walker (Eds), Ventilation of agricultural structures (Chapter 10). American Society of Agricultural Engineers, St. Joseph, Michigan 49085, ASAE Monograph no. 6, 372 pp.

Andersson, N.E., 1991. Spectral properties of shading materials. Tidskrift for planteavl 95(3): 345–351.

ASAE, 1984. Heating, ventilating and cooling greenhouses. ASAE Standards 1984. American Society of Agricultural Engineers, St. Joseph, Michigan 49085, ASAE EP406: 397–400.

Baille A. & B. Von Elsner, 1988. Low temperature heating systems. In: C. Von Zabeltitz (Ed.), Energy conservation and renewable energies for greenhouse heating. CNRE guideline No. 2, FAO, Rome, REUR Technical Series 3: 149-165.

Balemans, L., 1989. Assessment of criteria for energetic effectiveness of greenhouse screens. Rijksuniversiteit Gent, Gent, 157 pp.

Baron, J.J. & S.F. Gorski, 1986. Response of eggplant to a root environment enriched with CO_2. HortScience 21: 495–498.

Bell, G., J. Jung & H. Dehnz, 1990. CO_2-Gewinnung durch Nutzung von Rauchgasen aus mit Rohbraunkohle gefeuerten Heizkesselanlagen. *(CO_2 from coal fired boilers)*. Gartenbau 37(3): 77–78 (in German).

Berlinger M.J., 1993. The effect of greenhouse screens on the presence of western flower thrips: a preliminary study. In: J.C. Van Lenteren (Ed.), Contributions working group meeting. Pacific Grove, California, p. 13–16.

Bolas, B.D. & R. Melville, 1935. The effect on the tomato plant of carbon dioxide produced by combustion. Annals of Applied Biology 22: 1–15.

Bot, G.P.A., 1983. Greenhouse climate: from physical processes to a dynamic model. PhD Dissertation, Wageningen Agricultural University, Wageningen, 240 pp.

Breuer, J.J.G., 1990. Jaarbelastingduurkromme en energiebesparing voor de Nederlandse glastuinbouw: een studie naar relevante invloeden. *(Heat duration curve and energy saving in Dutch glass house cultivation)*. IMAG-DLO, Wageningen, IMAG-DLO rapport 90: 67 pp. (in Dutch).

De Graaf, R., 1985. The influence of thermal screening and moisture gap on the transpiration of glasshouse tomatoes during the night. Acta Horticulturae 174: 57–66.

De Jong, T., 1990. Natural ventilation of large multi-span greenhouses. PhD Dissertation, Wageningen Agricultural University, Wageningen, 116 pp.

De Jong, T., 1992. Alternatieve koeler voor gesloten ruimten. *(Alternative heat exchanger for closed spaces)*. Klimaatbeheersing 21(5): 123–128. (in Dutch).

De Jong, T., N.J. Van de Braak & G.P.A. Bot, 1993. A wet plate heat exchanger for conditioning closed greenhouses. Journal of Agricultural Engineering Research 56: 25–37.

Döring, K., 1987. CO_2-Düngung: mehr Qualität, weniger Kosten? *(CO_2-fertiliser: more quality, less costs?)*. Zierpflanzenbau 17: 649. (in German).

Enoch, H.Z. & J.M. Olesen, 1990. Neue Studien zu Carborain: Sprühen und Giessen von Pflanzen mit

CO_2-angereichertem Wasser. (*New study of Carborain*). Zierpflanzenbau 12: 501. (in German).

Goebertus, T.M., 1989. The twofold effect of an aluminium screen: a temperature increase in winter and a temperature decrease in summer. Acta Horticulturae 245: 470–475.

Goeijenbier, P., 1986. Gebruik recirculatie-ventilatoren. (*Use of recirculation fans*). Tuinderij 66(24): 46–47. (in Dutch).

Goeijenbier, P. & G.P.A. Van Holsteijn, 1986. Vochtafvoer bij schermen vraagt nauwkeurige regeling. (*Vapour release with screens requires accurate control*). Vakblad voor de Bloemisterij 41(4): 62–65. (in Dutch).

Hand, D.W., 1982. CO_2 enrichment: the benefits and problems. Scientific Horticulture 33: 14–43.

Hand, D.W., 1986. CO_2 sources and problems in burning hydrocarbon fuels for CO_2 enrichment. In: H.Z. Enoch & B.A. Kimball (Eds), Carbon dioxide enrichment of greenhouse crops. Volume I: status and CO_2 sources. CRC Press Inc., Boca Raton, Florida, pp. 99–121.

Hand, D.W., 1990. CO_2 enrichment in greenhouses: problems of CO_2 acclimation and gaseous air pollutants. Acta Horticulturae 268: 81–101.

Hand, D.W. & M. Hannah, 1990. Noxious side-effects of flue gases. Grower 113(6): 15–17.

Heijna, B.J., 1985. Warmtebesparing door bed- en stralingsverwarming in kassen. (*Heat saving by bed and radiative heating in greenhouses*). IMAG-DLO, Wageningen, IMAG-DLO rapport 72: 35 pp. (in Dutch).

Holländer, B. & H. Krug, 1991. Wirkungen hoher CO_2-Konzentrationen auf Gemusearten I: Symptome, Schadbereiche und Artenreaktionen. (*Effect of high CO_2-concentrations on vegetables*). Gartenbauwissenschaft 56: 193–205. (in German).

Huijs, J.P.G., 1988. Total energy system for assimilation lighting. Chronica Horticulturae 28(2): 20–21.

Jacobi, U., M. Zschoche, W. Recker & A. Vierig, 1990. Einsatz von CO_2-Generatoren zur CO_2-Düngung auf der Basis von Erd-, Bio- und anderen Gasen. (*Use of CO_2 generators*). Gartenbau 37(3): 74–77. (in German).

Kiel, A.J., 1990. CO_2 enrichment with natural gas fired hot-air heaters. Acta Horticulturae 268: 111–120.

Knies, P., 1991. Drie kasverwarmingssystemen voor restwarmte. (*Three greenhouse heating systems for waste heat*). IMAG-DLO Rapport 91–15, Part 1: 68 pp, Part 2: 127 pp. (in Dutch).

Knies, P., N.J. Van de Braak & J.J.G. Breuer, 1983. Infrared heating in greenhouses. IMAG-DLO, Wageningen, IMAG-DLO research report 83–7: 16 pp.

KNMI, 1982. Climatological data of stations in The Netherlands No. 10: normals and standard deviations for the period 1951–1980. KNMI, De Bilt, 118 pp. (in Dutch).

Kooiman, A.J., 1990. GIVEG approval requirements for hot-air guns suited for CO_2 enrichment. Acta Horticulturae 268: 121–125.

Kozai, T. & S. Sase, 1978. A simulation of natural ventilation for a multispan greenhouse. International Society of Horticulture Science, Wageningen, Acta Horticulturae 87: 339–349.

Kozai, T., S. Sase & M. Nara, 1980. A modelling approach to greenhouse ventilation control. International Society of Horticulture Science, Wageningen, Acta Horticulturae 106: 125–136.

Kwantitatieve informatie voor de glastuinbouw 1990–1991. (*Quantitative information for greenhouse horticulture 1990–1991*). Informatie en kennis centrum akker- en tuinbouw, Ministerie van Landbouw, Natuurbeheer en Visserij, 107 pp. (in Dutch).

Langer, K.-H., S. Schmidt & W. Dietrich, 1990[a]. Einrichtung und Nutzung eines bodennahen Verteilungssystems für CO_2 in Gewächshäusern. (*Distribution system for CO2 in greenhouses*). Gartenbau 37(3): 78–79. (in German).

Langer, K.-H., G. Hofmann, H.-H. Bath & W. Dietrich, 1990[b]. CO_2-Erzeugung durch Stoffumwandlung bei der Verbrennung von Erdgas mit Sauerstoffüberschuss. (*CO_2 from combustion of natural gas with O_2 excess*). Gartenbau 37(3): 71–72. (in German).

Meerman, H.J., 1989. Koeling in kassen. (*Cooling in greenhouses*). Klimaatbeheersing 18(9): 322–325. (in

Dutch).

Meinders, H., H.W. Vahl & C.A.J. Middendorp, 1984. Technical developments in thermal screening systems. Acta Horticulturae 148: 443–452.

Meijndert, J., 1982. Rookgascondensors echt nuttig gebruiken. (*Efficient use of flue gas condensers*). Vakblad voor de bloemisterij 37(4): 86–89. (in Dutch).

Meyer, J., 1982. Bewertung von beweglichen Energieschirme im Hinblick auf den Wärmeverbrauch von Gewächshäusern. (*Validation of movable energy screens against the background of energy consumption of greenhouses*). ITG, Hannover, Gartenbautechnisch Information Heft 11, 143 pp. (in German).

Mortensen, L.M. & R. Moe, 1992. Effects of selective screening of the daylight spectrum and of twilight on plant growth in greenhouses. Acta Horticulturae 305: 103–108.

Müller, G.J., 1987. Energieschirme unter Praxisbedingungen, Bewertung und Optimierung im Hinblick auf Energieverbrauch und Klimaführung. (*Energy screens in practice, validation and optimisation against the background of energy consumption and climate control*). ITG, Hannover, Gartenbautechnisch Information Heft 28, 181 pp. (in German)

Nederlandse Gasunie N.V., 1988. Physical properties of natural gas. Groningen, The Netherlands, 254 pp.

Nijeboer D.J. & G.P.A. Van Holsteijn, 1981. Perspectief voor gewasverwarming bij jaarrond chrysanten. (*Prospects for crop heating for year-round cultivation of chrysanthemums*). Vakblad voor de Bloemisterij 35(7): 28–33. (in Dutch).

Nijskens, J., 1986. Radiation transfer through covering materials, solar and thermal screens of greenhouses. Agricultural and Forest Meteorology 35: 229–242.

Okada, M., 1985. An analysis of thermal screen effect on greenhouse environment by means of a multi-layer screen model. Acta Horticulturae 174: 139–144.

Philips, 1987. Tuinbouw en kunstlicht. (*Horticulture and artificial light*). Philips Lighting Division, Eindhoven, 40 pp. (in Dutch).

Plaisier, H.F., 1990. Putting aluminized screen to work for you. Grower talks 53(12): 50–54.

Plaisier, H.F., 1992. Energy saving and climate improvement with thermal screens. Acta Horticulturae 305: 63–64.

Post, M.L. & R.H.M. Maaswinkel, 1984. Met een scherm betere horizontale temperatuur verdeling. (*Better horizontal temperature distribution with screens*). Tuinderij 64(22): 24–27. (in Dutch).

Richardson, G.M., 1985. The design of film-plastics clad buildings for horticulture. Plasticulture 67: 32–41.

Rijsdijk, A., 1989. Effect schadelijke gassen op gewas. (*Effect noxious gases on crop*). Groenten en Fruit 44(7): 30–31. (in Dutch).

Standard Committee 351 37, 1978. NEN 3859 Tuinbouwkassen, constructieve eisen. (*Greenhouses, structural requirements*). 1st edition. NNI, Delft, 19 pp. (also available in English).

Standard Committee 351 37, 1985. NPR 3860 Tuinbouwkassen, aanbevelingen voor en voorbeelden van de constructieve uitvoering, gebaseerd op NEN 3859. (*Greenhouses, recommendations for and examples of constructional performance*). 1st edition. NNI, Delft, 32 pp. (also available in English).

Standard Committee 351 37, 1988. NEN 3859 Tuinbouwkassen, constructieve eisen. (*Greenhouses, structural requirements*). 2nd edition (draft), NNI, Delft, 23 pp. (in Dutch).

Starkey, N.G., 1985. The effect of secondary glazing and fixed screens on greenhouse environment and crop response of tomatoes. Acta Horticulturae 174: 331–339.

Stoffers, J.A., 1989. Tuinbouwtechnische aspecten van druppelprofilering bij kasverwarmingsbuis. IMAG-DLO, Wageningen, 24 pp. (in Dutch).

Tantau, H.J., 1983. Heizungsanlage im Gartenbau. (*Heating systems in horticulture*). In: Handbuch des Erwerbgärtners. Ulmer, Stuttgart. pp. 245–252. (in German).

Telle, M.G., P. Kaspers & V.J.M. Visser, 1987. Onderzoek lucht/water warmtepomp in de glastuinbouw. (*Research on air/water heat pump in glass house cultivation*). IMAG-DLO, Wageningen, IMAG-DLO rapport 83: 60 pp. (in Dutch).

Van Berkel, N., 1975. CO_2 from gas-fired heating boilers: its distribution and exchange rate. Netherlands Journal of Agricultural Science 23: 202–210.

Van Berkel, N. & J.B. Verveer, 1984. Methods of CO_2 enrichment in The Netherlands. Acta Horticulturae 162: 227–231.

Van de Braak, N.J. & J.J.G. Breuer, 1991. Ventilatie in kassen. (*Ventilation of greenhouses*). IMAG-DLO, Wageningen, IMAG-DLO rapport 91–14: 21 pp. (in Dutch).

Van de Muyzenberg, E.W.B., 1980. A history of greenhouses. IMAG-DLO, Wageningen, pp. 291.

Van der Post, C.J., J.J. Van Schie & R. De Graaf, 1974. Energy balance and water supply in glasshouses in the West-Netherlands. International Society of Horticulture Science, Wageningen, Acta Horticulturae 35: 13–23.

Van Holsteijn, G.P.A., 1987. Met energiescherm op weg naar kleinere temperatuurverschillen. (*With energy screen towards smaller temperature differences*). Vakblad voor de Bloemisterij 42(49): 22–23. (in Dutch).

Van Holsteijn, G.P.A., 1992. Hogere produktie door padregistratie. Groenten en Fruit/Glasgroenten 2(36): 22–23. (in Dutch).

Van Koten, H., 1974. Bepaling van de vormfactoren van enkele configuraties bedrijfsgebouwen. (*Coefficients for several shapes of agricultural buildings*). TNO-IBBC, Delft, report no. B-76-273/08.2.467, 5 pp. (in Dutch).

Vogelezang, J.V.M., P.A. Van Weel & J.M. Freriks, 1988. Toepassing van verschillende tablet- en vloerverwarmingssystemen. (*Application of bench and floor heating systems*). Proefstation voor de Bloemisterij, Aalsmeer, rapport 48: 42 pp. (in Dutch).

Vogelezang, J.V.M., 1993. Bench heating for potplant cultivation: analysis of effects of root and air temperature on growth, development and production. Research Station for Floriculture, Aalsmeer, mededelingen no. 101, 115 pp.

Von Zabeltitz, C., 1988. Greenhouses used in Europe. In: C. Von Zabeltitz (Ed.), Energy conservation and renewable energies for greenhouse heating. CNRE guideline No. 2, FAO, Rome, REUR Technical Series 3: 9–16.

Waaijenberg, D., 1984. Research on plastic greenhouse cladding materials. Acta Horticulturae 154: 57–64.

Waaijenberg, D., 1988. Greenhouse covering material. In: C. Von Zabeltitz (Ed.), Energy conservation and renewable energies for greenhouse heating. CNRE guideline No. 2, FAO, Rome, REUR Technical Series 3: 43–55.

Waaijenberg, D., 1990. Standard for film-covered greenhouses. Acta Horticulturae 281: 129–137.

Wilkin, A.L. & B.J. Bailey, 1985. A mechanism for rolling a greenhouse screen. National Institute of Agricultural Engineering, Silsoe, divisional note no. 1285, 20 pp.

Working Group ISHS, 1991. European Greenhouse Standard, draft 01: Greenhouses, design, construction and loading, 16 pp. (available at CEN, Brussels).

Yates, D.J., 1986. Shade factors of shadecloth materials. Agricultural engineering Australia 15: 22–32.

List of symbols

A	surface area (m^2)
c	concentration ($kg\ m^{-3}$)
C_f	Carnot factor (-)
C_p	specific heat of air ($J\ kg^{-1}\ K^{-1}$)
F_d	dilution factor (-)
h	height (m)
H	volumetric burning value ($MJ\ m^{-3}$)
n	air factor (-)
o	perimeter (m)
P	driving power (W)
q_h	heat flux (W)
q_v	volumetric flux ($m^3\ s^{-1}$)
q_c	massic CO_2 flux ($kg\ s^{-1}$)
Q_v	time integrated volumetric flux (m^3)
r	heat of vaporisation of water ($J\ kg^{-1}$)
R	reflectivity (-)
s	spacing (distance) (m)
S	global solar radiation ($W\ m^{-2}$)
T	temperature (K)
u	windspeed ($m\ s^{-1}$)
U	greenhouse heat transfer coefficient ($W\ m^{-2}\ K^{-1}$)
V	volume (m^3)
vv	rate of air infiltration (h^{-1})
w	width (m)
x	specific humidity of air ($kg\ kg^{-1}$)

Greek symbols

β	window opening angle (-)
Δ	difference (-)
η	efficiency (-)
τ	light transmissivity (-)
ρ	density ($kg\ m^{-3}$)

Subscripts

b	burner
c	CO_2
ex	outside
f	flue gas
g	greenhouse
gg	greenhouse ground
in	inside
l	latent
p	photosynthesis
s	sensible
t	storage tank
vnt	ventilation
w	window
x	noxious gas

List of abbreviations

COP	coefficient of performance
EVA	ethylenevinylacetate
FEP	fluorethylenepropylene
IR	infra red
MAC	maximum acceptable concentration
NEN	Nederlandse norm (Netherlands standard)
NPR	Nederlandse praktijk richtlijn (Netherlands practice guideline)
PC	polycarbonate
PE	polythene
PMMA	polymethylmethacrylate
PVC	polyvinylchloride
RDF	refuse derived fuel
UV	ultra violet

5 Greenhouse climate control

5.1 Introduction

G.P.A. Bot

Greenhouse climate control has developed very quickly in the past decades. Originally it was primitive: only extreme conditions were avoided and to do so heating and ventilation were operated manually. In the late fifties thermostats for temperature control were introduced as labour saving equipment, and were later replaced by analogue electronic controllers (Strijbosch & Van de Vooren, 1975). In the sixties automatic control of ventilation windows was introduced, and shortly after that, system performance improved through the implementation of parameters such as outside temperature and radiation (Bowman and Wearing, 1970; Bokhorst *et al.*, 1972) and more refined control procedures primarily based on the common practices of climate control performed by leading growers. After, when these means of control became widely accepted, the further implementation of extra features in analogue systems became so expensive that the introduction of computer technology in the mid seventies was economically justified (Gieling, 1980). This development was strengthened by the introduction of equipment such as screens, CO_2 enrichment, artificial lighting that had to be controlled in relation to indoor and outdoor conditions.

Control systems have evolved into complex systems in which a lot of know-how is implemented: control algorithms, instrumentation and various climate processes. In this chapter the state of the art of the relevant items is described and evaluated.

5.2 Sensors and measurement

Th.H. Gieling and K. Schurer

5.2.1 Introduction

The need for measurement

Greenhouse management requires a continuous flow of quantitative data from measurements of physical, chemical or physiological phenomena in and around the greenhouse as an input.

This implies the use of sensors with an electrical output, connected to a data acquisition system. Improved methods of measurement and newly emerging semiconductor technologies have not only resulted in more and better sensors, but also provide opportunities to adapt sensors to specific requirements for horticulture. Potential developments include improvements in classical methods, such as digital output, smaller dimensions of the sensor, lower costs and higher reliability, as well as the development of new types of sensors.

The sensor

In order to describe the device that performs a "sense" action, the word "sensor" has been created in American English. A sensor incorporates three different functions: selecting the information sought from an abundance of information offered, transducing the information to a measurable form, and detecting the signal (Figure 5.2.1).

Often, the relationship between sensor (V) and output signal (S) cannot be described by simple laws of physics. An empirical relationship has to be determined: the function $V = f(S)$ has to be found by calibration. Once established a relationship can change over time. To ensure lasting reliability of measured values, a timetable for calibration has to be set up, with due regard to the severe conditions imposed on equipment by the greenhouse climate. Calibrations have to be performed according to strict procedures by organisations certified for this activity.

The following is an example of such a sensor. In a sensor for infrared radiation a filter acts as the selector (1) and only admits the infrared part of the whole radiation spectrum to increase the temperature of a blackened transducer (2). The temperature rise is then converted into an electrical signal by a thermopile detector (3).

In horticulture sensors are used to measure quantities in fields such as: local meteorology, indoor greenhouse climate, water and nutrient supply and feedback from greenhouse appliances (ventilators, valves, screens, etcetera). The overall uncertainty of sensors used in practice is in the class of ±5% of full scale.

The position of the sensor is of great importance, especially when measuring climate conditions. Horizontal and vertical gradients are intrinsic to all climate variables, both inside and outside a greenhouse. The greenhouse climate is characterized by moderate temperatures, a high to very high humidity, an intense solar radiation and little air movement. This complicates the measurement of air temperature and humidity.

5.2.2 Air temperature sensors

In greenhouse practice temperature sensors of the resistive type are used. There are platinum sensors (Pt) according to DIN-IEC 751 and standard ceramic NTC sensors (thermistors). In research mainly Pt100, Pt500, Pt1000 sensors are used, where the numeric value indicates the resistance at the reference temperature (0 °C). Sometimes thermocouple sensors are employed.

With air temperature measurements the uncertainty of the sensors should not exceed ± 0,3 °C. For platinum resistors the optimum choice is a tolerance one third of that of DIN–IEC 751 class B.

To measure the sensor resistance properly, a four wire connection between each resistance sensor and the electronic measuring equipment has to be attached (Figure 5.2.2). One pair of wires is used to conduct the measuring current from a current source to the resistance sensor. The other pair of wires is used to connect the measured voltage across the resistance sensor to a differential instrumentation amplifier with a high input impedance at the input of the data acquisition system. The value of the measuring current is derived from the voltage across a reference resistor (which has a constant and accurately known resistance) in series with the temperature sensor.

The value selected for the measuring current is about 1 mA for Pt100 and Pt500, to keep self-heating of the sensor within the uncertainty limits. A current of 0,2 mA should not be exceeded for

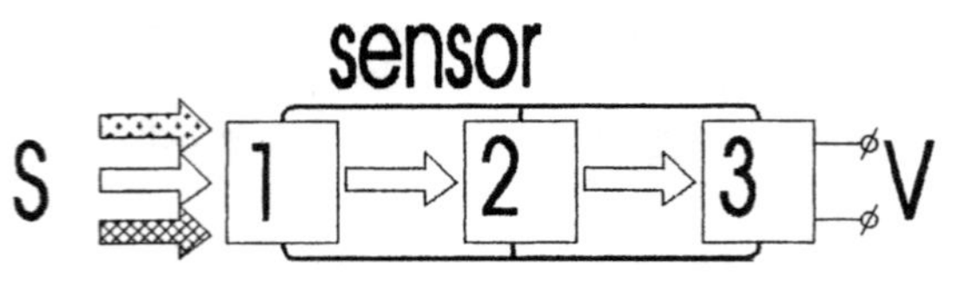

Figure 5.2.1 – Sensor.
S = Signal; V = Output; 1 = Selector; 2 = Transducer; 3 = Detector.

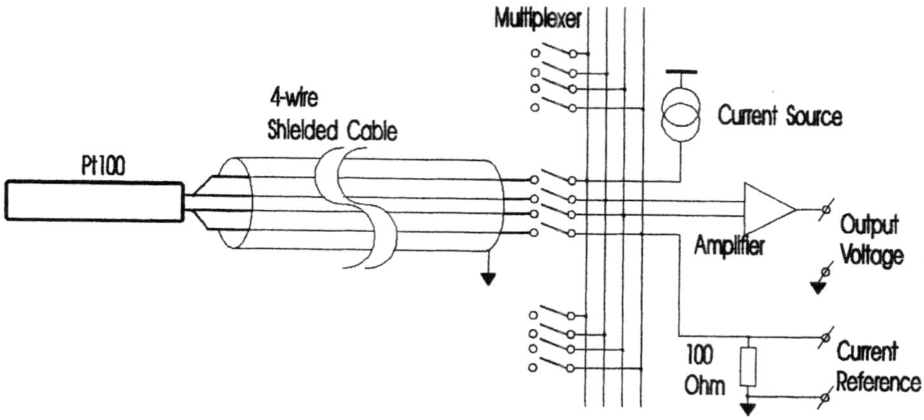

Figure 5.2.2 – Four wire connection to resistance sensors.

the high resistance NTC sensors. Often the current is just switched on for a few milliseconds during the actual measurement. This permits a higher value of the measuring current and hence a more reliable signal at the same level of self-heating.

The yearly maintenance should include a check on the calibration at two temperatures, for example at 0 °C and at ambient temperature.

5.2.3 Humidity sensors

Introduction

In section 3.4 various ways of defining humidity are presented, including relative humidity (RH) and vapour pressure deficit. If one humidity parameter is measured together with the air temperature the other parameters are amenable to computation.

Two types of humidity sensors are in common use in Dutch greenhouses: the dry- and wet-bulb instrument or psychrometer and the capacitive sensor.

For a representative RH value it is important to perform the measurements at a position that yields values characteristic for the conditions experienced by the crop. It is up to the user to decide where exactly that is: at the level of the growing points of the plants, near the bulk of the leaf-area or at some other position.

Dry- and wet-bulb psychrometer

The air temperature T and the temperature of a freely evaporating water-surface T_W are measured (see section 3.2.5) with a psychrometer. These temperatures are related to the water vapour pressure e over the psychrometer equation (Schurer, 1986):

$$e_W = e_s(T_W) - A \cdot P \cdot (T - T_W) \tag{1}$$

where the psychrometer coefficient A has a value between 6,2 10^{-4} and 6,6 10^{-4} K^{-1} and where P is the barometric pressure in the same pressure units as e (Pa or mbar). An accuracy of ±0,3 °C in dry- and wet-bulb temperature corresponds to ±5% RH.

From section 3.2.5 it can be understood that the psychrometer equation is based on the equilibrium of the convective heat fluxes to and from the wet surface. Heat transfer coefficients are largely dependent on the movement of the adjoining air. Proper operation of the wet-bulb thermometer therefore requires a minimum air speed of 2 to 4 m s^{-1}, depending on the diameter of the wet-bulb sensor (Wylie & Lalas, 1985). The same air movement is applied around the dry bulb sensor to ensure similar time constants for the two thermometers.

The wet-bulb is wetted by a wick immersed in a water reservoir. It should have an uninterrupted water supply. Insufficient water supply implies insufficient evaporation, high readings of the wet-bulb temperature and hence high RH values. The water reservoir should never run dry. The wick should be replaced when there are visible signs of contamination, if not more often. At installation a dry wick should be wetted completely to help it start transporting water.

Two more aspects should be incorporated into the design. The water in the reservoir is at ambient temperature. It should flow through a sufficient length of ventilated wick to allow it to cool to the wet-bulb temperature before it reaches the temperature sensor. Usually this requires about 2 cm of wick between the entrance in the airstream and the temperature sensor. Heat conduction to the sensor through the connecting wires and the sensor mounting capillary should also be kept low. To be sure of this, the wick should cover about two cm extra length of the capillary beyond the sensor.

Capacitive sensors

Sensors to replace the dry- and wet-bulb instrument should have a comparable accuracy, and should not need any maintenance for the duration of a growing season lasting some six to nine months. After that period calibration and occasional replacement is considered acceptable. Until quite recently tests invariably showed that electronic sensors did not meet these – reasonable – requirements (Visscher & Schurer, 1985). However, it is likely that in the not too distant future capacitive thin-film sensors will replace dry- and wet-bulb psychrometers in greenhouse control. Several sensor-manufacturers are now coming up with solutions for problems of poor long-term stability, drift at high RH and very slow response after wetting. Such problems have made it impossible to apply capacitive sensors in a humid environment over a long period. In recent tests sensors from twelve manufacturers have been exposed for six to twelve months to outdoor conditions (Visscher & Kornet, 1994). Most of the sensors were found to be capable of operating within an uncertainty of ±5% RH without calibration during the test period. These sensors can be expected to work equally well in a greenhouse and need no maintenance apart from a calibration before the start of each growing season.

To prevent the sensor from becoming contaminated a protective cap of a porous, inert material such as PTFE (teflon) or sintered metal, is required. The speed of response is somewhat reduced, but still satisfactory, and the long term reliability is greatly enhanced.

5.2.4 CO_2 sensors

Introduction

Instruments for the continuous monitoring of CO_2 concentration are relatively expensive. In research, it is therefore customary to use a multiplex sampling system and one analyzer. A steady flow of air is maintained in all sampling lines, so that a fresh sample is available the moment a line is connected to the analyzer.

Sampling lines are best made from a non-absorbing and gastight quality of tubing, such as nylon or high-density polyethylene. The diameter chosen should be such, that there will be no undue pressure-drop over the length of tubing required. Attention should be given to possible errors due to pressure differences between the sample in the analyzer and the greenhouse atmosphere. Though the result of the determinations is usually presented as a pressure independent volume fraction (expressed as volume parts per million, vpm), the actual measurement performed is a pressure dependent concentration measurement (in mol m^{-3} or sometimes kg m^{-3}).

Generally, infrared analyzers are used (Long, 1986). Alternatives, such as a conductometric or a photo-acoustic measurement have found little practical application in Dutch horticulture. Photo-acoustic instruments are in fact very recent; an adaptation for horticultural application has only just become available (Bicanic, 1992).

Infrared analyzer

Like all polyatomic molecules, carbon dioxide exhibits some strong absorption lines in the infrared range. Instruments based on this phenomenon measure the absorption over the length of a gas-cuvette (Long, 1986). A second cuvette is often used as a reference, filled with the same air matrix, but with a known CO_2 content. Sometimes, water vapour can cause interference. In that case a dryer is used at the entrance of the analyzer.

The absorption measured is nearly proportional to the number of absorbing molecules in the light-path, i.e. to the carbon dioxide concentration. The measuring range can be from 0 to 1000 vpm and higher. The instrument needs to be calibrated every three to six months. For more exacting work a calibration should involve at least three different concentrations.

5.2.5 The measuring box

The high levels of solar irradiation and the low air speeds inside the greenhouse will have adverse effects on the measurement of temperature and humidity. It is therefore common practice in Dutch greenhouses to mount the sensors in a ventilated box.

An example of a popular measuring box is shown in Figure 5.2.3. The diameter and height are 0.2 m and 0.3 m respectively. A fan is mounted in the top of the box, providing an air speed in excess of 2 m s^{-1} along the sensor (section 5.2.3). The air stream is upwards, thus ensuring that the air sample is not heated by the power dissipated by the fan. Sometimes a dust-filter is used at the inlet. Such a filter should be replaced before clogging occurs.

The air should not be allowed to reach the dry-bulb after it has been in contact with the wet-bulb. The box has a double wall and has a layer of reflective material on the outside. Thus, radiation is shielded effectively and variations in local air movements do not affect the measurements.

Capacitive sensors do not require a minimum air speed, but air stagnation has to be prevented. However, ventilation of the box should ensure a sufficient speed of response. Moreover, it should provide an adequate cooling of the irradiated surfaces.

The box can easily be positioned at a representative height, for example the growing point of the crop or the point where the leaf area index (LAI) as a function of height is largest. It can be used both inside and outside the greenhouse.

5.2.6 Wind and rain measurement

Wind detectors

Outdoor wind speed is measured with a cup anemometer. Various detectors are used such as a tachometer or a switch giving an on-off signal for every revolution (Hanan, 1984). Nowadays, a frictionless, interrupted light-beam switch is generally used.

Many anemometers show non-linearity errors at low wind velocities and a threshold value due to friction at near zero wind velocity. This is no problem for a windspeed signal that is used for climate control. If it is used as an input variable for models, it could cause considerable and inadmissible errors. Then, special types with a low starting speed have to be used.

The calibration of a cup anemometer should be checked yearly. Special attention has to be paid to

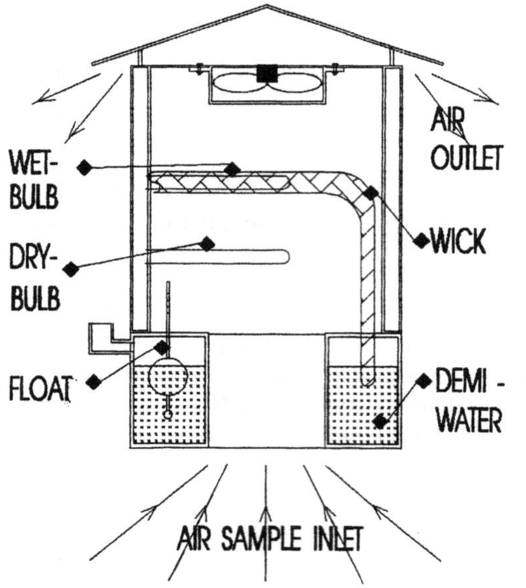

Figure 5.2.3 – Measuring box.

changes in starting speed, stopping speed and linearity.

Wind direction is measured with a wind-vane. The angle is determined from the measurement of a variable resistance or from a rotating coded disc (Meteorolical Office, 1969).

An interesting problem occurs when wind direction has to be filtered by a computer algorithm for smoothing a rapidly changing signal. When the signal is changing around north, e.g. between 350 degrees and 10 degrees (-10 degrees and +10 degrees around north), a normal smoothing algorithm would give the incorrect result: (350 + 10) / 2 = 180 degrees, so it would indicate south.

An example of a smoothing algorithm that always will yield the correct wind direction is given in Box 5.2.1.

Rain detectors

Most rain detectors consist of two comb-shaped, interlaced, gold-plated electrodes on a printed circuit board. When exposed to rain, raindrops will connect the two electrodes and generate a yes/no signal for rain (Meteorological Office, 1969). A small electrical heating element is mounted below the board to accelerate the evaporation of the raindrops. The electrodes should be cleaned now and then, to remove the salt crystals left behind after the water has evaporated.

In research tipping-bucket instruments are used for the quantitative measurement of the amount of precipitation.

5.2.7 Radiation sensors

Introduction

Electromagnetic radiation of optical wavelengths provides the driving force for the processes in the greenhouse (Bickford & Dunn, 1972; Chapter 2 and 3).

Four different quantities are measured more or less regularly (see also Table 3.1):

```
C = | D – DF_N–1 |
    IF (C < 180) THEN
        IF (D ≥ DF_N–1) THEN C = –C ENDIF
    ELSE
        C = 360 – C
        IF (D < DF_N–1) THEN C = –C ENDIF
    ENDIF
    IF (FC ≠ ) THEN
        DF_N = D + C * (FC – 1)/FC
        IF DF_N ≥ 360) THEN DF_N = DF_N – 360 ENDIF
        IF DF_N < 0) THEN DF_N = DF_N + 360 ENDIF
    ENDIF
```

C = Change, D = Unfiltered Wind Direction, DF = Filtered Wind Direction, FC = Filter Constant (FC ≥ 1), N = This Cycle, N –1 = Last Cycle.

Box 5.2.1 – Smoothing algorithm for wind direction.

a) Total global shortwave radiation (300 to 2500 nm) (W m^{-2})
b) Photosynthetically Active Radiation (PAR, 400 to 700 nm) (µmol m^{-2} s^{-1});
c) Net radiation (300 to 25,000 nm) (W m^{-2});
d) Light intensity (380 to 760 nm) (lx).

5.2.7.2 Measuring total shortwave radiation

The common instrument for measuring short wave radiation has a broadband thermal receiver, covered by one or two glass domes to protect the receiver from the weather and to provide a thermally stable environment. The glass domes transmit approximately the wavelength-band mentioned earlier. The instrument is known as a pyranometer or a solarimeter. It measures the total shortwave energy-flux density or irradiation (W m^{-2}) in the plane of the receiver. The usual position is horizontal. The angular response should follow a cosine function.

The total uncertainty of the shortwave irradiation measurement comprises errors due to calibration, non-linearity, angular response and positioning. The uncertainty can be kept well within ±5%, if the outer dome is cleaned monthly and the instrument is recalibrated every two years.

Some manufacturers offer silicon-cells with a response flattened from 450 to 1050 nm as "solid-state" pyranometers. These sensors must be equipped with a diffuser to ensure a good angular response. Such instruments measure between 70 and 90% of the total shortwave radiation, depending on cloudiness and solar height. Linearity, signal-level and price of a silicon pyranometer are comparable to those of a thermal instrument. The significantly increased uncertainty associated with limited wavelength range of the silicon pyranometer makes it an inferior substitute for the thermal instrument.

Sometimes it is useful to have a separate measurement of the diffuse sky radiation only. This can be achieved by a pyranometer with a shadow ring in an equatorial mount. The elevation of the ring should be adjusted every three to five days (Robinson, 1966; Meteorological Office, 1969). Tables are available to account for the interception of part of the diffuse radiation by the shadow ring.

PAR sensors

Green plants can utilise radiation between 400 and 700 nm for photosynthesis (PAR). Since the process is driven by the absorption of specific photons rather than by the total energy, the action spectrum of a crop reflects the photon flux between 400 and 700 nm (Figure 5.2.4). The corresponding SI-unit is mol m^{-2} s^{-1}. Commonly, the submultiple µmol m^{-2} s^{-1} is used.

PAR sensors consist of a silicon photocell with a diffuser and an optical filter (Biggs, 1986). They have good linearity and stability. Recalibration once every two years suffices.

Due to the spectral response of the PAR sensor (Figure 5.2.4) and the variations in the spectral distribution of shortwave radiation under different meteorological conditions, it is not possible to give a single conversion factor from total shortwave radiation to PAR. The instrument to be used should be tailored to the problem at hand: PAR as input in a plant growth model and total shortwave for temperature control and water supply.

Net radiometers

Net radiation is the balance of the downward and upward fluxes of short-wave and long-wave radiation together (section 3.2).

A net radiometer comprises two thermal radiation receivers, mounted back to back. The receivers have been prepared so that they have an equal sensitivity to both short-wave and long-wave radiation. Both receivers are covered with a thin polyethylene cap (transmissive to both short and long wave radiation) that can be lightly inflated to preserve their shape.

The measurement of net radiation is less straightforward than that of a shortwave flux, but the result comprises two extra terms of the energy balance: the reflected short-wave flux and the net long-wave flux. At present the net radiometer is still a research instrument, rather than one for everyday use. It needs a recalibration once a year.

Luxmeters

The sensitivity of the human eye covers a range very similar to that of photosynthesis. Yet the shape of the sensitivity-curve is very dissimilar. The eye has a low response in the blue and the red portions of the spectrum and shows a rather steep maximum in the green. From a maximum at 555 nm sensitivity drops to virtually zero at 380 nm and 760 nm (Figure 5.2.4). The range between these wavelengths

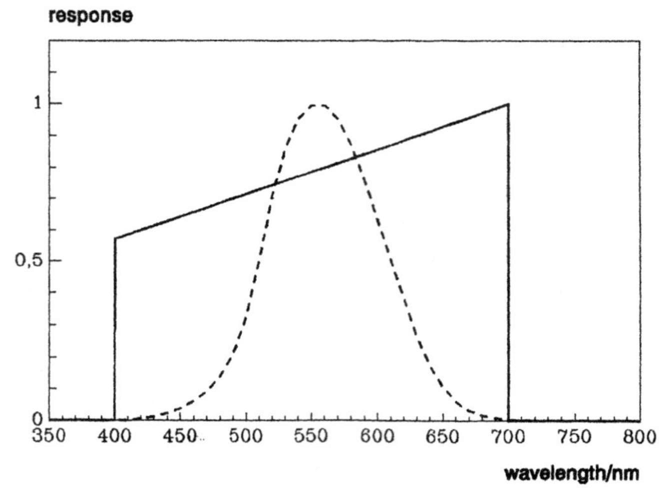

Figure 5.2.4 – Relative response. ——— = PAR; ----- = human eye.

is often referred to as the visible. Radiation weighted for the sensitivity curve of the eye is called "light". The SI-unit for light flux is the lumen (lm), for flux received per area (illuminance) the lux (lx). Due to the same reasons as given for PAR, there is no unique relation between illuminance (lx) and energy flux density (W m^{-2}) or between illuminance and PAR. Modern luxmeters comprise a silicon photocell with a diffuser and optical filters. Recalibration is needed every two years.

The widespread availability of luxmeters has led to their use in horticultural lighting. It is common practice to state the level of additional (artificial) lighting in a greenhouse in lx, rather than in mmol m^{-2} s^{-1}. Unfortunately, this has led to gross overestimation of the photosynthetic effect of radiation from high pressure sodium lamps in comparison to daylight (by about 50%). Modern lamps have been developed to be efficient sources of light rather than of PAR.

5.2.8 The "weather-station"

To incorporate outdoor weather in greenhouse climate control, a set of meteorological instruments is mounted near most greenhouses including an air temperature sensor, a radiation sensor, a precipitation sensor and sensors for wind direction and wind speed. In some cases a sensor is installed for the outdoor CO_2 level as well. In practice an outdoor humidity sensor is usually not installed due to problems with maintenance and freezing of the wet-bulb wick.

This so-called "weather-station" ranges from a yoke-like support on top of the ridge of the greenhouse roof, to a sophisticated weather-station on a tower. Care has to be taken that the sensors are easily accessible for preventive maintenance (cleaning of sensors).

For climate control in the greenhouse outdoor conditions should be measured at the height of the ventilators, i.e. at roof height. International standard heights for meteorological measurements (WMO, 1975) are not relevant in this respect. In positioning the sensors disturbing influences should be minimised. Ventilators or the boiler chimney can cause a local temperature rise, the CO_2 concentration will be higher near the boiler chimney, and nearby trees or buildings may shade the radiation sensor.

5.2.9 Root-zone measurements

Hydroponics

Hydroponics are widely used in the Netherlands (Verwer, 1976; Van Os, *et al*,, 1991) and worldwide (Collins & Jensen, 1983; Savage, 1985).

In nutrient supply systems so-called A-B tank systems are widely used in The Netherlands. These systems use four tanks, labelled A, B, Z and L. They contain respectively all the calcium salts (A tank), the phosphates and sulphates (B tank), an acid (Z tank) and a lye (L tank).

In a closed loop system, the water returned is pumped into a fifth tank, the buffer tank. Fresh water and samples from the A and B tank are added to the buffer tank to control electrical conductivity (EC). Samples from the Z or L tank are added to control pH.

Nutrients are supplied from the buffer tank to the plants through a valve and a set of trickle irrigation hoses. In closed loop systems, the excess water is caught in a gully. For each set of irrigation hoses an EC sensor is used in the corresponding gully to indicate the availability of water to the crop.

Chemo-sensors

Chemo-sensors provide an electrical signal in relation to the concentration of particles in fluids or gases. These particles can be atoms, molecules or ions (Hauptmann, 1990). If biological substances such as enzymes, bacteria or whole cells are part of a chemo-sensor, the word bio-sensor is used.

In horticulture chemo-sensors are mainly used in the root environment. In hydroponics EC and pH sensors are common practice. ISE (Ion Selective Electrode) (Albery *et al.*, 1986) and ISFET (Ion Selective Field Effect Transistor) (Bergveld, 1970) sensors are slowly gaining interest (Bailey *et al.*, 1988; Gieling *et al.*, 1988; Hashimoto *et al.*, 1989; Van den Vlekkert, 1992; Van den Vlekkert *et al.*, 1992).

ISE sensors are available for most macro nutrients, including K^+, Ca^{2+}, NO_3^-, SO_3^{2-}, NH_4^+ and ions detrimental to growth, such as Na^+ and Cl^-. Since the logarithm of activity is measured, the overall uncertainty of ISE measurements is not very good (Heinen, 1992). For instance, for a Ca^{2+} ISE sensor, a measuring accuracy of 1 mV implies an uncertainty in the ion activity of 8%.

The output impedance of the sensor is very high, which means that special precautions have to be taken with respect to connecting cables, signal grounding and electrical input impedance of the signal amplifiers.

ISFETs (Ion Selective Field Effect Transistor) are members of the ISE sensor family. Thus, problems related to uncertainty where ISE sensors are concerned, are also valid for ISFET sensors.

ISFET sensors were first conceived at Twente University in The Netherlands (Bergveld, 1970). An ISFET consists of a field effect transistor with an ion selective membrane on top of the gate. The principle of operation of ISFETs is the surface field effect. The conductance just beneath the surface of a piece of semiconductor material is affected by a perpendicular electrical field, generated by the potential difference over the ion selective membrane. The membranes used are insulators such as Si_3N_4, Al_2O_3 and Ta_2O_5 for pH, silicates which are sensitive to pNa or pK, or polymeric and solid state ion selective membranes for a variety of other ions.

ISFET sensors have some distinct advantages over conventional ISE sensors. The ISFET is a low cost, mass producible semiconductor component. It can be produced as a small but reinforced device, such as a dip-stick or a flow-through cell. Smart-sensors can be built by integrating ISFETs for various ions with electronics for amplification and data-handling on the same sensor body (silicon chip). The ISFET sensor does not use polluting reagents and it does not need excessive maintenance.

The commercial breakthrough for ISFETs is still limited by factors such as accuracy of measurement and life-expectancy (Van den Vlekkert, 1992). These problems are the subject of research in a project for the development of ISFETs for application in horticulture. (Van den Vlekkert *et al.*, 1992).

Electrical conductivity and pH sensors

In horticulture, EC (Electrical Conductivity, unit milli-Siemens per cm, mS cm^{-1}) and pH sensors are the simplest form of chemo-sensors.

EC is measured using three ring-shaped electrodes, mounted inside the water transport pipe at equal distances. An AC voltage of approximately 1 V is applied between the central electrode and the two interconnected and grounded end electrodes. The temperature of the fluid is measured and is used to modify the value of the AC voltage applied. The AC frequency may range from 400 Hz to 50 kHz. AC is used to avoid polarisation.

The EC is derived from the total current between the two end-electrodes and the central electrode. The total current ranges from 0,1 mA to 10 mA. The two currents are summed to eliminate parasitic effects such as that caused by the flow of the supply water.

Grounding both end-electrodes allows the sequential or parallel use of more EC electrodes in one water supply system. In horticultural practice two distinct EC electrodes are used in parallel, thus enabling a check on the functioning of both EC sensors against each other. EC normally ranges from 2 to 10 mS cm^{-1}.

In horticulture pH is measured with standard pH combination sensors. The life of the sensors is about one year. Here too, two sensors are used, so that one sensor checks the other. As a further check the values obtained from the EC and pH sensors are compared with the results of a bi-weekly lab analysis.

Measuring and controlling ion concentrations

Control of the supply of nutrients to the root environment shows two new developments.

The first development concerns replenishing nutrient ions to keep the concentration of each ion at a desired level. A method still widely used is to add nutrients in the form of salt crystals to the A and B tanks. Recently on-line liquid diluting systems have been developed. These systems dispense highly concentrated fluids from up to 14 tanks, each of which contain one particular electrolyte. These machines produce A and B fluids or add concentrated nutrients directly into the water supplied to the plants.

The second development concerns information feedback for control of the concentration of individual ions. Plants use water and individual ions. This process of selective uptake depends on variables including: plant activity, seasonal influences, climate variables and vegetative – generative phase. To avoid pollution of the environment more and more closed loop nutrient supply systems are being introduced into horticulture. EC and pH will only suffice as feedback signals for control if the excess water supplied to the plants runs off and is not used again. In a closed loop system the composition of the drainage water is hard to predict.

To be able to control the ion concentration in the nutrient solution, the relative concentration of each individual ion in the water returned should be measured (Gieling *et al.*, 1988; Kupers *et al.*, 1992). Van den Vlekkert *et al.* (1992) reported on the application of ISFET sensors and Albery *et al.* (1986) on the application of ISE sensors for control of ion concentration in the circulating nutrient solution.

Monitoring of soil moisture

Greenhouse economy and protection of the environment both require the close monitoring of soil moisture. We shall use the word soil moisture in the more general sense of substrate moisture, since the same reasoning refers to any kind of substrate. In the closed-loop systems now being developed, the main objective of a moisture monitoring system is to guarantee an optimum water supply to the rooting area.

The driving force for water uptake by roots is the difference in water potential (Schurer, 1986) between the substrate and the root-tissue. For any well-defined substrate there is a unique relationship between water potential and water content. Measurement of either of these is sufficient for greenhouse climate control. Current methods of measurement comprise deriving water content from EC of the bulk nutrient solution and the substrate, from the dielectric properties of the substrate in the time domain (TDR) or in the frequency domain, and from the attenuation and scattering of a beam of thermal neutrons. Water potential can be determined from dielectric measurements and from measurements with a hydraulic tensiometer.

Modern growing methods such as nutrient film technique (NFT) and aeroponics have no substrate in which measurements can be made. In these methods watering is controlled by the detection of excess water at the end of a drain. An EC sensor is suitable for this purpose.

Dielectric measurement

Methods to determine water content from the dielectric constant of the substrate are rapidly gaining ground. In a calibration procedure the relationship between dielectric constant and water content for each specific soil type has to be established.

In time domain reflectometry (TDR), a short electrical pulse is sent into a pair of electrodes and the time dependent reflected signal is analyzed to give the water content of the medium between the electrodes (Werkhoven, 1992). The method is still experimental, but completely automated instruments are emerging.

In an alternative approach the complex impedance between two electrodes is measured at one, carefully chosen (high) frequency (Hilhorst *et al.*, 1992). This method can give both water content and

EC. It is more easily automated and miniaturised than TDR, but it also needs a lot of further development.

A miniature version of the sensor can be incorporated into a well-defined substrate, for which the relation between water content and water potential is known. When this system is brought into a soil or substrate, an equilibrium will develop, in which the water potential in both media will be the same. Thus, a measurement of the water content of the known medium can be translated to the water potential of the soil.

Hydraulic tensiometer

A porous cup filled with distilled water can be used for high (near-zero) water potentials. When the cup is brought in contact with the soil an equilibrium will develop in which the tension inside the cup equals the suction of the soil or the substrate (Slavik, 1974). The tension inside the cup is measured with a pressure transducer. The method works for suction pressures down to a level of -80 kPa. At lower pressures there is a risk of air entering the cup. Errors will arise when contact between cup and soil is lost. Hydraulic tensiometers are available commercially.

5.2.10 Signals from appendages and greenhouse appliances

In glasshouses in The Netherlands the functioning of features such as windows, screens and mixing valves in heating systems is controlled by a computer. The effect of each control action is measured to be used as feedback.

Position is measured for windows, screens and valves. Most position sensors are variable resistances or potentiometers. The value of the potentiometer resistance is usually 39 ohm at 100%, in series with a fixed resistance of 100 ohm. In this way a Pt100 channel can be used to measure the resistance value (a Pt100 changes from 100 to 139 ohm over a range of 0 to 100 °C).

In some cases the position is not determined by a sensor, but by calculation. The position is the result of adding (subtracting) all the time periods the actuator is activated for opening (closing), in relation to the time it takes for the actuator to change from 0 to 100%. Every time the actuator reaches either the position for 0% or 100%, a switch for minimum or maximum resets the summation for this value.

The potentiometer sensor for valve position is coupled mechanically to the rotating shaft of the valve. The same mechanical coupling connects the shaft to the minimum and maximum position switches.

In the case of a ventilation window, the operation of the potentiometer depends on the kind of construction used for the window itself. Sometimes the shaft of the potentiometer sensor is operated from an arm with a tracking wheel mounted on the end. The wheel is pressed against the window pane by a lever. In some cases the potentiometer is connected directly to the rotating shaft that opens the windows.

The position of screens is usually determined by the method of integrating the activation time.

5.2.11 Shielding against RFI and LEMP interference

Introduction

Present-day measurement and control involve the application of electronic circuits, which are generally sensitive to RFI (Radio Frequency Interference) and LEMP (Lightning Electro Magnetic Pulse). The circuits can be designed either with discrete electronic components or with microchips. The trade-off for microchip technology is between low cost and good reproducibility versus increased electrical vulnerability (Clark & Povey, 1985).

Historical data on damage to crop and livestock from the files of a major Dutch insurance company show that considerable damage to production in agriculture is caused by failures in control equipment (Gieling & Van Meurs, 1984). Furthermore, it was shown that lightning and lightning-induction is the main cause of damage to computer equipment (87%) in agriculture and horticulture. Of course, trivial causes such as a broken or disconnected sensor wire can be equally detrimental.

At first glance the problem only seemed significant when livestock were involved. Careful analysis, however, indicated that a correlation exists between the reaction time to equipment failure and the extent of damage to any type of production (Table 5.2.1).

Damage increases considerably where there is a shorter reaction time. The introduction of new and fast reacting cultivation techniques in horticulture decreases the overall response time of a crop to failures in equipment. Examples of these easily disturbed cultivation techniques are: hydroponics (e.g. nutrient film technique, growing on rockwool, aeroponics) or critical climatic circumstances caused, for example, by measures for energy saving or cultivation in closed systems. Extreme summer or winter outdoor climatic conditions also present an increased risk.

RFI and LEMP shielding

Most preventive measures taken against electro-magnetic interference concern lightning and lightning-induction, as these are the main cause of equipment failure. For years and years sensitive analogue and digital equipment have been applied to control critical processes in the petrochemical industry. Here good results with respect to preventive measures against LEMP and RFI are shown in the literature (Högberg *et al.*, 1985). These measures are easily applicable to agriculture (Hasse *et al.*, 1985).

Preventive measures include:

- Filters on incoming cables, to short-circuit all transient voltages to a central ground electrode (e.g. mains power, telephone, terminals to other buildings);
- An efficient potential equalisation. All electrically conducting metallic parts should be connected to each other and to the central ground electrode (e.g. all piping and heating installation appliances). So-called "clean" grounding electrodes must also be connected to the central ground electrode;
- Cables to sensors and actuators should have double shielding. The inner shield is only connected to the mass terminal of the computer electronics as a shield against RFI noise. The outer shield is connected to ground on both ends of the cable to function as a shield against lightning-induction;
- The greenhouse construction can act as a so-called "Faraday cage" and offers some extra shielding against RFI and LEMP noise. For this purpose all construction parts (columns, gutters, glazing bars) should have a galvanic interconnection and be connected to the central ground electrode.

Table 5.2.1 – Estimation of worst-case reaction time as a result of equipment failure for agricultural processes.

Product loss	Pigs for breeding	Pigs for meat production	Tomato grown in soil	Poultry	Tomato grown soilless
-10%	60 min	30 min	40 min	10 min	2 min
-50%	120 min	60 min	120 min	20 min	4 min
-100%	480 min	120 min	360 min	40 min	8 min

5.3 Control principles

J. BONTSEMA

5.3.1 Input-output systems

In classical control, processes or systems are mainly considered as input-output systems. The main problem then is how to choose the input if the output of the process has to follow a prescribed path. The internal process variables are not considered. This can be represented schematically as in Figure 5.3.1.

In this setting there may be more than one input and/or output. The inputs can be divided into two classes: the control inputs and the exogenous inputs (or disturbances). For the outputs two classes can also be considered: the measured outputs and the outputs to be controlled. In classical control it is usually assumed that the output to be controlled can also be measured. Schematically this is given in Figure 5.3.2.

For a greenhouse the above can be represented as in Figure 5.3.3. In this case CO_2 supply, heat supply and the window opening are the control inputs, outdoor temperature, outdoor humidity, outdoor CO_2-concentration, wind speed, wind direction and global radiation are the exogenous inputs or disturbances. The outputs of the greenhouse climate process are indoor temperature, the indoor CO_2-concentration and the relative humidity. To include, for instance, global radiation as a disturbance may cause some confusion. Photosynthesis in plants is not possible without radiation. Since radiation can not be affected from a control point of view, it is considered as a disturbance. It affects for instance the indoor temperature and via the photosynthesis it influences the CO_2-concentration.

In the sequel we assume that the greenhouse process can be divided in several subprocesses which only have one control input, one disturbance and one output.

5.3.2 Models for input-output systems

In practice it is not always necessary to model a process in order to control it. A good example of this case is controlling the temperature in a living room with a heater. Nobody uses a model for this and still everybody is able to create a comfortabel temperature in the room by switching the heater on and off. The normally used temperature controllers in houses are based on the same principle. In practice trial and error is still used to tune controllers. However, for refined tuning, a model of the process to be controlled is necessary. Then the question arises what kind of models should be used. For the greenhouse climate for instance very complicated models exist (Bot, 1983), described by high order non-linear differential equations. This kind of model is not developed to tune standard controllers, although it can be very useful for designing the control configuration and for simulation of the controlled behaviour of the greenhouse climate. In control practice one commonly uses linear models, since the controller should force the process to stay near a desired trajectory. It is supposed to be designed in such a way that the deviations from the desired trajectory are small. Then it is reasonable that the non-linear process is approximated by a linear one.

For several reasons these models are not normally represented by differential equations, but by their Laplace transformations.

The Laplace transformation of a (smooth) time signal $y(t)$ is given by (Doetsch, 1974):

$$Y(s) = \int_0^\infty y(t)\, e^{-st} dt \qquad \text{(Eq. 5.3.1)}$$

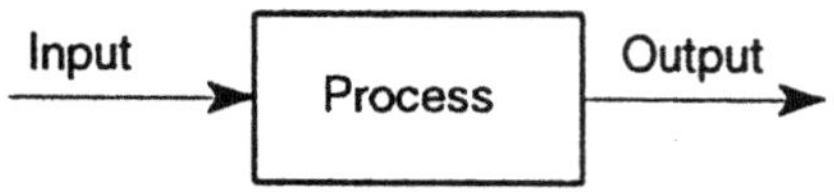

Figure 5.3.1 – Input-output systems.

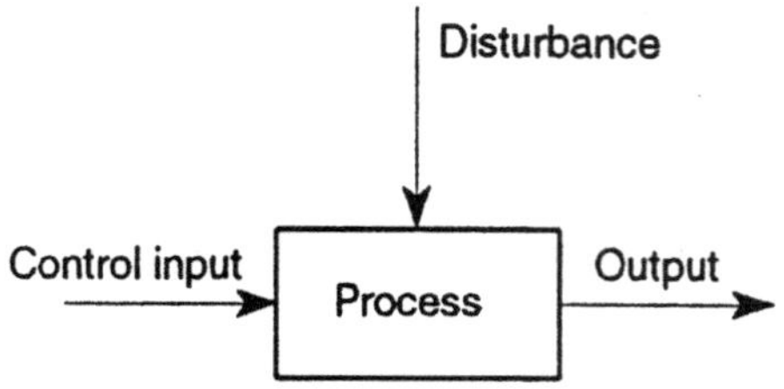

Figure 5.3.2 – Input-output system with disturbance.

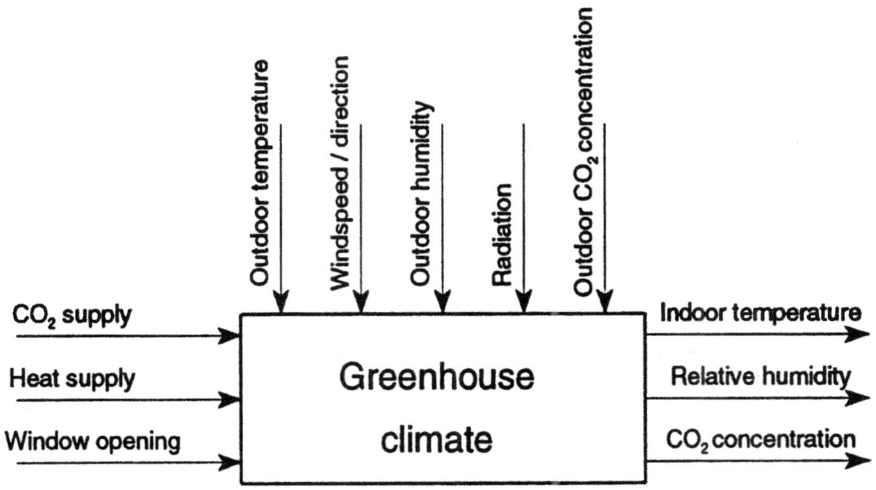

Figure 5.3.3 – The inputs and outputs for the greenhouse climate.

The process is then given by its transfer function, which is the Laplace transform of its impulse response. For single input, single output systems the transfer function $H(s)$ is the ratio of the Laplace transforms of the output $Y(s)$ and the input $U(s)$:

$$H(s) = \frac{Y(s)}{U(s)} \qquad \text{(Eq. 5.3.2)}$$

For a first order system the transfer function is given by:

$$H(s) = \frac{K_p}{\tau s + 1} \qquad \text{(Eq. 5.3.3)}$$

Here K_p is the static gain of the process and τ is the time constant of the process (s).

For a pure delay or dead time system the transfer function is given by:

$$H(s) = e^{-t_d s} \qquad \text{(Eq. 5.3.4)}$$

Where t_d, the dead time, is the time before the process responds to a change in the input.

It turns out that the transfer function of the majority of processes can be approximated by the transfer function of a first order process combined with a pure time delay:

$$H(s) = \frac{K_p e^{-t_d}}{\tau s + 1} \qquad \text{(Eq. 5.3.5)}$$

In section 5.3.5 on controller tuning there is a discussion about how to determine K_p, τ and t_d.

5.3.3 Feedback systems

Consider the schematic representation of the greenhouse climate according to Figure 5.3.2 and suppose that the climate has to be controlled. This can of course be done by choosing a control input by intuition. This kind of control is known as open loop control. If, however, the disturbances acting on the process change or if the process itself changes then the output will change and will be no longer equal to the desired output as before. In order to cope with this control problem a feedback configuration is used. Then the measured output is compared with the required output and the difference is fed back into the process in a certain way. A schematic representation of this is shown in Figure 5.3.4.

The output is measured by the measuring device and this measured output is then compared with the desired output or setpoint (this does not need to be a constant, but may be a time varying signal). This difference is transformed by the controller to deliver a control signal for the final control element. This final control element in its turn produces the input for the process.

The greenhouse climate is a process with many variables. The three control inputs, CO_2 supply, pipe temperature and ventilation all act simultaneously on the climate variables, CO_2, temperature and humidity. However, here the greenhouse climate is considered as consisting of three single loop systems, i.e. systems with one input and one output. The interaction in the total system is neglected. In the greenhouse we then have the following situation. For the climate variable CO_2 the process describes the dynamics between the CO_2-supply and the actual CO_2-concentration. The measuring device is a CO_2-meter and the final control element is a valve in the CO_2-supply.

For the climate variable temperature the measuring device is a thermometer e.g. a Pt100, and the final control element is the heating system. The process input in this case is the pipe-temperature. Note that in this case the final control element itself is a dynamic process. Also the measuring device has its own dynamics, but in general these dynamics are so fast that they can be ignored.

For the climate variable relative humidity the process input is the ventilation. The final control element is the motor which opens or closes the windows.

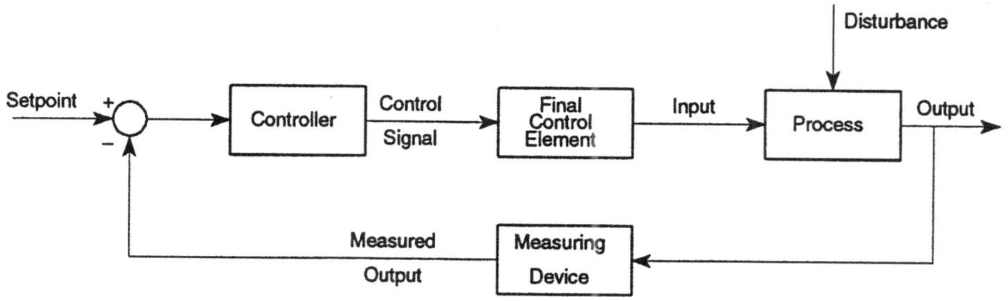

Figure 5.3.4 – Feedback configuration of process and controller.

5.3.4 Controller types

The most simple controllers are the on-off controllers. For temperature for instance, if the measured greenhouse temperature is higher than the desired temperature, the heating is switched off and if the measured temperature is lower, the heating is switched on. In practice there will an upper and lower bound for the desired output to prevent the actuation system from switching on and off too often. For the tuning of the controller one has to give the desired output value and a certain bandwidth around this value. The actual output will always oscillate around the desired value.

The most widely used controller types are the so called proportional, integrating and differentiating controllers; the PID type controllers. If one only uses a P-controller the error between setpoint and actual output is multiplied by the proportional gain of the controller and fed back into the process. This configuration will in general lead to a so-called offset: the actual output will never reach the desired value, since if the error between setpoint and output becomes constant and therefore also the input to the process, the output of the process will not change any more. In order to get rid of this offset an integrating factor can be introduced in the controller. The error will be integrated in time and even if the error is constant the input will grow, so the error will finally become zero. The integrating action has the disadvantage that it leads to less damping in the controlled system and it gives rise to oscillatory behaviour. These disadvantages can be reduced by introducing a differentiating element. The general formula for a PID-controller is:

$$u(t) = K_c \{ e(t) + \tau_i \int_0^t e(t)dt + \tau_D \frac{de(t)}{dt} \} \qquad \text{(Eq. 5.3.6)}$$

here $u(t)$ is the input for the process and $e(t)$ is the difference between setpoint and actual output of the system, K_c is the proportional gain, t_I is the integral time constant and t_D is the derivative time constant.

5.3.5 Controller tuning

After choosing a certain controller type the values of the parameters of the controller have to be selected. In case of PID controllers the first decision is which action is needed, so one has to choose between a P, PI, PD or PID controller and then to select a value for K_c, t_I and t_D. In order to select the type of controller and the best values of these parameters performance criteria are needed, such as small maximum error, short settling time, minimal integrated error, minimum or no overshoot, small rise time, desired decay ratio and so on.

In order to select the controller type the following rules are used (Stephanopoulos, 1984): apply simple P-control if possible, if overshoot is undesirable use PI-control and choose a PID-control if the speed of the closed loop process has to be increased. PID-control is also used if the PI-control lacks robustness.

For the choice of the parameters the Ziegler-Nichols method can be used, which is known as the process reaction method (Palm, 1986). In steady state a step of magnitude A is applied to the input of the uncontrolled process and the output of the process is recorded. This output is then approximated by a signal which is the output of a first order system with dead time. (equation (5.3.5)).

The parameters in the transfer function, static gain K_p, dead time t_d and time constant τ can easily be obtained from the recorded response (see Figure 5.3.5):

$$K_p = \frac{\textit{steady state output}}{\textit{steady state input}} = \frac{B}{A}$$

$$\tau = \frac{B}{S} \qquad \text{(Eq. 5.3.7)}$$

t_d = time until the approximate system responds.

Minimising the integral of the absolute value of the error between setpoint and output, Ziegler & Nichols (1942) have found by calculation and experiments that the parameter values of the controllers should be chosen as follows:
for proportional control

$$K_c = \frac{1}{K}\frac{\tau}{t_d} \qquad \text{(Eq. 5.3.8)}$$

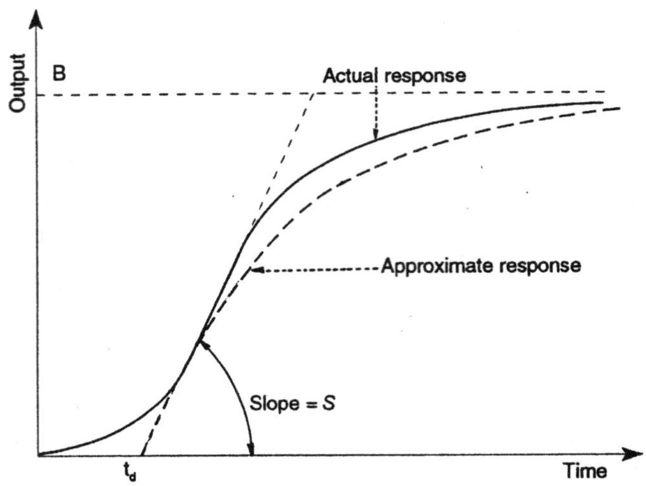

Figure 5.3.5 - Approximation of the response of a first order system with dead time.

for proportional-integral control

$$K_c = 0.9 \frac{1}{K} \frac{\tau}{t_d} \qquad \text{(Eq. 5.3.9)}$$

$$\tau_I = 3.3 t_d$$

and for proportional-integral-derivative control

$$K_c = 1.2 \frac{1}{K} \frac{\tau}{t_d} \qquad \text{(Eq. 5.3.10)}$$

$$\tau_I = 2\, t_d$$
$$t_D = 0.5\, t_d$$

From these controller settings it can be seen that for the integral control the controller gain is smaller than for only proportional control. The reason for this is the destabilising effect of the integral part. If a derivative part is added, the controller gain can be increased, due to the stabilising effect of this action. Ziegler & Nichols (1942) found that this parameter setting gives satisfactory behaviour; for instance there is enough damping and the second overshoot is less than 25% of the first overshoot.

5.3.6 Other control configurations

Feedback control systems are quite satisfactory, but sometimes slow. For this reason some other control structures can be developed. Here we will only discuss feedforward control and cascade control.

Feedforward control is based on the principle that if the disturbances in the process can be measured this information can be used to reduce the effect of the disturbances. The feedforward structure is shown in Figure 5.3.6.

The setpoint element is used to ensure that the output equals the setpoint. The feedforward controller is used to minimise disturbance. Since in general the setpoint element or the feedforward controller can not be implemented exactly, the output will not follow the setpoint exactly where there are setpoint changes and/or changes in the disturbance. For this reason in practice the feedforward control is used in combination with a feedback controller, see Figure 5.3.7.

Cascade control is used when the process consists of two sub-systems and a measurable disturbance which is acting on the main process, see Figure 5.3.8.

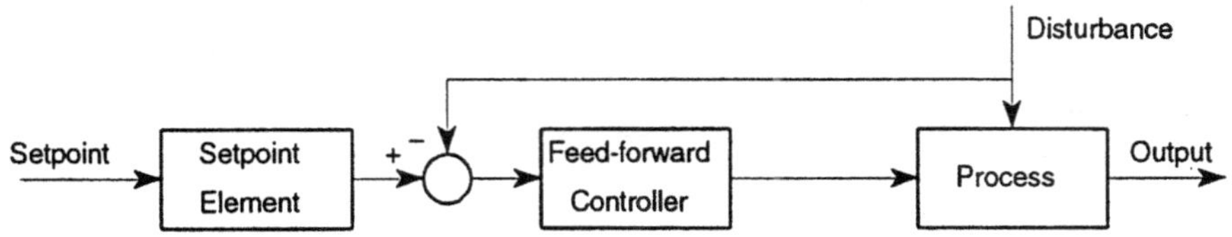

Figure 5.3.6 – Feedforward control structure.

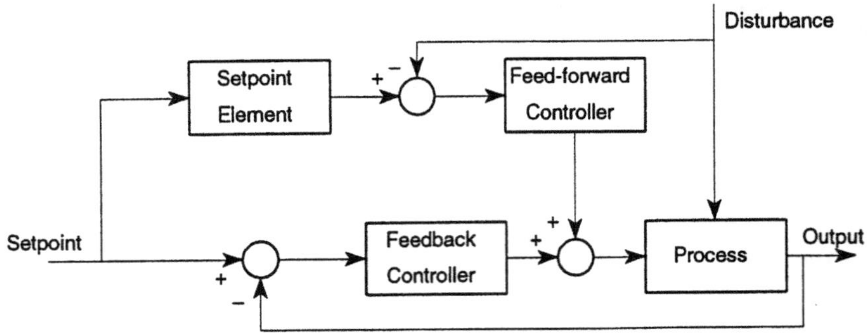

Figure 5.3.7 – The feedforward/feedback control structure.

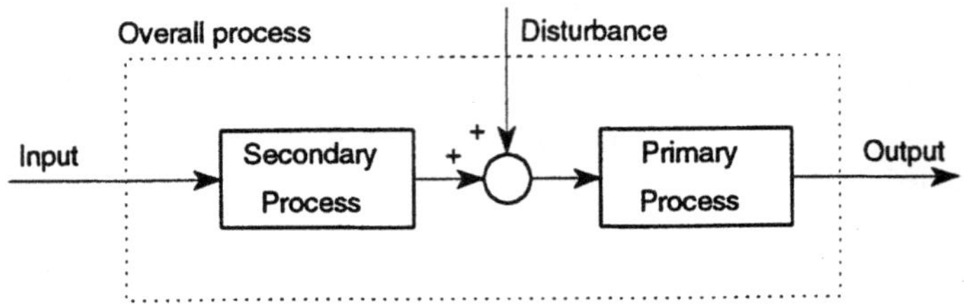

Figure 5.3.8 – Process with disturbance

For instance, in the temperature control of the greenhouse the secondary process is the heating system and the disturbance s the solar radiation.

The conventional feedback configuration is shown in Figure 5.3.9. If the main process is slow it takes a while before the controller will react to a change in the disturbance. For instance, if the solar radiation increases in the greenhouse system, after a while, due to the dynamics of the greenhouse finally, the temperature of the greenhouse will increase. Then the controller will decrease the pipe temperature, but this whole process takes some time.

Cascade control reduces this effect by a direct feedback to the secondary process, in this case the heating system. The cascade control configuration is shown in Figure 5.3.10.

The output of the secondary process together with the disturbance is compared with the output

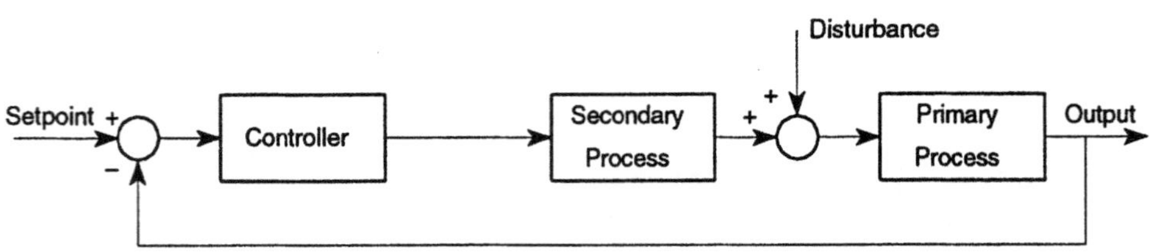

Figure 5.3.9 – The conventional feedback structure.

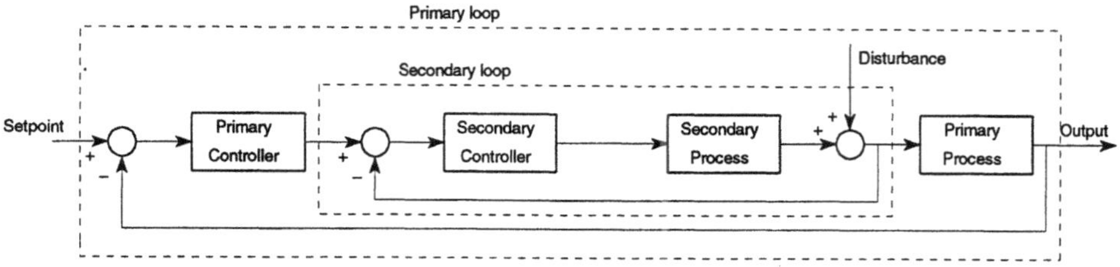

Figure 5.3.10 – The cascade control configuration

of the primary controller and fed back via the secondary controller. In this way the influence of the disturbance can be corrected very quickly. Notice that the setpoint of the secondary loop is determined by the output of the primary controller. The secondary loop does not affect the stability of the overall loop, and therefore high gains can be made in the secondary loop. In our greenhouse example this means that if the solar radiation increases, this effect is corrected by the secondary controller and the pipe temperature is decreased almost directly. The tuning of the primary and secondary controllers can be done in the same way as before. First the secondary controller is tuned and then the tuning of the primary controller is determined while the secondary controller is already set up.

5.3.7 Practical considerations

So far only the principles of some controller types and control configurations have been considered. In order to apply these controllers in practice more work is necessary. For implementation in climate computers controllers have to be digitalized and have to be adjusted to avoid input saturation. The interaction in the greenhouse process also has to be taken into account. Some of these aspects will be discussed in section 5.4.

5.4 Current implementation of hardware and software

W.TH.M. VAN MEURS

5.4.1 Hardware

Plant production can only be indirectly controlled by climate control. As indicated in section 5.1 the hardware of these control systems has evolved from analogue to digital.

In 1974, the first climate control computer system appeared on the Dutch market. This was a stand-alone system, controlling an arbitrary number of greenhouse compartments. In this set-up any sensor or actuator could be connected to the central computer.

The introduction of the single-board microprocessor resulted in two hardware developments. On the one hand a small, single board, dedicated measurement and control processor was developed, able to control one compartment with a limited number of control loops. A decimal keyboard and a LED display on the front of the box enabled the user to communicate with the system (Gieling, 1980). On the other hand a distributed system was developed, equipped with a central host computer connected to local measurement and control processors (Van Meurs, 1980). On the host computer man-machine interfacing, data handling, graphics and alarms were performed together with all calculations of the control algorithms and the Input/Output (I/O) with the distributed small front-end computers. Generally the small front-ends only measured the sensor signals and activated the relays for the valves and ventilators.

Nowadays the grower can choose from a variety of commercial systems with a centralised or distributed set-up.

The first computers started with 8–16 Kb EPROM for the programmes and 4 to 8 Kb memory. To satisfy the growers' demands with respect to controlling, data storage, graphics and alarms, the memory has been extended to 128 Kb (or more) EPROM and as many RAM.

Nowadays the climate computer can be linked in an internal network to the computer controlling the nutrient solution and the management computer. The last one can be connected by modem to the auction and the bank.

5.4.2 Software

The first computer programme for climate control was a direct translation of the actions of the analogue control units, i.e. without any connections between the different control loops. All the necessary control actions were written in one main programme without any sub-routines. In the late seventies and the early eighties, much work was done to upgrade the programmes. The main programme has been divided into smaller modules, one for each function. A computerised system offers the opportunity for more sophisticated climate control methods by offering, for example, more on-line calculations and the implementation of models.

The first programmes were written in assembler language for fast execution to reduce the need for expensive memory boards. The introduction of personal computers has reduced the price of the hardware significantly. Nowadays, memory space is no longer a limiting factor. For the development of control programmes more modern languages such as Pascal or C are used.

The variables to be controlled in a greenhouse are: air temperature, relative humidity, CO_2 and, optionally, radiation and irradiation. For those purposes, greenhouses are installed with a heating system, ventilators, a CO_2 installation, screening and artificial lighting. The present climate computers are equipped with software algorithms to control these installations, making allowance for the interdependences. In the next sections the different control items will be discussed in a general way.

5.4.3 Temperature control

5.4.3.1 Heating and ventilation setpoints

Originally two greenhouse air temperature setpoints for heating were set: one for the night and one for the day period. The values depended on the crop in the greenhouse. Nowadays the growers require a more flexible set-point choice. Therefore most systems offer adjustment for 4–8 independent periods. However, the principles can be understood from a two period set-up.

The temperature set point for ventilation is needed to prevent too high temperatures. Therefore this is a fixed value or a calculated one around the set-point for heating. However, under most conditions ventilation is needed to prevent too high humidity. In this case the principles of humidity control have to be followed (section 5.4.4), overruling those for temperature control.

A typical profile for the greenhouse heating and ventilation setpoints is given in Figure 5.4.1.

The setpoints for temperature and ventilation are calculated with regard to sunrise and sunset. As an option a fixed clock time can be used for the day/night and night/day changes. Obviously, the setpoint has to change at sunrise in such a way that the greenhouse air is heated by solar radiation. Due to rapidly increasing transpiration and slowly increasing air temperature, the dew-point will increase. This causes humidity problems with condensation on fruits or flowers, resulting in disease and negative effects on quality. Therefore the starting point A (Figure 5.4.1) of the set point increase is calculated, dependent on the time of sunrise, and the slope (1–2 °C h^{-1}) of line AB as preamble to sunrise at B. Likewise, point D depends on slope CD and the starting point C which is related to sunset.

In most systems the setpoints are recalculated every minute while a check of the control loops is completed every 15 seconds, including measurements of all sensors.

From research and practical experience growers have developed heuristic rules that at higher light levels temperature has to be adjusted in order to achieve better crop growth. Therefore the temperature setpoints that are set are proportionally dependent on the radiation level (light-dependent control; Bokhorst *et al.* 1972).

Variable outside conditions (e.g. solar radiation, wind speed) act as disturbances on climate control. The effects are described in Chapter 3. This means that comprehensive climate management has to consider the external weather conditions as well as the greenhouse climate itself. In most control systems outside weather data are used in algorithms designed to compensate for the disturbing effects.

5.4.3.2 Heating systems

The heating system in a typical Dutch greenhouse is described in section 4.3. The air temperature control is performed as a closed feedback control loop, reducing the offset of the air temperature. The heating system can be represented by approximating it as a first order transfer function with a delay (section 5.3). The principle of the control algorithm to maintain the air temperature is a proportional-integral (PI) approach in a master/slave system (Figure 5.4.2).

The master system controls the greenhouse temperature, and the slave system controls the hot-water temperature in the pipes. Maximum and minimum temperatures of the heating system have to be defined. These levels are dependent on the greenhouse installation, lay out of the heating pipes, and the crop.

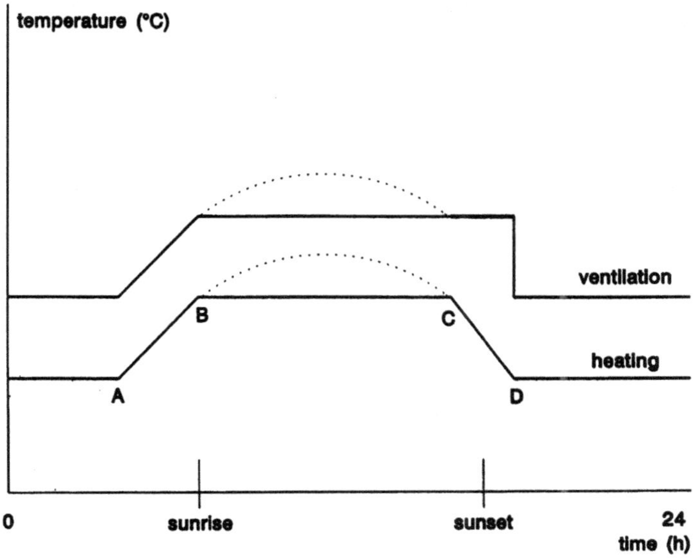

Figure 5.4.1 – Typical profile of temperature and ventilation setpoints over 24 hours.

In the case of two heating pipe systems the control algorithm needs to be expanded with a split range controller. This controller calculates out of the total hot water setpoint, the setpoint of each individual system, taking into account the adjustable maximum and minimum temperatures of each system. The split-range algorithm in particular, controls the smooth changeover of heat demand of one heating system to two systems. (Valentin & Van Zeeland, 1980).

The basic relation for PI control (section 5.3) in discrete form is given by the equation:

$$u(k) = K_c e(k) + K_i t_s \sum_{j=0}^{k-1} e(k-j) \qquad \text{(Eq. 5.4.1)}$$

where the proportional action is:

$$u'(k) = K_c e(k) \qquad \text{(Eq. 5.4.2)}$$

and the integral action is:

$$u''(k) = K_i t_s \sum_{j=0}^{k-1} e(k-j) \qquad \text{(Eq. 5.4.3)}$$

In these equations the controller input $u(k) = u(t)$ at time $t = kt_s$, where t_s is the sampling time interval. K_c is the proportional gain and K_i ($= K_c \times \tau_i$) the integral gain (in section 5.3 the product of proportional gain and integral time constant). The error $e(k)$ [°C] is the difference between the setpoint value $T_{sp}(k)$ and the measured greenhouse air temperature $T_i(k)$. Methods to derive K_c and K_i from the system properties are derived in section 5.3.

Setpoints are not changed stepwise but smoothly along lines AB and CD (Figure 5.4.1). Beside the reasons already mentioned in section 5.4.3, this is also done to achieve smooth control. If heat demand is represented by a stepwise change of the temperature setpoint for the morning, the central boiler cannot meet the demand, so the heating pipe temperature will not reach the desired value. Saturation occurs in the actuating signal and the integral part u"(k), (equation (5.4.3)), will go to infinity. This effect is known as "winding up". As a consequence it causes an undesired greenhouse temperature rise. A sloping change in the setpoint diminishes this effect of saturation. Maximum and minimum values for u"(k) have to be adjusted to keep the saturation between limits. An improved control method to diminish winding up is the anti-wind up approach, which keeps the control input u(k) within limits of the realized pipe temperature (Udink ten Cate & Van Zeeland, 1981).
Equation (5.4.1) can be rewritten as:

$$u(k) = u(k-1) + \Delta u(k) \qquad \text{(Eq. 5.4.4)}$$

where

$$\Delta u(k) = K_c [\, e(k) - e(k-1) + K^*_i e(k) \,] \qquad \text{(Eq. 5.4.5)}$$

with $K^*_i = K_c K_i$ and the extra condition:

$$T_h(k-1) - c \leq u(k-1) \leq T_h(k-1) + c \qquad \text{(Eq. 5.4.6)}$$

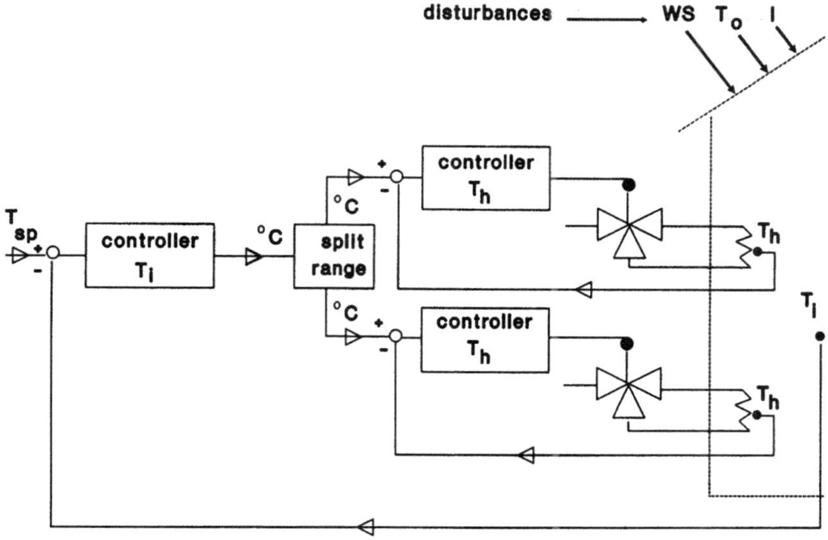

Figure 5.4.2 – Greenhouse temperature control by a hot-water piping system, including split-range control for two systems (feedback control).

c is a constant (≈ 5°C), $T_h(k-1)$ (°C) is the measured pipe temperature and u (°C) is the setpoint temperature for the heating system.

Several other modifications have been developed in research to improve the controllers, such as adaptive control (Udink ten Cate, 1983; Verwaayen, 1988). A finely tuned controller has an accuracy of better than +/- 0.2 °C and over- and undershoots less than 25% of the temperature step.

Another method is the combination of feedback and feedforward control. This combination increases the stability of the control loop and the accuracy of control. The feedforward control reduces beforehand the offset in expected changes of the inside conditions, due to changes outside. One method is to calculate the heat load of the greenhouse dependent on the temperature difference between inside and outside, the wind speed and the incoming radiation. A provisional setpoint for the water temperature T'_h can be calculated as

$$T'_h = T_h\,min + \frac{T_o max - T_o}{T_o max - T_o min}\,(T_h max - T_h min) + aWS - bI \qquad \text{(Eq. 5.4.7)}$$

in which $T_o$max and $T_o$min are the outside temperatures at which the pipe temperature is respectively minimum $T_h$min and maximum $T_h$max. T_o is the measured outside temperature, *WS* is the wind speed and *I* is the global radiation inside. The coefficients *a* and *b* have adjustable values. Other methods have been developed by Heyna (1980) and Tantau (1984). The PI controller of inside air temperature gives an addition to this setpoint for the water temperature. An example of this control principle is given in Figure 5.4.3.

As a general remark it can be stated that a control system cannot be more accurate than the sum of the resolution and accuracy of each element in the loop. This involves the sensor itself, the transducers, the computer calculations and the actuators.

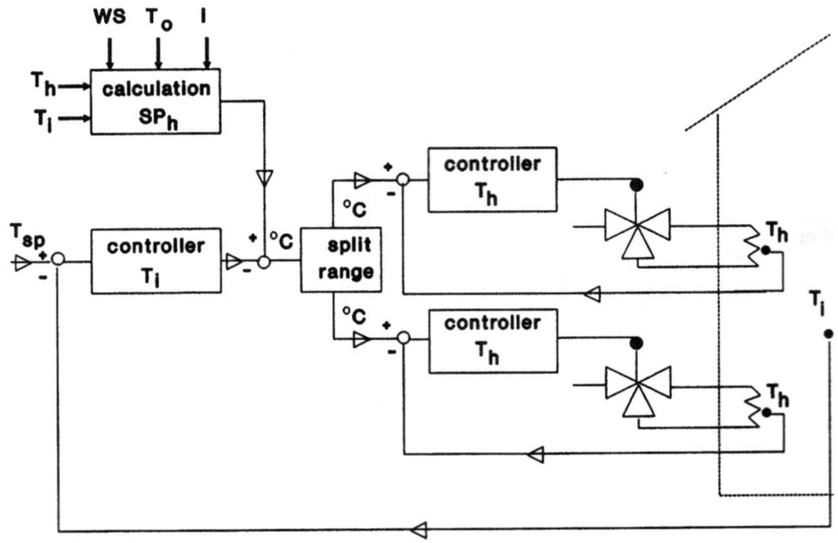

Figure 5.4.3 – Greenhouse temperature control by a hot-water piping system, including split-range control for two systems (feedforward-feedback approach).

One must keep in mind however, that even if the accuracy is very good, it does not apply to the temperature at every spot in the greenhouse. The air temperature is measured in only one place. Nearly all greenhouses exhibit a vertical and/or horizontal gradient in temperature and also in humidity.

Heated concrete floors are in use as stand-alone systems or in combination with a pipe system. As the response time from the heated water of the concrete floor to the air is in the order of 5 to 8 hours, it is practically impossible to control the air temperature in this way. For botanical and human welfare, the floor temperature is not allowed to rise above about 26 °C (realised at ≈ 45 °C water temperature). Therefore, it is possible to use the concrete floor as a slowly controlled base heating system, while the pipe system in the greenhouse is used as a fast supplementary heat source controlling above all the air temperature. The efficiency of the heated floor is very low.

Some greenhouses are equipped with air heaters (section 4.3). The control of these heaters is on/off, giving rise to a hysteresis of 1-3 °C, dependent on the difference between the setpoint and the measured greenhouse temperature. If there is more than one heater in a compartment, there will be a sequence control of these. Fluctuations, inherent to this type of control, are in the order of some degrees Centigrade.

5.4.3.3 Temperature control by ventilation

Ventilation is used to prevent the temperature in the greenhouse exceeding the setpoint value. The apertures of the ventilators, installed in the roof of the greenhouse, are controlled by a proportional algorithm (Figure 5.4.4).

In a closed loop system, the ventilator position is measured. In a semi-closed loop system, the position of the ventilators is calculated every control run and used in the control algorithm as the "measured" value. The calculated values are used when the measurements of the ventilator positions by potentio-

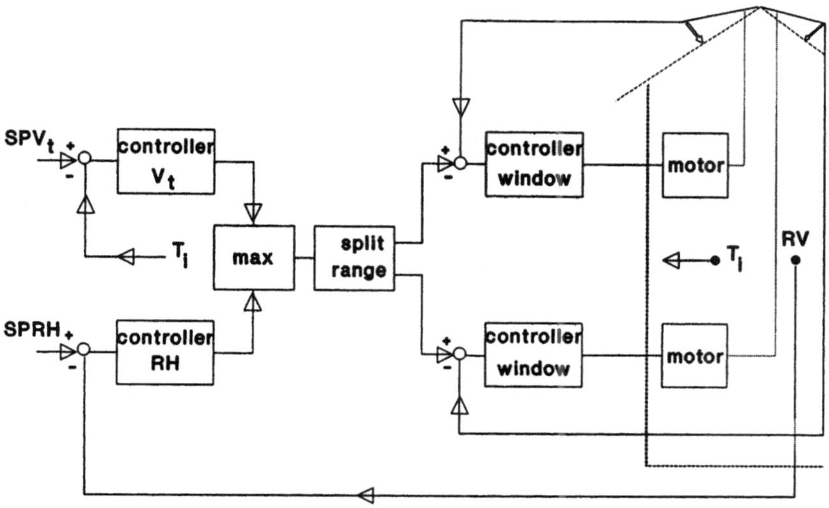

Figure 5.4.4 – Greenhouse temperature and humidity control by ventilation.

meters are inaccurate due to mechanical or temperature effects. In this case the ventilator positions have to be verified a few times a day, mostly in the extreme positions. Automatic calibration has to be done when switching from hand to automatic position.

The ventilation rate of a greenhouse depends on the ventilator apertures, the wind speed and the temperature difference between inside and outside (section 3.3.4). Accurate temperature control could be achieved by direct control of the ventilation rate. As the ventilation rate is hard to determine on-line for commercial greenhouses, the effect on the ventilation rate is accounted for in the temperature control by adjusting the proportional band dependent on wind speed and temperature difference. A second possibility is a correction of the primary calculated ventilator position, based on the same factors.

Measurements of the wind direction are used to decide whether leeward side or windward side opening is appropriate. The programme decides on which side the ventilator is opened first (commonly the leeward side).

Maximum and minimum ventilator apertures are defined for storm, frost and rain both for day and night.

5.4.4 Humidity control

Ventilators are used for humidity control as well as temperature control. In a greenhouse with closed ventilators the humidity will increase due to transpiration of the crop, even at night. Although a high humidity is not always a problem as far as plant growth is concerned (Bakker, 1991), growers try to prevent high humidity because of the increased risk of diseases. In wintertime condensation against the cold roof will reduce humidity. In summer this will be less because of the decreased temperature difference between roof and inside. Therefore humidity can be controlled by ventilation. As pointed out in section 5.4.3, relative humidity can also rise when the system changes from day to night temperature setpoint.

On the other hand humidifiers can be installed to humidify the air. The installation and control programme is also used for crop cooling at high greenhouse temperatures (section 4.4).

From the above considerations it is evident that the grower has to determine a day and a night setpoint for the relative humidity (RH) or the humidity deficit (Δx).

If no humidification is used, there are two approaches in the control algorithm. First, the ventilation setpoint will increase in the case of a low RH value (high Δx) and will go down where there is a high RH value (low Δx). The second approach is to control the ventilator opening proportional to the difference between the RH setpoint and measured value, in the range between the temperature setpoint (or just below) and the ventilation setpoints. Above the ventilation set point the largest calculated ventilation opening for temperature or relatie humidity is taken.

Because the influence of a ventilator opening is much greater on the relative humidity than on the temperature, the value of the proportional band for humidity control is larger than for temperature control.

The combination of ventilating and heating is used to stimulate transpiration, particularly on very cloudy days when the temperature difference between inside and outside is relatively small and crop transpiration is low. To activate crop transpiration the setpoint of the pipe temperature is put at a minimum value (minimum pipe temperature). In this case the greenhouse temperature may rise above the temperature setpoint for ventilation so the ventilators will be forced to open. Energy transfer from pipes to plants due to long wave radiation, air movement and decreased RH will activate transpiration (Chapter 3). The minimum pipe temperature will be set even in summer.

Simultaneous heating and ventilation results in a considerable increase in energy consumption by the greenhouse.

5.4.5 CO_2 control

W.TH.M. VAN MEURS AND E.M. NEDERHOFF

A strategy for the enrichment of CO_2 from a central heater is described here. As mentioned before, there are several variations possible and in use. The algorithm calculates the instantaneous setpoint dependence on the heat demand, the radiation, the wind speed and ventilator position (Figure 5.4.6).

It uses a number of settings, among others also three CO_2 levels. The high CO_2 level (line A in Figure 5.4.6) is taken as setpoint as long as heating is required independent of the ventilation and radiation. The medium level (line B, Figure 5.4.6) is the setpoint where no heat is required and the radiation exceeds a preset level. Line C represents the low level setpoint when no heat is required and the radiation is below a preset level. A minimum value of the CO_2 is set when the greenhouse is ventilated to a certain extent. The transition from a higher level to the minimum value is proportional to the ventilator positions. A higher wind speed shifts line B and C to the left.

In addition, growers can increase the heat demand (e.g. setting a higher minimum heating pipe temperature) or delay the ventilation, by choosing other settings for the greenhouse air temperature or humidity control. Both are indirect measures to enable the maintenance of a higher CO_2 concentration. If a heat storage tank is present, the utilisation of heat should also be considered in the CO_2 strategy. There is one system commercially available that optimises the supply of CO_2 during the day and takes into account the expected heat demand during the night, on the basis of the local weather forecast.

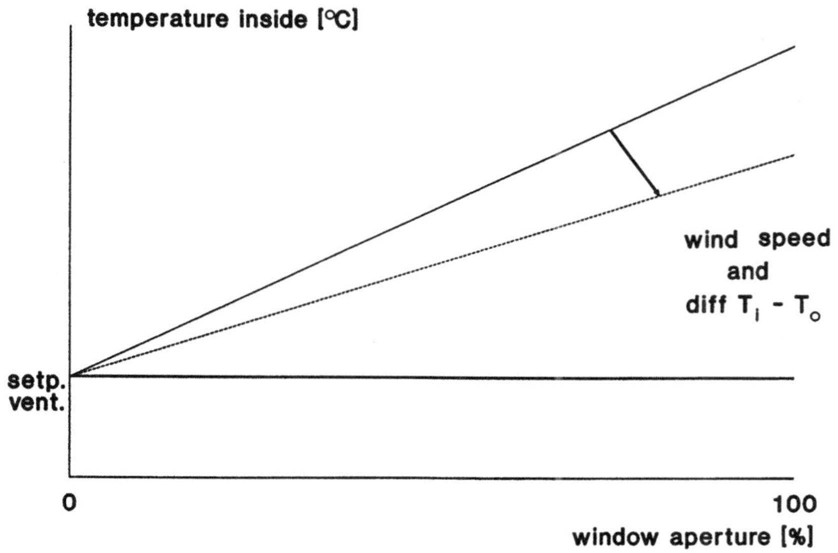

Figure 5.4.5 – Window aperture dependent on inside temperature (proportional control). The slope is dependent on wind speed and inside-outside temperature difference

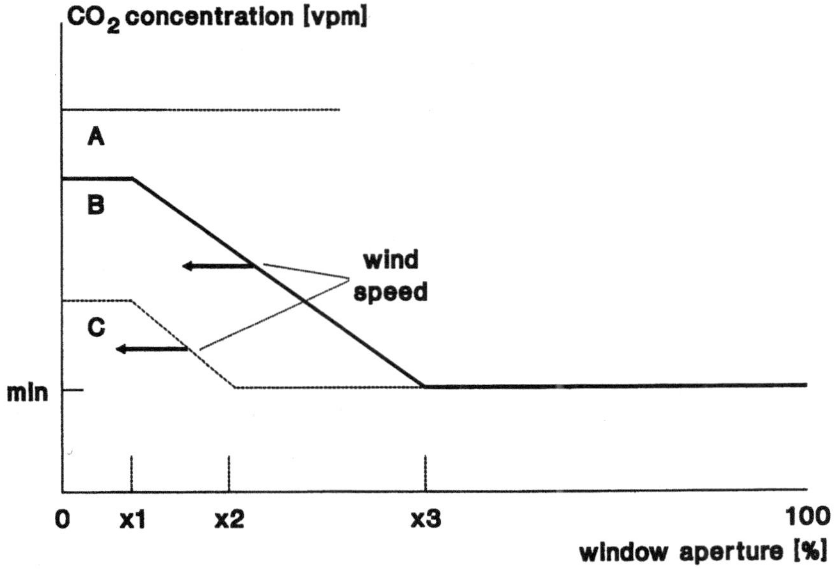

Figure 5.4.6 – Setpoint of CO_2 enrichment dependent on heat demand (A), and the solar radiation level (A and B) versus window aperture.

5.4.6 Artificial light control

W.TH.M. VAN MEURS

For assimilation lighting, the high pressure sodium lamp is commonly used (Chapter 4). To prevent a short lamp life, these lamps should not be subjected to frequent switching on and off. A minimum on period of 20 minutes is recommended. A minimum off period of 10 to 15 minutes is required to cool the lamp down before it is switched on again.

The electric power consumption of a group of lamps is measured and logged. This is done firstly as a check on control function, and secondly to check whether any lamps have broken down.

Cogenerators are commonly used, the generator being driven by a gas-fired engine. The electricity generated is consumed by the lamps and the engendered heat by the engine for heating the greenhouse. As heat and light are not always required simultaneously, a buffer for heat storage is required. From the point of view of control, this means that the algorithm has to decide if the central boiler, the buffer or a combination of both will be the supplier of the necessary hot water for the greenhouse. A substantial part of power consumed by the lamps is converted into heat (≈ 78%) which reduces the heat demand of the greenhouse.

From the point of view of efficiency, it makes no sense to switch the total energy installation on and off very frequently, and therefore a minimum "on-time" has to be set.

Lamps for day length control are used for *Chrysanthemum* and some pot plants such as *Kalanchoë blossfeldiana*. The aim is to regulate the flowering time of the plants. In this way it is possible to produce flowers (e.g. chrysanthemums) throughout the year.

The day length is controlled through a combination of artificial lighting and black screens. The lamps are used to simulate a longer day, whereas the dark (black) screens shorten the natural daytime. Both are controlled by clock adjustments for on and off time. In many applications cyclic lighting is used. Here the duration of illumination is divided into successive periods of, for example, 10 minutes lamps on and 20 minutes lamps off.

For day length light control incandescent lamps are used. The electric power consumption of the lamps is measured and logged.

5.4.7 Screen control

Thermal screens

Thermal screens (section 4.5) are parked during the day and unrolled at night. A control strategy is chosen, unrolling the screen for the dark period, when either the temperature difference between inside and outside exceeds a preset value or the temperature of the heating pipes exceeds a certain value. The screen is closed in one operation. If the humidity exceeds an allowed value, the screen can sometimes be opened between 1 and 30 cm, to get rid of the humidity by condensation at the cold roof or by opening the ventilators a little. After the humidity has decreased, the screen is closed again.

A second option is not to control humidity on the basis of a small opening but to park the screen completely and not to use it during the night period. In the morning the screen is opened at a time related to the moment of sunrise or to a certain level of daylight, whichever comes first. The screen is opened first in small steps, to prevent a sudden decrease in greenhouse air temperature (cold-fall).

When screens are unrolled, fans can be used, to force a small air movement under the screen.

Shading screen

The decision to close this screen depends on the time of the day and on two light levels, one for opening and one for closing. If the screen is closed, the temperature in the greenhouse can increase. In that case the screen is set to a small opening to allow more ventilation inside the greenhouse.

Black screen
As mentioned before, the dark screens are used to shorten the natural day length. The screen has to cover the whole greenhouse area. Parking and unrolling of the screen is controlled at adjusted times depending on the desired day length.

5.5 Conclusions

G.P.A. BOT

As described in this chapter classical automatic control usually starts with simple feedback control schemes. The user specifies a setpoint or setpoint trajectory for one or more variables of interest, and the task of the controller is to match this setpoint as closely as possible by manipulating the controls. In greenhouse control, this basic idea has been extended and modified for several reasons. First, the greenhouse is, in fact, a multivariable system. In the simplest set-up there are two variables, humidity and temperature, and two control actuators, the heating system and the windows or ventilators. The problem with single loop controllers is that for limiting the humidity in the greenhouse the windows may need to be opened, while the heat losses to the environment may lead to a demand for heating, and thus for closing the ventilators. Also, when the solar radiation input exceeds the heat losses, so that cooling is needed, the ventilators must be used for temperature control as well as humidity control. Thus, the separate control loops for temperature and humidity both lead to a desired window setting, and some way of combining the outcome of each is necessary (section 5.3.2). This is generally achieved by some heuristic combination rule. A further heuristic development is the introduction of desired bands, rather than setpoints, the forward coupling of temperature setpoints to incoming solar radiation and the introduction of CO_2 enrichment.

Thus, the historic development of automatic control systems for greenhouse climate control has naturally led to a system that effectively consists of some simple proportional (P) or proportional-integral (PI) controllers which are linked together through a set of quite complicated rules. It is only by the introduction of greenhouse climate computers in combination with direct digital control (DDC) that the implementation of these rapidly expanding operation rules is made possible. Moreover, the computer memory enables the detailed specification of desired setpoint trajectories, such as day and night temperatures, starting time and slope of the temperature setting around sunrise and sunset, specification of the desired tube system temperature, numerical values of forward adaptation parameters, and so on. Also, reporting facilities can be built in, as well as a set of alarm and diagnosis facilities. Consequently, modern greenhouse climate control systems not just automate the short term immediate control action in response to outside influences, but also partly automate the implementation of a longer term strategy, formulated by the grower. Although it cannot be denied that today's systems have several advantages, the current greenhouse control computer systems could be described cynically as "a modern complex version of a digitalised collection of analogue controllers".

Despite the tremendous success of computerised greenhouse control systems, their heuristic development also entails some serious drawbacks.

Firstly, the system has become extremely complicated to operate. A total number of 300 or 400 settings for a moderate greenhouse nursery is not uncommon. The physical and practical meaning of these settings is not always clear to the user. In practice mistakes occur easily; for instance the undesired setting of conflicting requirements. Although the software can be adapted to help and guide the grower, the principle drawback remains that settings have to be specified that bear only indirect meaning for the ultimate goal of the grower: to make money. The present systems provide little information about the consequences of the chosen settings for product yield, risk, energy costs, vulnerability to ambient factors and discharges to the environment. Some commercial management tools however contain information on photosynthesis.

Secondly, the intricate problem of interacting control loops has not been solved in a systematic way. This is not to say that present heuristic rules have not been set up with a lot of knowledge, good common sense and experience. It is expected, however, that a more fundamental approach should do better.

Thirdly, present control systems concentrate on the physical environment. Information about the plant growth as a function of physical climate variables is only used in an indirect way through generic blueprints of desired setpoint trajectories which are sometimes modified from day to day depending on the crop performance. If the knowledge on plant physiological and physical processes could be incorporated, fundamental improvements in the diurnal climate control could be achieved (Challa, 1990). Optimisation of plant production in this way is much more than the present simple control of actuators and requires an integrated approach of the disciplines described in the previous chapters: plant physiology, physics, horticulture, construction, equipment and control.

The factors above call for new, more intelligent control systems. These systems should be based on fundamental rather than heuristic approaches to greenhouse control. Also, the control issue is crucial in new advanced equipment designs, such as power-heat cogeneration systems and heat storage systems, which cannot be applied properly without a fundamental and integrated control framework. The possibilities, advantages and state of the art of this integrated approach are highlighted in the next chapter.

References

Albery, W.J., B.G.D. Haggett & L.R. Svanberg, 1986. The development of electrochemical sensors. In: W.G. Gensler (Ed.), Proceedings of the NATO Advanced Study Institute on "Advanced Agricultural Instrumentation". Il Ciocco, 1984. Martinus Nijhoff, Rotterdam, pp. 349–392.

Bailey, B.J., B.G.D. Haggett, A. Hunter, W.J. Albery & L.R. Svanberg, 1988. Monitoring nutrient film solutions using ionselective electrodes. Journal of Agricultural Engineering Research 40: 129–142.

Bakker, J.C., 1991. Analysis of humidity effects on growth and production of glasshouse fruit vegetables. PhD thesis, Wageningen Agricultural University, Wageningen, 155 pp.

Bergveld, P., 1970. Development of an ion sensitive solid state device for neurophysical measurements. IEEE Trans.Biomed.Eng. BME–17, 70 pp.

Bicanic, D., P. Torfs, M. Lubbers & A. Tam, 1992. Horticultural sensing by photoacoustics and thermal lensing. Acta Horticulturae 304: 29–41.

Bickford, E.D. & S. Dunn, 1972. Lighting for plant growth. The Kent State university press, Kent, Ohio.

Biggs, W., 1986. Radiation measurement. In: W.G. Gensler (Ed.), Proceedings of the NATO Advanced Study Institute on "Advanced Agricultural Instrumentation". Il Ciocco, 1984. Martinus Nijhoff, Rotterdam, pp. 3–20.

Bokhorst, D., A. Van Drenth & G.P.A. Van Holsteijn, 1972. Lichtafhankelijke klimaatregeling voor kassen. IMAG-DLO, Wageningen, ITT Publ. 74, 80 pp. (in Dutch).

Bot, G.P.A., 1983. Greenhouse climate: from physical processes to a dynamic model. PhD thesis, Wageningen Agricultural University, Wageningen, 240 pp.

Bowman, G.E. & G.S. Weaving, 1970. A light-modulated greenhouse control system. Journal of Agricultural Engineering Research 15(3): 255–264.

Calvert A. & G. Slack, 1976. Effect of carbon dioxide enrichment on growth, development and yield of glasshouse tomatoes. II. The duration and daily periods of enrichment. Journal of Horticultural Science 51: 401-409.

Challa, H., 1990. Crop growth models for greenhouse climate control. In: R. Rabbinge, J. Goudriaan, H. van Keulen, F.W.T. Penning de Vries & H.H. van der Laar (Eds), Theoretical production ecology: reflections and prospects. Simulation Monographs 34. Pudoc, Wageningen, pp. 125–145.

Clark, O.M. & M.A. Povey, 1985. Integrated protection for computer systems. In: Conference proceedings of the 18th conference on lightning protection (ICLP-85). VDE-Verlag GmbH, Berlin, Offenbach, pp. 325–331.

Collins, W.L. & M.H. Jensen, 1983. Hydroponics: a 1983 technology overview. National Science Foundation, no 82-SP-1009.

De Jong, T., 1990. Natural ventilation of large multi-span greenhouses. PhD thesis, Wageningen Agricultural University, Wageningen, 116 pp.

Doetsch, G., 1974. Introduction to the Theory and Application of the Laplace Transformation. Springer Verlag, Berlin.

Gieling, Th.H., 1980. Commercial greenhouse computer systems. Acta Horticulturae 106: 59–66.

Gieling, Th.H. & W.Th.M. Van Meurs 1984. Damage to climate control equipment and crop caused by lightning. Acta Horticulturae 148: 297–303.

Gieling, Th.H., E. Van Os & A. de Jager, 1988. The application of chemo-sensors and bio-sensors in soilless cultures. Acta Horticulturae 230: 357–361.

Hanan, J.J., 1984. Plant environmental measurement. Bookmakers Guild Inc., Longmont, Colorado, 326 pp.

Hashimoto, Y., T. Morimoto, T. Fukuyama, H. Watake, S. Yamaguchi & H. Kikuchi, 1989. Identification and control of hydroponic system using ion sensors. Acta Horticulturae 245: 490–497.

Hasse, P. & J. Wiesinger, 1985. Schutz von Rechenanlagen und -Geräten gegen Blitzstörungen. In: Conference Proceedings of the 18th conference on lightning protection (ICLP-85). VDE-Verlag GmbH, Berlin, Offenbach, pp. 299–304. (in German).

Hauptmann, P., 1990. Sensoren, Prinzipien und Anwendung. Carl Hanser Verlag, München, Wien, 182 pp. (in German).

Heinen, M. & K. Harmanny, 1992. Evaluation of the performance of ion-selective electrodes in an automated NFT system. Acta Horticulturae 304: 273–280.

Heyna, J.B., 1980. Een nieuwe K-waarde voor kasverwarmingsbuizen. IMAG-DLO, Wageningen, IMAG-DLO Xnr 6882/80-11-21. (in Dutch).

Hilhorst, M.A., J. Groenwold & J.F. de Groot, 1992. Water content measurements in soil and rockwool substrates: dielectric sensors for automatic in situ measurements. Acta Horticulturae 304: 209–218.

Högberg, R., E. Lötberg & V. Scuka, 1985. Lightning protection of electronic installations: design considerations. In: Conference proceedings of the 18th conference on lightning protection (ICLP-85). VDE-Verlag GmbH, Berlin, Offenbach, pp. 291–298.

Kupers, G., J. Van Gaalen, Th.H. Gieling & E.A. Van Os, 1992. Diurnal changes in the ion concentration of the supply and return water of a tomato crop grown on rockwool. Acta Horticulturae 304: 291–300.

Long, S.P., 1986. Instrumentation for the measurement of CO_2 assimilation by crop leaves. In: W. Gensler (Ed.), Advanced Agricultural Instrumentation. Martinus Nijhoff, Rotterdam, pp. 39–91.

Meteorological Office, 1969. Handbook of meteorological instruments, I, Instruments for surface observations. Her Majesty's Stationary Office, London.

Palm, W.J., 1986. Control Systems Engineering. Wiley, New York, 695 pp.

Robinson, N., 1966. Solar Radiation. Elsevier Publishing Company, Amsterdam, 347 pp.

Savage, A., 1985. Hydroponics worldwide: proceedings international conference on hydroponics. Honolulu, 194 pp.

Schurer, K., 1986. Water and plants. In: W.G. Gensler (Ed.), Proceedings of the NATO Advanced Study Institute on "Advanced Agricultural Instrumentation". Il Ciocco, 1984. Martinus Nijhoff, Rotterdam, pp. 429–456.

Slavik, B., 1974. Methods of studying plant water relations. Springer Verlag Berlin, Heidelberg, New York, 449 pp.

Stephanopoulos, G., 1984. Chemical Process Control. Prentice Hall, Englewood Cliffs, 696 pp.

Strijbosch, Th. & J. Van de Vooren, 1975. Developments in climate control. Acta Horticulturae 46: 21–22.

Tantau, H.-J., 1984. Adaptive control of greenhouse climate. Acta Horticulturae 148: 251–258.

Udink ten Cate, A.J. & J. Van Zeeland, 1981. A modified PI-algorithm for a glasshouse heating system. Acta Horticulturae 115: 351–358.

Udink ten Cate, A.J., 1983. Modelling and (adaptive) control of greenhouse climates. PhD thesis, Wageningen Agricultural University, Wageningen, 159 pp.

Valentin, J. & J. Van Zeeland, 1980. Adaptive split-range control of a glasshouse heating system. Acta Horticulturae 106: 109–115.

Van den Vlekkert, H.H., 1992. Ion sensitive field effect transistors. Acta Horticulturae 304: 113–126.

Van den Vlekkert, H.H., J.P.M. Kouwenhoven & A.A.M. Van Wingerden, 1992. Application of ISFETs in closed loop systems for horticulture. Acta Horticulturae 304: 309–320.

Van Meurs, W.Th.M., 1980. The climate control computer system at the IMAG. Acta Horticulturae 106: 77–84.

Van Os, E.A., M.N.A. Ruijs & P.A. Van Weel, 1991. Closed business systems for less pollution from greenhouses. Acta Horticulturae 294: 49–57.

Verwaayen, P.W.T., 1988. Regeltechnisch onderzoek in kassen: het gebruik van twee warmtebronnen bij toepassing van twee buisverwarmingssystemen. IMAG-DLO, Wageningen, IMAG rapport 102, 73 pp. (in Dutch).

Verwer, F.L., 1976. Growing horticultural crops in rockwool and NFT. IMAG-DLO, Wageningen, IMAG-DLO Publ. no. RR76-5.

Visscher, G.J.W. & K. Schurer, 1985. Some research on the stability of several capacitive thin film (polymer) humidity sensors in practice. In: Moisture and humidity 1985, Measurement and control in science and industry. Instrument Society of America. Research Triangle Park, North Carolina, pp. 515–323.

Visscher, G.J.W. & J.G. Kornet, 1994. Test of air humidity sensors. In preparation.

Werkhoven, C., 1992. Sensors for irrigation scheduling of cultures in the field. Acta Horticulturae 304: 259–264.

WMO/World Meteorological Organisation, 1971. Guide to meteorological instrument and observing practices. World Meteorological Organization, Geneva, WMO no 8, TP3-4th edition.

Wylie, R.G. & Th. Lalas 1985. Accurate psychrometer coefficients for wet and ice-covered cylinders in laminar transverse airstreams. In: Moisture and humidity 1985, Measurement and control in science and industry. Instrument Society of America. Research Triangle Park, North Carolina, pp. 37–56.

Ziegler, J.G. & N.B. Nichols, 1942. Optimum Settings for Automatic Control. ASME Transactions, Vol. 64, 8: 759.

List of symbols

A	psychrometer coefficient (K^{-1})
e	water vapour pressure (Pa or mbar)
e	tracking error (-)
H	transfer function (-)
I	global radiation inside ($W\ m^{-2}$)
K	integral gain (-)
K_c	controller gain (-)
K_p	process gain (-)
P	barometric pressure (Pa or mbar)
S	Signal (-)
s	Laplace variable (-)
T	temperature (K)
t	time (s)
t_d	dead time (s)
t_s	sampling time interval (s)
U	Laplace transform of u (-)
u	input (-)
V	sensor output (-)
WS	wind speed ($m\ s^{-1}$)
Y	Laplace transform of y (-)
y	output (-)

Greek symbols

Δ	difference (-)
τ	time constant (s)
τ_D	derivative time constant (s)
τ_I	integral time constant (s)

Subscripts

o	outside
s	saturated
w	water-surface

Superscript

*	saturation value

List of abbreviations

DDC	direct digital control
DIN EC	EC according to Deutsche Industrie Norm
EC	electrical conductivity
EPROM	Erasable programmable read only memory
ISE	ion seclective electrode
ISFET	ion selective field effect transistor
Kb	kilobyte
LAI	leaf area index
LEMP	lightning electro magnetic pulse
NFT	nutrient film technique
P	proportional
PAR	photosynthetic active radiation (400–700 nm)
PI	proportional integral
PD	proportional derivative
PID	proportional integral derivative
PTFE	polytetrafluoroethylene
RAM	random access memory
RFI	radio frequency interface
RH	relative humidity
TDR	time domain reflectometry

6 Towards integration

6.1 Introduction

G. VAN STRATEN

In the previous chapters various biological and technical aspects that are relevant to greenhouse cultivation, and its relation with greenhouse climate control, have been explained and discussed. Greenhouse cultivation is subject to continuous change due to technical, horticultural, social, legal and other developments. Thus, the greenhouse of the future will not be the same as the greenhouse of today. Technical and horticultural developments have to be framed within shifting pressure from society towards energy conservation and environmental protection. Moreover, there is a desire to meet market demands for total product quality and for a wider spectrum of produce.

The aim of this chapter is to integrate the achievements of the previous chapters by providing an outlook on new technological developments in greenhouse design, and new possibilities for advanced control, while taking these developments into account. In this chapter the expected development in the design of greenhouses in The Netherlands is described first. The use of modern materials, ad-vanced construction methods, and the installation of heat storage (section 4.6) and cogeneration units (section 4.7) can be foreseen. Subsequently it outlines a generalized framework for greenhouse operation, based on the principles of model based optimal control, in line with the ideas presented in the previous chapters. This framework can be used not only to improve the operation of present day greenhouses, but also offers an integrated prospect for the more complex structures of the future. Advanced control schemes aiming at economy and ease of operation, such as the one described here, together with knowledge based management information systems, will largely contribute to the competitive power of greenhouse production systems.

6.2 Greenhouse construction and equipment

N.J. VAN DE BRAAK

As stated previously in Chapter 4, present greenhouse constructions are standardized to a large extent. The same can also be said of the cultivation methods of the various crops, as well as the equipment in the greenhouse. This, of course, will hamper the penetration of entirely new greenhouse concepts. Yet, there is a large amount of research under way that will gradually lead to renewal.

6.2.1 The dilemma of greenhouse cover design

The dilemma of greenhouse cover design is that it should constitute a barrier to heat loss by convection, while at the same time it should give free access for solar radiation to the plants. The reduction of energy consumption in greenhouses is becoming more and more important in the fight to diminish global environmental pollution i.e. carbon dioxide emission. This is why both greenhouse builders and researchers are looking for greenhouse constructions and covering materials that will lead to energy conservation on the one hand, without reduction of light transmittance and without detrimental effects to greenhouse climate on the other. Although glass is a bad conductor of heat, the panes used in greenhouses are so thin that the overall heat transfer is still considerable. Other aspects that have to be considered in finding the most suitable compromise are costs, life span, uniformity, vulnerability, and labour costs for cleaning and handling.

Experiments with foamglass, an insulating glass material, are being conducted for special applications, but this material seems to be particularly sensitive to dirt. Moreover, in spite of the high production and yield levels achieved in the horticultural industry, the material is far too expensive for greenhouses.

Double glazing reduces the heat loss considerably. However, due to reflection of light at each surface, the light transmission of these panes is less than that of single glass, which until now has been a major drawback. As the energy costs account for roughly 12 to 25% of the total production costs (section 1.3.4) and 1% light reduction generally reduces yield by 1%, at least 4% energy should be saved for each percent of light loss to be able to economically justify the installation of double glass.

Recently Out & Breuer (1994) reported that special coatings on glass can reduce the reflection of photosynthetic active radiation (PAR). This coating can be used in combination with double window panes or with glass with a low emissivity coating. A low emissivity coating suppresses heat losses by radiation, so that the heat transfer either by convection or by radiation can be diminished without affecting light transmission.

Another way to improve the light transmittance of greenhouses is to reduce the amount and size of construction parts which cause shading, or to apply a highly reflective coating, e.g. white paint, to construction parts in order to cut down the total light interception (section 4.5.3). During the last decade the size of glass panes has gradually increased and the dimensions of gutters have decreased.

6.2.2 Energy conserving greenhouses

Waayenberg & Freney (1993) reported a new greenhouse construction consisting of plastic film suspended on a new type of polymer cables, providing a minimum of shading parts. Though lifetime and optical properties of the covering material still have to be improved, this construction forms a good starting point for the development of energy conserving greenhouses.

Energy may also be saved by employing screens. Future developments in the area of screen technology will mainly result in better control of the screen position, further integration of the screen in the greenhouse construction, and better prevention of droplet condensation on screen materials. Research is in progress to provide guidelines for the control of screens in order to optimize energy consumption and crop production.

In the field of heating systems much attention will be given to the introduction of cogeneration units in order to increase the overall efficiency of energy consumption. The development and introduction of burners with a low NO_x emission (low NOX-burners) will contribute to decreasing pollution of the environment. Also, the treatment of flue gases of cogenerators (section 4.6) in order to make them suitable for CO_2 supply can be foreseen.

Concerning ventilation the trend is to mount windows on both sides of the ridge which overlap, in order to increase the potential ventilation capacity.

6.2.3 Conclusion

The effects of energy saving measures on the greenhouse climate are a major concern, as the greenhouse climate has an immediate impact on the production process. In general these measures will reduce the removal of moisture from the greenhouse, resulting in higher relative humidities. Although this could be solved by increased ventilation, this would also lead to extra loss of energy and CO_2. So, the development of future climate control equipment for greenhouses will have to be directed towards the separation of the functions of the present ventilation systems, i.e. cooling, dehumidification and control of the CO_2-level. Such separation would also enhance the possibility of entirely closed greenhouses. Investigations by De Jong (1993) have shown, however, that at present entirely closed greenhouses are not economically feasible in The Netherlands, due to high electricity prices and investment costs.

The use of new materials, advanced construction methods and the installation of additional equipment such as heat storage devices, cogenerators, dehumidifiers and mechanical ventilation, will further increase the complexity of greenhouse climate control. This emphasizes the need for a new approach to the control of the greenhouse climate.

6.3 Greenhouse climate control systems

G. VAN STRATEN AND H. CHALLA

6.3.1 Requirements for intelligent climate control systems of the future

Climate control is one of the tools used in manipulating greenhouse production, and should thus be considered as a part of the overall management rather than as an isolated activity. Management may be defined as the collection of activities directed to reach certain goals. One of the goals of a grower, as an entrepreneur, in general is to maximize his profit.

Often, three levels of management are considered: strategic, tactical and operational. At the strategic level decisions on capital investments for equipment determine the technical possibilities for climate control. At the tactical level, before the start of a new cultivation, the grower decides what crop and cultivar to cultivate, when to plant or sow the crop, and how to make best use of human resources. Connected with the tactical plan is an expectation of average climatic conditions, prices that the grower will receive for his product, and associated with this a "blueprint" of how the crop will grow, develop and produce as a function of time. The tactical plan should provide the framework for the operational level, i.e. control. The control system is the instrument in the hands of the grower that enables him to follow the tactical plan, and moreover, allows him to modify the original plan in response to deviations from the original assumptions, such as the actual weather conditions, the behaviour of the crop, or unexpected developments in the market. The climate control system, therefore, is a tool of operational management. This is not to say that the control system should merely be a device to follow preset setpoint trajectories as closely as possible. Instead, control should rather be cast in the frame of optimal steering on the basis of a prescribed goal function derived from criteria formulated at the tactical level, which then, ideally, results in desirable and realisable trajectories of the climate variables of interest. Before such a model based, goal-oriented control strategy can be developed it is necessary to analyze the criteria which must be taken into account in order to formulate the goal function.

In relation to the goals of climate control the following criteria are significant: physical yield (in kg, or numbers per m^2, section 2.3.2), crop quality (i.e. crop production capacity, see section 2.3.1 and 2.3.2), product quality (section 2.3.3), timing of the production process (section 2.3.1), production costs and production risks (Challa & Van Straten, 1993). These criteria will often give rise to conflicting climate requirements (e.g. yield versus quality, yield versus costs). For example, when yield increase requires extra economic inputs, as in the case of pure CO_2 enrichment, additional yield and associated extra costs have to be compared (Challa & Schapendonk, 1986). Also, seemingly attractive cost savings that do not affect short-term yield, for instance by lowering the night tempera-ture as far as possible, may have negative implications for the crop's long-term production capacity. The tactical and operational management system has to provide for balanced solutions to these conflicts in a transparent way. The criteria will now be briefly reviewed.

Physical yield

Yield is strongly affected by the climate conditions and as such it is a major criterion for climate control. Photosynthesis is the primary driving process, but the allocation of dry matter to harvestable product is also of crucial importance (Chapter 2). In addition it should be noted that manipulating short-term yield may have long-term implications on the plant's production capacity (section 2.3.2).

Crop production capacity

For crops with a long growing period it is particularly important for the grower to keep his crop in good condition for production. The internal balance between vegetative and generative growth is an important criterion (section 2.3.2) in maintaining productivity of generative crops. Temperature is the major climatic factor controlling this balance. Other phenomena affecting crop quality are physiological disorders and pests and diseases (see "risk prevention" below).

Product quality

Quality is a concept with a wide scope (section 2.3.3). It is not just influenced by climatic conditions, but also by the nutrient supply and the water balance of the plant (section 2.2.2). The external quality (e.g. size, weight, shape, colour) and the absence of visible injury are particularly relevant for the selling price. With respect to internal quality, keeping-quality and taste are also influenced by the climatic conditions during cultivation, but the relations between climatic factors and product quality are highly crop specific, and often not available in the form of quantitative relationships (section 2.3.3).

Timing of the production process

The market may show predictable patterns with some crops. Known examples are Christmas (Poinsettia), Easter, Valentines Day, Mother's Day, etc. In these cases timing is extremely important. Timing, besides, is also crucial for the cost of production in relation to labour requirements and space utilization in greenhouses (e.g. pot plants). After establishment of the culture the production process can be advanced or delayed to a certain extent through temperature and daylength (section 2.3.1).

Production costs

Part of the production costs can be directly attributed to climate control, e.g. heating, CO_2-enrichment, electricity consumption for supplementary lighting (Chapter 4). In addition there are indirect effects of climate control due to, for example, cost of labour, or length of the production cycle.

Risk prevention

During cultivation there is a continuous risk of damage to the crop and the product, due to pests, diseases, physiological disorders and environmental stress (sections 2.2.2 and 2.3.3). Humidity and temperature, but sometimes also radiation, have to be kept within certain limits in order to prevent acute problems. Beside these instantaneous reactions there are long-term adaptations of the crop to the climatic conditions which determine its sensitivity to pests, diseases and environmental stress (Levitt, 1980). So, adequate climate control is a significant tool in the integrated control of pests and diseases, and in risk management in general.

A characteristic of many of the criteria mentioned is the absence of an exact standard and the difficulty of quantification in economic terms. There is a notion of 'ideal' and of unacceptable situations, but in between there is often a gradual range. Moreover, the criteria have a number of climate factors in common, which makes decoupled climate control impossible. All this has lead to the present rule based controllers, which are largely built upon empirical knowledge and experience (Chapter 4). They require specification of a large number of settings, which is difficult to do and frequently leads to errors. Also a direct link between the climate settings and the consequences for the ultimate goals of the grower is missing (Chapter 5).

6.3.2 Design specifications for intelligent climate control systems

A new control system should meet the requirements defined according to the criteria above, while at the same time removing the drawbacks and shortcomings of present controllers. This could be achieved by a system that automatically handles in an integrated way the parts that can be described by established quantitative relations, i.e. models, and in addition provide means by which the grower can interact with the system to incorporate factors that cannot be quantified on the basis of present knowledge (Challa *et al.*, 1994).

Reviewing the requirements set out above, the design specifications of an ideal intelligent climate control system can be summarized as follows:

1. The system should be such that information that can be formulated in formalized quantitative models no longer needs to be processed by the grower. The grower should not have to bother with technical details of the control system that cannot be interpreted in terms of the ultimate goals and criteria.
2. The system should allow the specification of information that cannot be quantified, but nevertheless is important. This encompasses state constraints necessary for risk prevention (e.g. humidity bounds), constraints arising from the requirement to maintain long-term production capacity (e.g. temperature integrals, or constraints to dry weight distributions, see section 2.4), and the possibility to override the automatically generated climate path in case of disease risk or occurrence. A decision support system may be helpful to facilitate the translation of the grower's knowledge expressed in the criteria above into information needed for the control system.
3. The system should give the grower ample opportunity to inform the system about required and observed changes in timing, product prices, energy costs, environmental constraints, changes in crop status due to diseases, and risk assessment.
4. As at present, the system should have facilities for alarms and diagnosis in response to equipment failures.
5. The system should have a user interface that allows the grower to communicate in terms of the decision criteria. The low level details should not normally be accessible to the grower. The system will provide projections and forecasts in terms of crop yield and timing, energy consumption, operation costs and environmental burden as a function of the grower's settings of the tactical goal parameters of the system.
6. Ideally, the system should be flexible enough to encompass new quantitative information as soon as it becomes available from further research.

6.3.3 Improvements in parts

Several solutions have been proposed to improve climate control in the direction of the requirements formulated above, by providing partial solutions (Challa *et al.*, 1988). The first steps were taken by economic optimization of temperature control (Challa & Van de Vooren, 1980), later followed by optimization of CO_2 control (Challa & Schapendonk, 1986; Seginer *et al.*, 1986; Nederhoff, 1990) and that of supplementary lighting (Heuvelink & Challa, 1989). These early attempts were followed by others, which can be categorized as follows (without being complete).

Multivariable control schemes

These schemes tackle the problem of mutual loop interaction, by treating the control in a multivariable fashion. Van Henten (1989) has described a standard linear quadratic control (LQ) approach. Young *et al.* (1993) developed a digital multivariable PI controller, based on a non-minimal state space interpretation of an experimentally obtained input-output model, combined with LQ inspired tuning rules. These controllers can give good performance in terms of set-point tracking properties, but they do not handle the problem of set-point generation itself.

Controlling processes rather than state variables

An example here is transpiration control (Van Meurs & Stanghellini, 1989). The idea is that requirements of crop quality can be interpreted in terms of desired transpiration rates (or transpiration/crop growth ratio, Aikman & Houter, 1990). Thus, by controlling transpiration rate (inferred by detailed transpiration models) a more direct goal oriented control is achieved than in the case of humidity control. The ideas around the "speaking plant" concept (Takakura *et al.*, 1974; Hashimoto, 1989) also fall into this category.

Partial optimization

The intimate relationship between photosynthetic yield, solar radiation and CO_2-concentration has inspired developments aiming at an optimal CO_2 supply during the day (Seginer *et al.*, 1986; Chalabi, 1992; see also section 5.3). In one commercially available controller, suitable for a greenhouse with a heat storage tank and direct use of CO_2 contained in the flue gas of the boiler, the burner of the boiler is manipulated in such a way that during the day CO_2-production is optimally coordinated by the demand of the plant for photosynthesis, while ensuring that by the end of the day the storage tanks are filled for the night. The controller uses a weather forecast for the whole day, to calculate the optimal time pattern of heating and associated CO_2 supply.

Seasonal optimization

Optimization using plant models, while maintaining the "classical" computerized controllers has been proposed by Tantau (1991, 1993). Here, the goal is to calculate the settings that, given the plant dynamics and the dynamics of greenhouse plus controller, give maximum yield. This approach does not question the heuristic nature of present day climate controllers, which makes incorporation in current systems relatively easy.

A similar approach, advocated mainly by Seginer *et al.* (Seginer, 1991; Seginer & Sher, 1993), does not incorporate the greenhouse dynamics and the fast actual weather changes, but tries to find the optimal seasonal settings, by characterising the greenhouse physics as a momentarily reacting system, and by taking smoothed expected weather patterns. The implicit assumption here is that the control problem can be decomposed into a slow, seasonal optimization problem, and a fast momentary control problem, and that the latter does not influence the former. It should be noted that due to the greenhouse dynamics and the actual weather variations, this approach does not provide clear-cut solutions to the momentary control problem. Yet, the principles are similar to those outlined below, and the possibility of a hierarchical decomposition is worth further analysis.

6.3.4 Towards integrated optimal climate control

6.3.4.1 Basic structural elements

The basic components of the integrated optimal operational controller for greenhouse management and control that fulfils the requirements of section 6.3.1 can now be described. The system should contain the following intrinsic parts:

1. Descriptive dynamic models for the greenhouse physics (Chapter 3), crop photosynthesis and evaporation, and crop growth as a function of the control variables and the external influences (Chapter 2);
2. A user interface to allow the grower to set goal parameters in economic terms (price expectations, energy and CO_2-costs, desired timing), and to formulate constraints and limitations required by those aspects that cannot be properly handled with quantitative models (risk prevention, crop and product quality etcetera, as described before in this chapter);
3. A dynamic optimization tool to calculate the optimal pathways of control variables and state variables in open loop,
4. A procedure for automatic calibration and adaptation of the models to observations;
5. A control law or procedure to implement the optimal control and to accommodate deviations from the ideal path by introducing feedback;
6. A facility to evaluate the effects of changes in the goal function and constraint specification made by the grower, to help the grower in making the proper decisions.

Elements 4 and 6 are not absolutely necessary in a basic version of the controller, but are needed for any practical application of these advanced control concepts. Since the optimization plays a central part, the next paragraphs will be devoted to a description of the basic principles.

6.3.4.2 Systems dynamics introduction

In general, a dynamic system can be described by a set of difference or differential equations. In their most general form these are partial differential equations, to cope with spatial variability. Such equations can be turned into ordinary differential or difference equations by spatial discretization. Also, often an assumption is made about homogeneity over certain volumes in space, again leading to ordinary differential or difference equations.

Assuming ordinary differential equations in the state variables x the dynamics of the greenhouse and the crop can be described in very general terms as

$$\dot{x} = f(x,u,d) \qquad \text{(Eq. 6.3.1)}$$
$$y = g(x,u,d)$$

where $x(t) \in \mathbb{R}^{n_x}$ is a (column) vector representing the n_x state variables of the system, $y(t) \in \mathbb{R}^{n_y}$ are the n_y observed outputs, $u(t) \in \mathbb{R}^{n_u}$ the n_u manipulated inputs (control variables), and $d(t) \in \mathbb{R}^{n_d}$ the n_d inputs from outside that cannot be manipulated (disturbances). The functions f and g are vector valued, possibly non-linear, functions of the state, output and disturbance vectors, with dimensions n_x and n_y, respectively. They also contain the parameters of the model, left out here for brevity; ideally, parameters are time-invariant.

State variables are defined by the choice of the particular model, which, in turn, is dictated by the ultimate use. The main state variables relevant to greenhouse climate control are the air temperature, the CO_2-concentration, the air moisture content, the non-structural dry weight of the crop (assimilates) and the structural dry weight of the crop, possibly with subdivisions over the various parts of the plant. In a mathematical sense knowledge of the present values of the state variables together with knowledge on future inputs completely defines the system's behaviour in the future, i.e. no knowledge of the past is needed. In practice, there is a differential or a difference equation for each state variable.

The output variables are variables accessible to observation. In other words, they communicate information about the process to the outside world. In general they are static, possibly non-linear, functions of the states and the inputs. Sometimes, states can be observed directly, e.g. the air temperature, so the output coincides with the state in these cases. However, in general, transformations of the state variables are in play, for example total biomass is an output variable which follows from the sum of the state variables describing structural dry weight and non-structural dry weight. Another example is relative humidity being an output variable which depends in a non-linear way on the moisture content and the temperature. Output variables can also be quite complex rate functions, for instance the plant water uptake rate (section 2.2.2), or the plant photosynthetic rate (section 2.2.1). The latter example shows that it may be useful to calculate output variables that are not directly accessible to observation, but are of interest from a physiological point of view.

The disturbances come from outside, and may or may not be measurable. Examples in greenhouse control are solar radiation, wind speed, outside air temperature, CO_2-concentration and moisture content. In most control systems the idea of the controller is to suppress the influence of the external disturbances as much as possible. This is also partly true in greenhouse climate control, e.g. to reduce the effect of outside temperatures, but the effect of radiation on photosynthesis is essential, and should therefore be exploited, rather than suppressed.

The control variables can be set manually, but in the case of greenhouse control they are often generated by a suitable controller (section 5.2). Typical control variables in greenhouse control are the heating valve opening, the opening of the ventilating windows, and the position of the CO_2-supply valve (Chapter 4). In fact, it would be possible to redefine these controls in terms of heat supply, ventilation flux and CO_2-flux, but in practical applications the calculation of these variables is part of the model. A consequence of calculating the heating valve opening rather than the heat supply is, that it becomes necessary to augment the model with an additional state variable for the average pipe temperature.

Systems described by equation (6.3.1), and similar equations as given in Chapter 2 and section 3.6, can be easily simulated with suitable simulation software if sequences $u(t)$ and $d(t)$ are given. The purpose of control is to generate values of the control variables $u(t)$ such that the system plus controller behave in a predefined manner.

6.3.4.3 The principle of optimal control

The basic idea of optimal control is to generate a sequence of controls $u^*(t)$ such that the associated path $x^*(t)$ given by equation (6.3.1) is in some sense optimal. A vast amount of theory is available for when the system equations are linear, and the goal function is a weighted quadratic sum of state deviations and control deviations. This so-called LQ-theory leads to a closed loop feedback control law, where the control action $u'(t)$ (relative to a base level) is directly proportional to the state deviations $x'(t)$ (relative to the desired values). Since u and x are vectors the control is multivariable, rather than a collection of single loop controllers. This kind of control is very suitable for tracking a predefined path, while suppressing the influence of disturbances.

In greenhouse climate control, however, the task is not to suppress disturbances, but to generate the path that is best adapted to the actual conditions prevailing. Also, the quadratic goal function is too restrictive, and not very realistic here. Fortunately, general dynamic optimal control theory allows for any kind of goal function (Bryson & Ho, 1975; Lewis, 1986). It is quite logical to take an economic criterion, e.g. the money made in selling the produce, minus the cost associated to the control. In general terms the dynamic optimization problem can be expressed as follows. Given the model

$$\dot{x} = f(x,u,d) \tag{Eq. 6.3.2}$$

with initial conditions

$$x(t_0) = x_0 \tag{Eq. 6.3.3}$$

a goal function

$$J = \Phi\,(x(t_f)) + \int_{t_0}^{t_f} L(x,u,d)dt \tag{Eq. 6.3.4}$$

where Φ is a value associated with the states at the final time t_f, L are the instantaneous benefits minus costs associated to the state and control trajectories, and constraints imposed on the final state

$$\psi\,(x(t_f)) = 0 \tag{Eq. 6.3.5}$$

find the time evolution $u^*(t)$ of the controls such that equation (6.3.4) is maximized, or expressed differently

$$u^*(t) = \arg\max J\,(\hat{x}\,(t), u(t)) \tag{Eq. 6.3.6}$$

while satisfying the terminal conditions of equation (6.3.5) and the state equations

$$\frac{d\hat{x}}{dt} = f(\hat{x},u,d) \tag{Eq. 6.3.7}$$

where the caret (ˆ) is used as a reminder of the fact that the controls can be calculated in advance only if *expected* future values of the disturbances d are used.

The solution of the problem stated above is not an easy task. A well known method is inspired by the Lagrange multiplier approach in static optimization, and is known as the Hamiltonian approach (Bryson & Ho, 1975; Lewis, 1986). First, the Hamiltonian is formed

$$H = L + \lambda^T f \tag{Eq. 6.3.8}$$

The column vector $\lambda(t)$ represents the so-called adjoint variables, or co-states, which are introduced to incorporate the system equation as a constraint in the optimization. There are as many co-states as state variables. The co-states are used as intermediate auxiliary variables, which allows the minimization of J, such that the optimal state and control trajectories, $x^*(t)$ and $u^*(t)$ also obey the systems dynamics equation (6.3.2). Note that the Hamiltonian is a scalar. By substituting the Hamiltonian in the goal function (equation (6.3.4)), and setting the increment of J equal to zero by zeroing each term separately, we find the necessary conditions for an optimum (Lewis, 1986):

state equation

$$\dot{x} = \frac{\delta H}{\delta \lambda} = f \qquad t \geq t_0 \tag{Eq. 6.3.9}$$

costate equation

$$-\dot{\lambda} = \frac{\delta H}{\delta x} = \left(\frac{\delta f}{\delta x}\right)^T \lambda + \frac{\delta L}{\delta x} \qquad t \leq t_f \tag{Eq. 6.3.10}$$

stationarity condition

$$0 = \frac{\delta H}{\delta u} = \frac{\delta L}{\delta u} + \left(\frac{\delta f}{\delta x}\right)^T \lambda \tag{Eq. 6.3.11}$$

and the final boundary condition

$$\left(\frac{\delta \Phi}{\delta x} + \left(\frac{\delta \Psi}{\delta x}\right)^T v - \lambda\right)^T \Bigg|_{t_f} dx(t_f) + \left(\frac{\delta \Phi}{\delta t} + \left(\frac{\delta \Psi}{\delta t}\right)^T v + H\right)\Bigg|_{t_f} dt_f = 0 \tag{Eq. 6.3.12}$$

The variable ν in the final boundary condition is an adjoint variable related to the fixed final value specified in equation (6.3.5). The final boundary condition formulated this way allows the specification of problems with a required final condition, defined by equation (6.3.5), while the final time does not have to be fixed. The latter is useful where there are timing problems. If the final time is fixed, the second term vanishes. If there is a free final state, then the final state constraint also disappears, and equation (6.3.12) reduces to

$$\lambda(t_f) = \frac{\delta \Phi}{\delta x}\Bigg|_{t_f} \tag{Eq. 6.3.13}$$

and thus specifies the final boundary condition to the costate equation (6.3.10).

The solution of the optimization above entails the forward calculation of the states and the backward calculation of the co-states, such that the initial and final values are satisfied, and is therefore a two-point boundary value problem (TPBV). The optimal control path follows from the stationarity condition, with the optimal state and costate satisfying the state and costate equation.

In cases where the Hamiltonian is linear in the controls, the stationarity condition equation (6.3.11) can no longer be used, because the control u vanishes. Then, resort must be made to Pontryagin's maximum principle, which states that at the optimal state and costate trajectory the control must be made that minimizes the Hamiltonian. In practice, this means the control takes on either its maximum or its minimum admissible value, whenever the so-called switching function derived from the Hamiltonian changes sign. The result is bang-bang control, or bang-singular-bang control, see Lewis (1986) for further details.

6.3.4.4 Applicability to greenhouse climate control

Does the optimal control framework according to the theory above fulfil the requirements of the greenhouse control system of the future?

Firstly, it is clear that the method relies on the models being developed. The more advanced the models, the more advanced the control can be.

Secondly, the method allows full specification of a goal function. With single harvest crops, the market price expectation appears in the final term ϕ in the goal function, and L, which will be negative, then represents the costs of heating and CO_2-supply. With crops which are harvested more or less continuously, the price expectations can best be incorporated as a benefit term in L, because they are totalled over the full season. The final state constraint allows the specification of a fixed crop weight, or other fixed characteristics at the end of the optimization period, if so desired, but it can be left out if the final state is free. The boundary condition allows the specification of minimum time problems, but also of problems with a fixed end time.

The specification of imponderabilia by formulating constraints, which is the third requirement, is a little bit more involved. State constraints could be accommodated by incorporating them in the Hamiltonian. Another, perhaps more practical possibility is to incorporate penalty functions in the objective function. This solution is particularly suitable if the constraints are not hard, such as humidity constraints.

Finally, the interactions with the grower can be manifold. A user interface can be constructed that transfers price expectations, energy and CO_2 costs in a form that is needed for optimization. The same holds for the constraints. After optimization, the optimal pathways have been calculated, and can be displayed to the grower, along with the value of the goal function, and the profit in economic terms. He can easily examine the sensitivity of the solution to constraint values he sets, and also to changes in expected weather conditions. Thus, maximum flexibility is achieved, and the responsibility of the grower himself as an entrepreneur is fully respected.

Two problems remain: the practical solution, and the safeguarding against inevitable errors in the models used.

6.3.4.5 Problems concerning time scales and unknown disturbances

Although the theory is conceptually very attractive there are a number of problems which need to be solved before a practical application can be implemented.

In greenhouse climate control dramatic differences in time scale of the dominant processes occur, as indicated in sections 2.1 and 3.6. The temperature, moisture content and CO_2 dynamics are very fast as compared to the dynamics of the crop biomass. The crop itself contributes to the fast dynamics of the physical parameters due to the almost instantaneous response of transpiration and photosynthesis. The existence of fast and slow dynamics complicates considerably the calculation of the full theoretical solution, even if the weather conditions can be assumed to be known. This is because the controls need to be calculated on the basis of short time intervals related to the fast processes, while the total calculation period is equal to a full growing season. This is a very heavy computational task.

The second problem is that the derivation above assumes that the disturbances only are known functions of time. However, in greenhouse climate control the disturbances are mainly due to the weather, and assuming the weather to be known over the full season in advance is not very realistic. Seasonal average expectations are known from statistical data, but in the short-term considerable deviations can occur. At best, weather forecasts may be reliable at the scale of a day or so.

The third problem is that dynamic optimal control in its pure form is essentially open loop, because the control actions are calculated in advance. This implicitly assumes that the model is correct over the entire horizon, an assumption which is nearly never justified. In other words, there is a need to introduce corrections to the model in case the observable outputs appear to deviate from the ones calculated from the theoretical optimal time paths x^* and u^*.

A potential solution to these problems is outlined below. First, the problem of various time scales is addressed. When looking at the overall model for the behaviour of greenhouse plus plant material, the greenhouse physical processes are relatively fast compared with the crop behaviour. Moreover, the crop itself is not directly influenced by the manipulated control variables, but only by the external variables and the physical states. So, a decomposition is possible in the following form

$$\frac{dx_g}{dt} = f_g(x_g, x_p, u, d) \qquad \text{(Eq. 6.3.14)}$$

$$\frac{dx_p}{dt} = f_p(x_g, x_p, d) \qquad \text{(Eq. 6.3.15)}$$

where x_g: the indoor greenhouse state variables, e.g. temperature, CO_2-concentration and humidity.

x_p: the plant state variables, e.g. total biomass, leaf area index, and biomass distribution over root and shoot, leaves and fruits,

u: the control inputs, e.g. heat supply, ventilation, CO_2-input flux.

d: the external inputs, e.g. solar radiation, outdoor temperature, outdoor humidity, outdoor CO_2, wind speed.

Photosynthesis, respiration and transpiration are not directly visible in this generic representation, but appear as important terms within both equations. These terms respond in an immediate fashion to indoor climate and radiation, and are therefore controlled by the greenhouse climate states. Photosynthesis and respiration form the major input to the crop growth model (Chapter 2). In general, the vector function f_g in equation (6.3.14) has relatively fast dynamics, while the vector function f_p in equation (6.3.15) has slow dynamics. This has led several researchers who have studied the optimal control problem over the season to treat the first equation as pseudo-static, and perform the optimization over the season with the second equation. This procedure results in optimal "setpoint" paths of the climate variables temperature, humidity, CO_2-concentration, and so on, which are assumed to be realized by an ideal local controller (Seginer, 1991). Van Henten & Bontsema (1992), and Van Henten (1994) have developed a formal procedure, using the method of singular perturbations, which allows the optimization of the seasonal problem, using average weather expectations and pseudo-instantaneous greenhouse response. Van Henten (1994) has shown that for a vegetative crop such as lettuce the co-state patterns associated with structural and non-structural dry weight are very similar for various weather patterns. Since the co-states represent a kind of "shadow price", this information can be used to subsequently solve the problem of the minute-by-minute optimal control, where now the momentary weather variations can be taken into account (Tap *et al.*, 1993; Tchamitchian *et al.*, 1993; Van Henten, 1994).

The second problem is the problem of the unknown weather. As outlined above, the results of the seasonal optimization do depend on the actual weather – there are obviously "bad" and "good" years – but the time pattern of the associated co-states are quite similar. So, for the long term, the known statistical expectation of the weather is sufficient. To put it differently: if the long term seasonal weather would be known in advance, better results could be achieved, but in practice these "dream" patterns can never be realized. For the daily momentary control, however, the situation is quite different. The question arises what advantages can be taken from using weather forecasts in the process of optimization. It can be postulated that the horizon over which the forecast will have an effect is in the order of the time constant of the greenhouse physics. This stimulates the idea of using a predictive control approach, where the short-term optimization is performed over a limited horizon, using short term forecasts. In the next section some preliminary results on the validity of this concept will be presented. The coupling with the long term season is then achieved through the co-state values obtained over a seasonal optimization using average weather patterns.

Finally, the problem of robustness of the solution has to be discussed. The expectation is that the proper solution can be found by expanding the predictive control concept to an approach known as receding horizon predictive control (RHPC). The idea is that the optimization is calculated over the forecasting horizon (24 hours maximum, but usually less, see below). Only the first value of the calculated optimal control sequence is actually applied to the greenhouse. The system behaviour is then observed, and after the computation interval – typically 1 minute – the optimization is repeated starting with the observed state, again over the same forecasting length, i.e. the horizon is receding. Once again, the newly calculated value is applied, and the procedure is repeated. This RHPC procedure thus introduces feedback, because the state is observed, and the optimization is corrected, should any deviation from the expected values occur. Also, the procedure uses the best information available at any time. The only requirement is that the optimization must be performed within the control sample interval, but since the deviations from the previous time instant will be relatively small, it seems that this can be performed.

6.3.4.6 Some preliminary results

Van Henten (1994) has compared the performance of a real grower, growing lettuce, to simulated results of seasonal optimal control of the greenhouse climate. The latter was done in two ways. The first is full optimal control, where the criterion contains a penalty function for violation of lower and upper humidity levels. A sample of the resulting control patterns is shown in Figure 6.3.1. It was clearly shown that this method yields a larger harvest weight (some 20%), at lower CO_2-costs (about 8%), and with considerable energy savings (about 30%). It should be noted that these advantages could be obtained by exploiting knowledge about the weather of the coming day. A remarkable result is that the optimal control scheme does not give a heat pulse, as the grower did, in the morning. The ventilation patterns calculated by the optimal control are also quite different from those of the grower. The grower may have had reasons to operate the ventilators the way he did, for instance in the hope of maintaining better long term crop quality. Therefore a second optimization was performed by accepting the grower's ventilation pattern, assuming that this would accommodate potential effects related to long term crop quality. With this approach, the crop weight was about the same as that of the grower, but at far lower CO_2-costs (some 45%), and still with 30% savings on energy.

The feasibility of the receding horizon approach in a lettuce cultivation in a greenhouse without a heat storage tank was investigated by Tap (personal communication). As before, the criterion function consists of two parts: the direct economic profit, i.e. the difference in value of the increment in crop and the energy and CO_2 supply costs, and a penalty if the humidity constraints are violated. Figure 6.3.2 shows the profit, the penalty and the total criterion value for a particular day when the prediction horizon is gradually increased. It is interesting to note that the results were obtained by taking the present weather as the "best" forecast for the weather over the entire horizon. The dashed lines indicate the values obtained from a full 24 hour optimization, using the *a posteriori* actual weather. It can be seen that when the horizon is too short, solutions are not optimal. However, a horizon of 60–100 minutes is enough to achieve the optimum in this case. In other words, a prediction horizon of about 1 hour gives solutions that are very close to the optimal "dream" solution had the weather been known over 24 hours. Moreover, it turns out that using as a forecast that the next hour weather will be the same as the present weather, and updating this forecast every minute, gives

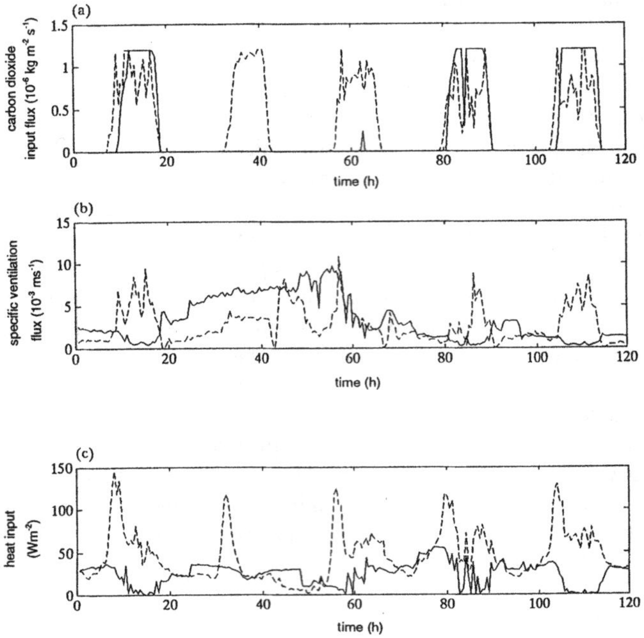

Figure 6.3.1 – Control of carbon dioxide supply (a), ventilation (b), and heating (c), according to the grower (dashed line), compared to optimal control strategies with humidity constraints, based on known weather inputs and a validated greenhouse dynamics plus lettuce crop growth model (solid line). (Courtesy, Van Henten, 1994).

almost optimal results. This is very important, because it means that instantaneous measurements can be used. In systems with larger physical time constants, longer range forecasts might bring further profits, but as weather forecast quality deteriorates over the longer horizon, the difference from the "dream" pattern occurring when everything would be known in advance has to be accepted.

Preliminary calculations have also been performed to compare, in simulation, the actual behaviour and performance of a modern widespread commercial rule-based controller, using standard blueprint control, and the optimal control concept outlined above. Again, the same lettuce model was used. One interesting observation is that the rule based controller with its usual settings cuts off carbon dioxide supply much earlier than the optimal controller. It became clear that considerable gains in crop yield can be achieved by increasing the CO_2 content much further. This, of course, can be done in the classical controller by changing the settings. The merit of the optimal controller is that it generates these solutions automatically, and therefore does not depend on the awareness of the grower. The comparison also made it clear that the constraints imposed on the optimization have a clear impact on the results. Since in the optimal controller excess humidity above the limit value is punished by a steep penalty function, the optimal controller sometimes tries to reduce humidity at night by heating. In the rule based controller this behaviour is not observed, at the expense of serious violations of the humidity limits. So, while it seems that the optimal controller leads to higher energy consumptions on those days, this is solely due to the weight put on violations of the constraints. The

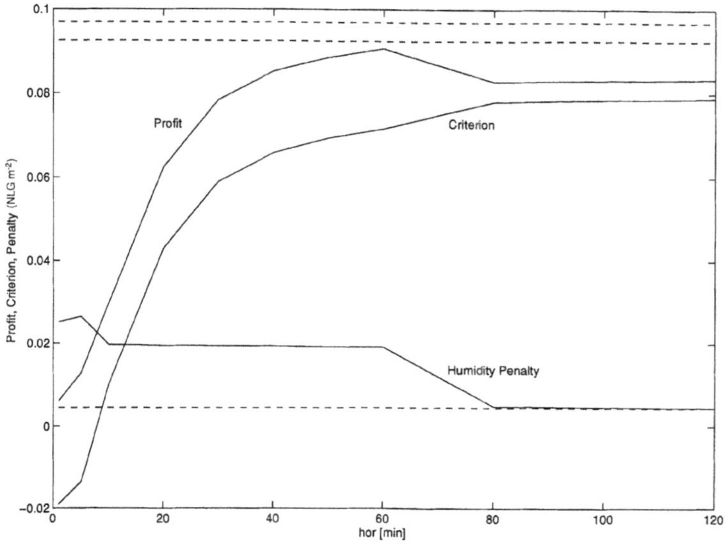

Figure 6.3.2 – Effect of the optimization horizon on profit, humidity penalty and total criterion value. Lettuce crop model, weather data of 6 August, 1992. Greenhouse without heat storage tank. Weather forecast over the horizon assumed equal to present weather ("lazy man's method"). Dashed lines: full 24 hours optimization. (Unpublished results, K.F. Tap).

advantage, however, is that it is very easy to see the economic consequences of the setting of the constraints, a piece of information that is lacking in the rule based controller. Of course, overall, the optimal controller provides the best performance, simply because it is optimal, and no better solution exists.

6.4 Further developments

G. VAN STRATEN

The dry matter distribution and development is very much species dependent, and much research is needed to specify quantitative models for each economically interesting crop. Lack of these models does not hamper the use of optimal control methods *per se*, because for the time being, qualitative knowledge about these phenomena can be incorporated by specification of constraints. Yet, in the future, improved models for the distribution of assimilates over the various plant parts, such as those described in section 2.3.2, will have to be modified for control purposes and incorporated within the optimization methods as presented here. Practitioners have also found that the assumption of homogeneous climate conditions within the greenhouse is not fulfilled (Bakker & Van Holstein, 1989). This suggests that more attention should be paid to the spatial distribution within greenhouses.

A matter of considerable theoretical and practical interest is the issue of the robustness of the optimal control to model structure and parameter deviations. Although the receding horizon approach probably alleviates this problem, on-line state estimation and model validation methods are needed to enhance robustness and ease of implementation.

By virtue of the models used in optimal controllers, part of the task that is currently represented in the form of blue prints is taken over by the climate system. However, there are a number of decisions that can not be based on quantitative models. In the system outlined above, these aspects are treated by setting constraints to the optimal controller. It is not an easy matter to decide on these constraints. Therefore, in the future, a knowledge based system may help to perform this task (Martin-Clouaire *et al.*, 1993). The problem may also be cast in the form of a constraint satisfaction problem (Martin-Clouaire, 1993). As an alternative, a combination of optimal and fuzzy control may be thought off. A fuzzy controller would rely heavily upon extracting knowledge from human operators. This can be a very practical approach for the short term, where it is known that one grower performs much better than the other. If the behaviour of the "good" grower is incorporated in a fuzzy controller, good performance might be attainable. On the other hand, all advantages of quantitative models in this approach would be lost. Therefore, a combined approach keeping the advantages of both, might be the way to proceed.

The principles developed for a standard greenhouse can be extended for more complex greenhouse equipment of the future. Internal heat storage, energy screens, combined heat-power units, use of reject heat from central or regional power plants, each of these will profit from advanced control systems based on the same principles.

As was stated in the introduction in section 1.1, the increasing awareness of the fundamental shortcomings of present greenhouse climate control systems, the gain in insight into the functioning of crops, the growing availability of models for the greenhouse-crop system, and the low price of computing power have invoked a revisiting of the greenhouse climate control problem. These developments have been accompanied by significant progress in the field of control theory, and together this has set the scene for a thorough reformulation of the problem. The concept of optimal control, using a economically founded goal function, provides a natural way to best exploit physiological, physical and control knowledge. The conviction is that a scientific foundation of this nature is extremely applicable, and will, in fact, be necessary in order to run the greenhouse of the future in the most efficient manner.

References

Aikman, D.P. & G. Houter, 1990. Influence of radiation and humidity on transpiration: implications for calcium levels in tomato leaves. Journal of Horticultural Science 65: 245–253.

Bakker, J.C. & G.P.A. Van Holstein, 1989. Horizontal temperature distribution in heated glasshouses: causes and effects. Acta Horticulturae 245: 226–231.

Bryson, A.E. & Y.C. Ho, 1975. Applied optimal control: optimisation, estimation and control. Hemisphere Publishing Corporation, New York, 481 pp.

Chalabi, Z.S., 1992. A generalized optimization strategy for dynamic CO_2 enrichment in a greenhouse. European Journal of Operational Research 59: 308–312.

Challa, H., G.P.A. Bot, E.M. Nederhoff & N.J. Van de Braak, 1988. Greenhouse climate control in the nineties. Acta Horticulturae 230: 459–470.

Challa, H. & A.H.C.M. Schapendonk, 1986. Dynamic optimalization of CO_2 concentration in relation to climate control in greenhouses. In: H.Z. Enoch & B.A. Kimball (Eds), Carbon dioxide enrichment of greenhouse crops. CRC Press Inc., Boca Raton, Florida, pp. 147–160.

Challa, H. & J. Van de Vooren, 1980. A strategy for climate control in early winter production. Acta Horticulturae 106: 159–164.

Challa, H. & G. Van Straten, 1993. Optimal diurnal climate control in greenhouses as related to greenhouse management and crop requirements. In: Y. Hashimoto, G.P.A. Bot, W. Day, H.-J. Tantau & H. Nonami (Eds), The computerized greenhouse: automatic control application in plant production. Academic Press, pp. 119–137.

Challa, H., E. Heuvelink & K.J. Leutscher, 1994. Modelling in the search for balance between inputs and outputs in greenhouse cultivation. P.C. Struik, W.J. Vredenberg, J.A. Renkema & J.E. Parlevliet (Eds), Plant Production on the Thres-hold of a New Century, Kluwer Academic Publishers, pp. 187–195.

De Jong, T., 1993. Ontwerp van klimaatbeheersingsapparatuur voor gesloten kassystemen. *(Design of an air conditioner for conditioning closed greenhouses)*. IMAG-DLO, Wageningen, IMAG-DLO Report 93–20: 35 pp. (In Dutch with English summary and annotations).

Hashimoto, Y., 1989. Recent strategies of optimal growth regulation by the speaking plant concept. Acta Horticulturae 260: 115–121.

Heuvelink, E. & H. Challa, 1989. Dynamic optimisation of artificial lighting in greenhouses. Acta Horticulturae 260: 401–412.

Levitt, J., 1980. Responses of plants to environmental stresses. Vol. 1: Chilling, freezing, and high temperature stresses. Academic Press, Orlando, Florida, 497 pp.

Lewis, F.L., 1986. Optimal Control. Wiley-Interscience, New York, 362 pp.

Martin-Clouaire, R., T. Boulard, M.-J. Cros & B. Jeannequin, 1993. Using empirical knowledge for the determination of climatic setpoints: an artificial intelligence approach. In: Y. Hashimoto, G.P.A. Bot, W. Day, H.-J. Tantau & H. Nonami (Eds), The computerized greenhouse: automatic control application in plant production. Academic Press, pp. 197–224.

Martin-Clouaire, R., 1993. Determination of climate setpoints by satisfaction of soft constraints. In: Preprints of the 12th World Congress IFAC, Sydney, Vol. I. The Institution of Engineers, Sydney, pp. 321–324.

Nederhoff, E.M., 1990. Technical aspects, management and control of CO_2 enrichment in greenhouses. Acta Horticulturae 268: 127–138.

Out, P.G. & J.J.G. Breuer, 1994. Meer licht, betere isolatie van kassen *(More light, better isolation of greenhouses)*. IMAG-DLO, Wageningen, IMAG-DLO Report, 44 pp. (In Dutch with English summary).

Seginer, I., A. Angel, S. Gal & D. Kantz, 1986. Optimal CO_2 enrichment strategy for greenhouses: a simulation study. Journal of Agricultural Engineering Research 34: 285–304.

Seginer, I., 1991. Optimal greenhouse temperature trajectories for a multi-state-variable tomato model. In: Y. Hashimoto & W. Day (Eds), Mathematical and control applications in agriculture and horticulture. Proceedings of the IFAC/ISHS workshop, Matsuyama, Japan (Sept. 30-Oct. 3). Pergamon Press, Oxford, pp. 73–79.

Seginer, I. & A. Sher, 1993. Optimal greenhouse temperature trajectories for multi-state-variable tomato model. In: Y. Hashimoto, G.P.A. Bot, W. Day, H.-J. Tantau & H. Nonami (Eds), The computerized greenhouse: automatic control application in plant production. Academic Press, pp. 153–172.

Takakura, T., T. Kozai, K. Tachibana & K.A. Jordan, 1974. Direct digital control of plant growth. I. Design and operation of the system. Transactions of the ASAE 17: 1150–1154.

Tantau, H.-J., 1991. Optimal control for plant production in greenhouses. In: Y. Hashimoto & W. Day (Eds), Mathematical and control applications in agriculture and horticulture. Proceedings of the IFAC/ISHS workshop, Matsuyama, Japan (Sept. 30–Oct. 3). Pergamon Press, Oxford, pp. 1–6.

Tantau, H.-J., 1993. Optimal control for plant production in greenhouses. In: Y. Hashimoto, G.P.A. Bot, W. Day, H.-J. Tantau & H. Nonami (Eds), The computerized greenhouse: automatic control application in plant production. Academic Press, pp. 139–152.

Tap, R.F., L.G. Van Willigenburg, G. Van Straten & E.J. Van Henten, 1993. Optimal control of greenhouse climate: computation of the influence of fast and slow dynamics. In: Preprints of the 12th World Congress IFAC, Sydney, Vol. X. The Institution of Engineers, Sydney, pp. 321–324.

Tchamitchian, M., L.G. Van Willigenburg & G. Van Straten, 1993. Optimal control applied to tomato crop production in a greenhouse. In: J.W. Nieuwenhuis, C. Praagman & H.L. Trentelman (Eds), Proceedings of the 2nd European Control Conference, Groningen (June 28–July 1, 1993), pp. 1348–1352.

Van Henten, E.J., 1989. Model based design of optimal multivariable climate control systems. Acta Horticulturae 248: 301–306.

Van Henten, E.J. & J. Bontsema, 1991. Optimal control of greenhouse climate. In: Y. Hashimoto & W. Day (Eds), Mathematical and control applications in agriculture and horticulture. Proceedings of the IFAC/ISHS workshop, Matsuyama, Japan, (Sept. 30–Oct. 3). Pergamon Press, Oxford, pp. 27–32.

Van Henten, E.J. & J. Bontsema, 1992. Singular perturbation methods applied to a variational problem in greenhouse climate control. In: Proceedings of the 31st IEEE Conference on Decision and Control, Tucson, Arizona, December 1992. IEEE Inc., pp. 3068–3069.

Van Henten, E.J., 1994. Greenhouse climate management: an optimal control approach. PhD dissertation, Wageningen Agricultural University, Wageningen, 329 pp.

Van Meurs, W.T.M. & C. Stanghellini, 1989. A transpiration-based climate control algorithm. Acta Horticulturae 245: 476–481.

Waayenberg, D. & J.W. Freney, 1993. Kunststofkas met tuiconstructies: ontwerp, uitvoering en toetsing van een prototype *(Plastic greenhouse supported by guy ropes: design, construction and testing of the prototype)*. IMAG-DLO, Wageningen, IMAG-DLO Report 93–24, 51 pp. (In Dutch with English summary and annotations).

Young, P.C., A. Chotai & W. Tych, 1993. Identification, estimation and true digital control of glasshouse systems. In: Y. Hashimoto, G.P.A. Bot, W. Day, H.-J. Tantau & H. Nonami (Eds), The computerized greenhouse: automatic control application in plant production. Academic Press, pp. 3–50.

List of symbols

d	the external inputs, e.g. solar radiation, outdoor temperature, outdoor humidity, outdoor CO_2, wind speed
f_g	rate of change of the greenhouse climate variables
f_p	rate of change of the plant crop variables
H	Hamiltonian
J	economical goal function (net profit)
L	instantaneous benefits minus costs
n_x	number of state variables
n_y	number of observation variables
n_d	number of disturbance inputs
n_u	number of control inputs
x	vector of state variables
y	vector of observation variables
d	vector of disturbance inputs
u	vector of control inputs
t	time (s)
u	the control inputs, e.g. heat supply, ventilation, CO_2-input flux
x_g	the indoor greenhouse state variables, e.g. temperature, CO_2-concentration and humidity
x_p	the plant state variables, e.g. total biomass, leaf area index, and biomass distribution over root and shoot, leaves and fruits

Greek symbols

λ	co-state vector (marginal value of the state variables)
ν	co-state vector associated to final state constraints
Φ	value of the states at the final time t_f
Ψ	final state constraint function

Subscripts

o	onset
f	final

Superscripts

*	optimal
$\hat{}$	expected
$\dot{}$	time derivative

List of abbreviations

PAR	photosynthetic active radiation (400–700 nm) (mmol m^{-2} s^{-1} or Wm^{-2})
RHPC	receding horizon predictive control
TPBV	two-point boundary value

Index

Editors

DR. IR. J.C. BAKKER
IMAG-DLO
formerly Glasshouse Crops Research Station

PROF. DR. IR. G.P.A. BOT
IMAG-DLO and Wageningen Agricultural University
Department of Agricultural Engineering and Physics

PROF. DR. IR. H. CHALLA
Wageningen Agricultural University
Department of Horticulture

IR. N.J. VAN DE BRAAK
IMAG-DLO

Authors

DR. J. BONTSEMA
Wageningen Agricultural University
Department of Agricultural Engineering and Physics

ING. J.J.G. BREUER
IMAG-DLO

DR. IR. A.N.M. DE KONING
Glasshouse Crops Research Station

IR. J. M. JACOBS
formerly Glasshouse Crops Research Station

ING. TH. GIELING
IMAG-DLO

IR. H. GIJZEN
Wageningen Agricultural University
Department of Horticulture
formerly AB-DLO

IR. E. HEUVELINK
Wageningen Agricultural University
Department of Horticulture

ING. J.P.G. HUJS
IMAG-DLO

ING. P. KNIES
IMAG-DLO

DR. IR. L.F.M. MARCELIS
AB-DLO

IR. D. MEIJAARD
LEI-DLO

DR. IR. E.M. NEDERHOFF
Glasshouse Crops Research Station

DR. K. SCHURER
IMAG-DLO

DR. C. STANGHELLINI
IMAG-DLO

IR. P.A.C.M. VAN DE SANDEN
AB-DLO

ING. G.P.A. VAN HOLSTEIJN
Glasshouse Crops Research Station

DR. IR. U. VAN MEETEREN
Wageningen Agricultural University
Department of Horticulture

ING. W.TH.M. VAN MEURS
IMAG-DLO

PROF. DR. IR. G. VAN STRATEN
Wageningen Agricultural University
Department of Agricultural Engineering and Physics

DR. IR. C. VONK-NOORDEGRAAF
Research Station for Floriculture

ING. D. WAAIJENBERG
IMAG-DLO

IR. G.W.H. WELLES
Glasshouse Crops Research Station

Addresses of Institutes

AB-DLO
DLO – RESEARCH INSTITUTE FOR AGROBIOLOGY AND SOIL FERTILITY
P.O. Box 14
6700 AA Wageningen
The Netherlands

GLASSHOUSE CROPS RESEARCH STATION
P.O. Box 8
2670 AA Naaldwijk
The Netherlands

IMAG-DLO
DLO – INSTITUTE OF AGRICULTURAL AND ENVIRONMENTAL ENGINEERING
P.O. Box 43
6700 AA Wageningen
The Netherlands

LEI-DLO
DLO – AGRICULTURAL ECONIMICS RESEARCH INSTITUTE
P.O. Box 29703
2502 LS The Hague
The Netherlands

RESEARCH STATION FOR FLORICULTURE
Linnaeuslaan 2-A
1431 JV Aalsmeer
The Netherlands

WAGENINGEN AGRICULTURAL UNIVERSITY
DEPARTMENT OF AGRICULTURAL ENGINEERING AND PHYSICS
Bomenweg 3
6703 HD Wageningen
The Netherlands

WAGENINGEN AGRICULTURAL UNIVERSITY
DEPARTMENT OF HORTICULTURE
Haagsteeg 3
6783 PM Wageningen
The Netherlands

www.ingramcontent.com/pod-product-compliance
Lightning Source LLC
Chambersburg PA
CBHW061203240926
51846CB00006B/1097
* 9 7 8 9 0 7 4 1 3 4 1 7 0 *